Green Chemistry in Agriculture and Food Production

Editors

Vinay Kumar

Department of Community Medicine
Saveetha Medical College
Saveetha Institute of Medical and Technical Sciences
Chennai, India

Kleopatra Tsatsaragkou

Independent Researcher
Melissia, Athens
Greece

Nilofar Asim

Solar Energy Research Institute
Universiti Kebangsaan Malaysia
Bangi, Selangor
Malaysia

CRC Press
Taylor & Francis Group
Boca Raton   London   New York

CRC Press is an imprint of the
Taylor & Francis Group, an **informa** business

A SCIENCE PUBLISHERS BOOK

Cover illustration reproduced by kind courtesy of Dr. Vinay Kumar (first editor)

First edition published 2023
by CRC Press
6000 Broken Sound Parkway NW, Suite 300, Boca Raton, FL 33487-2742

and by CRC Press
4 Park Square, Milton Park, Abingdon, Oxon, OX14 4RN

© 2023 Taylor & Francis Group, LLC

CRC Press is an imprint of Taylor & Francis Group, LLC

Library of Congress Cataloging-in-Publication Data (applied for)

ISBN: 978-0-367-25431-5 (hbk)
ISBN: 978-1-032-43375-2 (pbk)
ISBN: 978-0-429-28953-8 (ebk)

DOI: 10.1201/9780429289538

Typeset in Times New Roman
by Radiant Productions

Preface

This book titled "Green Chemistry in Agriculture and Food Production" is a comprehensive compilation of the applied aspects of green chemistry applications in agriculture and food production. It deals with the various aspects of green chemistry in agriculture. It discusses the green technologies used in crop-pest control; green fertilizers in agriculture, green chemistry in organic farming; eco-safe farming with microbes; marine algae as green agricultures; production of value-added products from waste processing using green technologies; green technologies in food processing; food chain and green chemistry; green technologies in reduction of toxins; green technologies in food analysis; green methods in agrochemical residues; functional food ingredients, and foodomics.

This book will guide students, scientists, teachers and researchers in understanding green technology's fundamentals and applications.

Vinay Kumar
Kleopatra Tsatsaragkou
Nilofar Asim

Contents

CHAPTER 1

Natural Product Chemistry in Agriculture

Divya Utreja,[1,*] *Komalpreet Kaur,*[1] *N.K. Dhillon,*[2] *Anupam*[2]
and *Sarbjit Singh*[3]

India is blessed with rich biodiversity grown in different agroclimatic zones. Natural products are widely employed for the synthesis and commercialization of bio-based products. These products have a profound impact on the pharmaceutical, food, cosmetics, and agricultural industry. Natural products originated from co-evolutionary interactions and several biosynthetic pathways due to which they can survive in a stressful environment. Natural products constitute several primary and secondary metabolites. These two classes of metabolites consist of lignins, flavonoids, terpenes, alkaloids, phenolic compounds, nitrogenous compounds, aldehydes, ketones, alkaloids, glycosides, glucosinolates, isothiocyanates, limonoids, quassinoids, saponins, phenolics, flavonoids, quinones, piperamides, polyacetylenes, and polythienyls (Anulika et al. 2016). These are involved in growth, reproduction, development, and ecological function. In fact, it will be worth saying that nature constitutes incredible complex products that are of worth to mankind. Today, pharmaceutical industries are interested in research on natural products to explore and design them and develop novel bio-based products. Commercial insecticides in the agrochemical industry have been found to exhibit negative effects, such as mortality in various microbial communities, ground water contamination, human carcinogenicity, bird toxicity, soil pollution, and hazardous effects due to residues led to their withdrawal from the market and leading to decreased options for the farmers subsequently (Gad 2014, Baidoo et al. 2017, Qiao et al. 2012, Khalil 2014, Bernard et al. 2017).

[1] Department of Chemistry, Punjab Agricultural University, Ludhiana - 141004, Punjab, India.
[2] Department of Plant Pathology, Punjab Agricultural University, Ludhiana - 141004, Punjab, India.
[3] Eppley Institute for Research in Cancer and Allied Diseases, University of Nebraska Medical Center, Omaha, Nebraska 68022, USA.
* Corresponding author: utrejadivya@yahoo.com

The ever-increasing populations, changing climatic parameters, the outbreak of pandemic diseases, food requirements, and protection of the global economy are some of the interesting challenges that are being faced globally (Tsygankova et al. 2016). The crop ravages caused by various biotic and abiotic stresses negatively affect the growth and productivity curve of crops and are prime threats for sustaining agriculture productions (Atkinson et al. 2013, Narsai et al. 2013). Abiotic stress factors wrap extreme salinity, floods, radiation, temperature, mineral toxicity, heavy metals, and drought in it, while biotic stress factors comprise loss by fungi, viruses, viroid, insects, arthropods, bacteria, oomycetes, nematodes, mollusks, weeds, and other competitive plants (Dresselhaus and Huckelhoven 2018, Madani et al. 2018, Gull et al. 2019). Pathogenic infestations pose a threat to food security with an annual loss of 60 billion dollars (10–40%) and sometimes these lead to epidemics (Sharma et al. 2016, Anamika et al. 2019, Serge et al. 2019, Cal et al. 2012). These biotic factors badly affect the plants' photosynthetic and transport systems as they develop a competitive atmosphere over attacked sites (Singla and Krattinger 2016).

Plant-parasitic nematodes are destructive endoparasites that can negatively target numerous agricultural crops, especially vegetables, fruits, field crops, ornamental plants, and human health (Kaur et al. 2018, 2019, Jain et al. 2019, Kaur et al. 2020, 2021). The immense damage caused has been estimated to range from the US \$80 billion to \$1.58 billion per annum (21.3%) (Jain et al. 2007, Singh 2015). The nematode population favors sedentary infestations with other soil-borne pathogens leading to multifold effects and crop damage (Elling 2013, Ralmi et al. 2016). The chapter intends to orient the research toward the management of nematodes environmentally and economically beneficial manner.

Brassicaceae family is reported with biofumigant properties and is widely investigated for its potential green manure technique to combat plant-parasitic nematodes. The family produce glucosinolate compounds, which on hydrolyzation by myrosinase in soil produces isothiocyanates containing -N=C=S group that in turn adheres to the target organism (Riga et al. 2004) and enhance fumigant activity. Isothiocyanates cause immediate cell death by uncoupling oxidative phosphorylation, interrupt energy metabolic pathways and aggregation of intracellular protein (Caboni and Ntalli 2014), interrupt coupling between electron transport and phosphorylation, inhibit ATP synthesis, and break -S-S- bridges that lead to cell death. Another important activity is the interaction of isothiocyanate with topoisomerase II resulting in DNA supercoiling and results in the removal of the knotting effect from the host genome. Glucosinolates such as epi-progoitrin (2(S)-2-hydroxy-3-butenyl-GLS), glucoerucin (4-methylthio-butyl-GLS) (-OH), Glucoiberin (3-methylsulfinylpropyl-GLS) (-S-), epi-progoitrin (-S=O) on hydrolysis by myrosinase boost their activity by 100%. These easily react with essential biological nucleophiles, such as thiol and amino groups of the nematode enzymes for the alkylation. The dead J_2 appeared straight, rigid with brown internal organs. It was reported that aliphatic isothiocyanate was more active in biofumigants than aromatic isothiocyanates. The allelochemicals secreted act as deterrents, repellants, oviposition inhibitors, and antifeedants (Caboni et al. 2012). Nematode from family can combat pathogens, such as *Meloidogyne chitwoodi, Meloidogyne hapla, Meloidogyne incognita*, as it contains six isolates of glucosinolates (Curto et al. 2006). The study reported the integrated practices to combat nematode by the combination of Brassica green manures as biofumigants with commercialized nematicides 1,3-dichloropropene (1,3-D) reduced 35% cost and did not affect non-pathogenic *Pseudomonas species*. The economical and eco-friendly practice effectively reduced the population of *Pratylenchus penetrans* and *M. chitwoodi, Paratrichodorus allius*. The combination is a boon for the free-living nematodes and *Pseudomonas* spp. (Riga 2011).

The term 'biofumigation' defines the suppressiveness exhibited by plants of the Brassicaceae family on noxious soil-borne pathogens and is related to the discharge of biocidal isothiocyanates (ITCs), particularly due to the glucosinolates hydrolysis (GSLs, thioglucosides) present in the residues of the crop, catalyzed by myrosinase (MYR, β-thioglucosideglucohydrolase) isoenzymes (Matthiessen and Kirkegard 2006, Motisi et al. 2010). Along with GSL hydrolysis products, the tissues of decomposing *Brassica* produce several volatile sulfur-containing toxins, viz. methyl sulfide, carbon disulfide, dimethyl disulfide, dimethyl sulfide, methanethiol, etc., which may participate in the process of biofumigation (Lord et al. 2011). The active sites of ITC mainly thiol and amine groups of various enzymes or other volatiles react with the biological nucleophiles of target nematodes (Avato et al. 2013). Charles et al. 2015 described the usefulness of formulations of *Brassica* against root-knot nematodes (*M. javanica*) in tomato crops. A glasshouse trial was conducted for evaluating the antinemic potential of varied sources of glucosinolate (radish, mustard, and cabbage) and *Brassica*-based formulations (cake, extract, and unmacerated) in decreasing population of *M. javanica* on tomato. The results revealed that mustard was the most influential brassica for managing nematodes, while other plant species cabbage and radish also reduced the population of *M. javanica* as compared to untreated control.

The higher plants constitute an important class of secondary metabolites known as terpenoids, which exhibit a broad spectrum of biological activities (Ohri and Pannu 2009). Terpenoids are a large group of compounds derived from the combination of several isoprene units. These volatile plant extracts are extracted through the hydro-distillation process. The volatile characteristic of the family poses them to act as attractants, which are explored for their wide application in the agricultural industry. These differ from each other through a varied number of functional groups present on them and carbon skeleton. These are classified into hemiterpenes, monoterpenes, sesquiterpenes, diterpenes, sesterpenes, triterpenes, tetraterpenes, and polyterpenes (Kaur et al. 2019). Isoterpenoids are the subclass of the prenylipids, especially terpenes, sterols, and phenylquinones derived from isoprene units. In the simple form of terpenoids, rishitin (phytoalexin from *Solanum tuberosum*), α-humulene (*Pinus massoniana*), odaracin (*Daphne odora*), aldehydes hemigossypol, and 6-methoxyhemigossypol (*Gossypium hirsutum*) were found to be highly nematotoxic as these induce resistance against *Ditylenchus dipsaci*, *Bursaphelenchus xylophilus, Aphelenchoides besseyi*, and *M. incognita*. Others include the alantolactone, solavetivone, odoratrin, and humulene which induce resistance against various pathogenic nematodes (Khalil 2014). Terpenoids generally induce the genotoxic effect which in turn activates the octapaminergic receptors (Enan

2001, Kostyukovsky et al. 2002) and interfere with the GABA receptors (Priestley et al. 2003) of the target organisms. Terpenoids being lipophilic in nature readily collaborates with the cytoplasmic membrane of the microorganisms and ultimately disrupts the structures of fatty acids, phospholipids, and depolarize the membrane of the mitochondria (Bakkali et al. 2008).

Monoterpenoids are formed by only two isoprene units having ten carbon chains, which are active components of 90% of the essential oils received from plants. Nematicidal activity of 22 monoterpenoids was evaluated against root-knot nematode *M. incognita* by Echeverrigaray et al. (2010). These prominently reduced egg hatching and induced J_2 mortality. Geraniol, carveol, citronellol, α-terpineol, borneol, citral, and terpinene-4-ol (Figure 1) were found to be the most effective monoterpenoids as efficient nematicidal agents (Kong et al. 2007). Geraniol is reported active against the juveniles of nematodes. These monoterpenoids decreased galling at even 100 mg/kg of dose. The studies have shown that the presence of hydroxyl and carbonyl groups is very essential for the compound to exhibit nematicidal character, whereas the presence of acetyl-esters and cyclic ethers decreases the effect.

Abdel-Rahman et al. (2013) studied the effect of 34 terpenoids on free-living nematode *C. elegans*. It was observed that the terpenoids bearing hydroxyl, carbonyl, and phenolic groups act as potent nematicidal agents, e.g., carvacrol, pseudoionone, eugenol and limonene (Figure 2) as compared to other compounds deprived of these functional groups. Among these, eugenol was reported to be effective against juveniles of *M. javanica*, *Tylenchus semipenetrans* and limonene was reported to inhibit reproduction of *M. incognita*.

The monoterpenoids, such as carvacrol and thymol, were reported to be very toxic against soil saprophytic nematode *C. elegans* with 100% mortality, even at 250 µg/ml of tested concentration followed by (-)-perillaldehyde, eugenol, geraniol, and menthol. The monoterpenoids are reported to exhibit more potent activity than

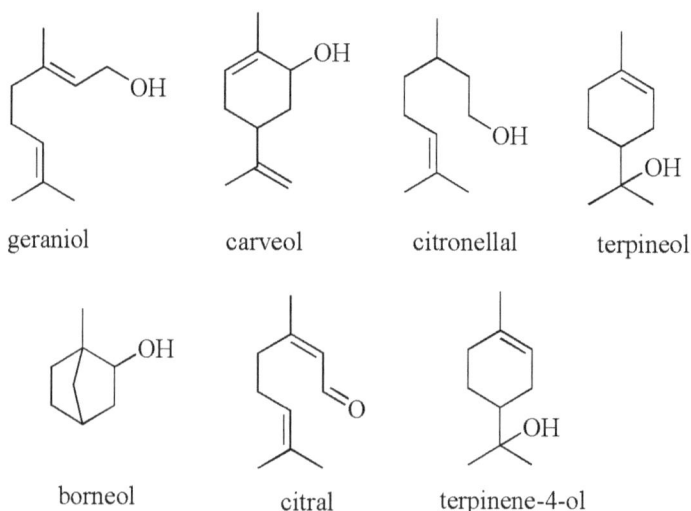

Figure 1. Monoterpenoids as nematicidal agents.

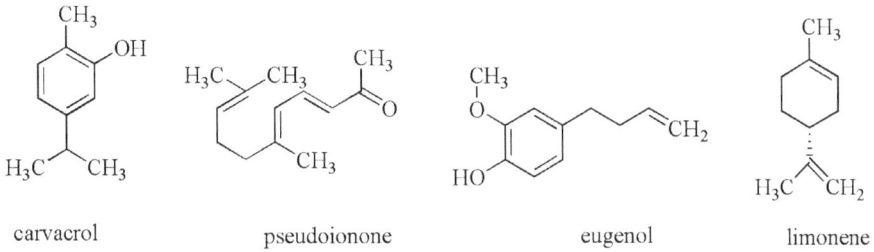

Figure 2. Terpenoids as nematicidal agents.

the oxamyl with 13.4% mortality even at a lower concentration of 250 µg/ml (Tsao and Yu 2011).

The five compounds from a perennial herb from *Peperomia japonica*-Pepermianone, Peperomialin, Peperomialin acetate, 4-hydroxy-2-[(3,4-methylenedioxyphenyl)undecanoyl]cyclohexane-1,3-dione and Proctorione had been isolated and evaluated for their nematicidal potential against *C. elegans* and *Heligmosomoides polygyrus* (Figure 3). The compounds were obtained through silica gel column chromatography and HPLC separation in methanol extract. Pepermianone and Proctoroine were found to be effective against larval stage 1 but not against infective larval stage 3 and showed LC_{50} value ranges from 10.1–16.7 µg/ml, respectively. The nematicidal potential of compounds is contributed by the presence of 1,3-benzodioxole, keto group, and alkyl chain. The presence of the triketone group connected to the alkyl chain positively affects the antiparasitic potential of the compound. But the remaining compounds had shown antinemic potential, even at 50 µg/ml of concentration.

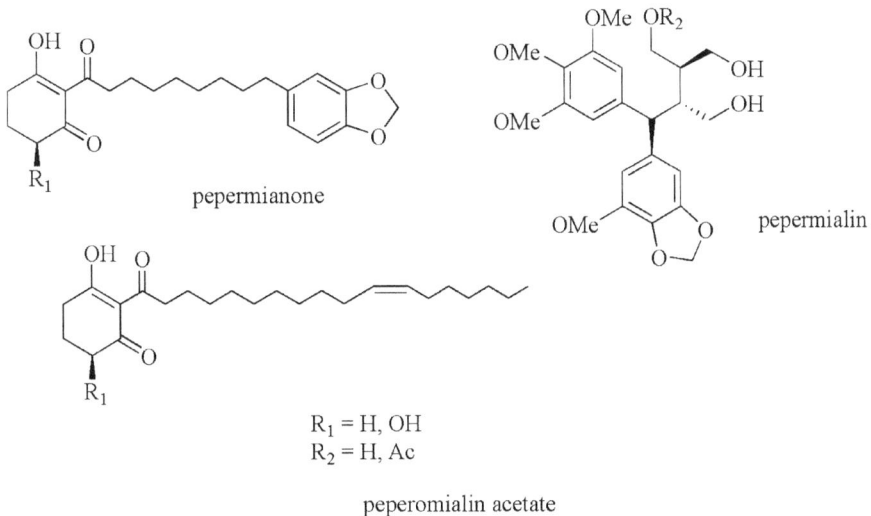

pepermianone

pepermialin

peperomialin acetate

R_1 = H, OH
R_2 = H, Ac

Figure 3. *Peperomia japonica* active compounds.

Fumaria parviflora (Fumariaceae) is a small, branched annual herb particularly found in wheat fields. The active components of *F. parviflora* are alkaloids, such as adlumidicien, copticine, fumariline, perfumine, protopine, fumaranine, fumaritine, paprarine, glycosides, saponins, steroids, flavonoids, and triterpenoids. The various extracts of stem and root were prepared in hexane, ethylacetate, chloroform, and methanol were studied for their antinemic potential. The root and stem extracts of *F. parviflora* possess a strong nematicidal character due to the presence of additional components flavonoids and steroids than other parts of the plant. All the extracts are reported to exhibit significant activity. But the *n*-hexane extract of root and stem exhibited 100% second stage of juvenile mortality followed by ethylacetate, which helps in the reduction of disease parameters and increases plant growth parameters because of tannins and glycosides that produce a synergistic effect (Naz et al. 2012). Tannins effectively bind to the protein as they bind glycoprotein enriched cuticles of the larvae leading to death. Another mode of action of *F. parviflora* bionematicide is by inhibition of formation of DNA by intercalating with it (Maqbool et al. 2004).

Caper (*Capparis spinosa* L.) is a deciduous spice bush found widely in the Mediterranean and West Asia. The major chemical classes of Caper, include alkaloids, indoles, lipids, flavonoids, and aliphatic glucosinolates. The hydrolysis of aliphatic glucosinolates by myrosinase produces isothiocyanates, nitriles, thiocyanates, glucose, and sulfate ion which imparts fumigant activity to the compound. The leaves, stem and buds' methanolic extract of the spice bush were studied against second-stage juveniles of *M. incognita*. It was reported that methanolic extract of the stem is more effective with EC_{50} 215 mg/l due to paralysis in *M. incognita* and fumigant activity (Caboni et al. 2012). The methanolic extract of stem comprised of methylisothiocyanate, furfural, 2-thiophenecarboxaldehyde, fosthiazte, 2-thiophenecarboxaldehyde oxime, 2-thiophenecarbonitrile, 2,5-thiophenedicarboxyaldehyde, 5-formylthiphene-2-carbonitrile, 2,5-thiophenedintrile, pyrrole-2-carboxaldehyde, N-methyl-2-pyrrolecarboxaldehyde, 2-pyrrolidone, indole-3-carboxaldehyde, indole-3-carbonitrile, 1*H*-indole-2,3-dione, 2-pyridinecarboxaldehyde, and 3-pyridinecarboxaldehyde (Figure 4). Working on the paralytic activity, it was observed that heteroaromatic aldehydes induce death of nematode in straight shape (Caboni et al. 2012), organophosphorus fosthiazate induce death in a coil shape, methanolic extract of clove in a straight or bent shape, and levamisole in semicircular or coil shape, respectively (Wiratno et al. 2009, Kong et al. 2006). The presence of high aromatic scaffolds with more than 2 hydrogen bond acceptors contributes toward the enhancement of nematicidal character, while unsymmetrical bond geometry, weak hydrogen bond acceptor, reduced electrophilicity, and different oxidation states contribute to reduced activity. The heteroaromatic aldehydes found in caper are highly reactive due to the presence of carbonyl functional group, which can easily react with primary amino groups and thiol groups that form products like Schiff bases hemithioacetals. These products through nucleophilic addition reactions damage the cuticle of the nematode and result in internal fluid leakage of nematode and death. Due to the structural similarity of heteroatomic aldehydes and avertin B, these are reported to be a potent inhibitors of vacuolar ATPase enzymes.

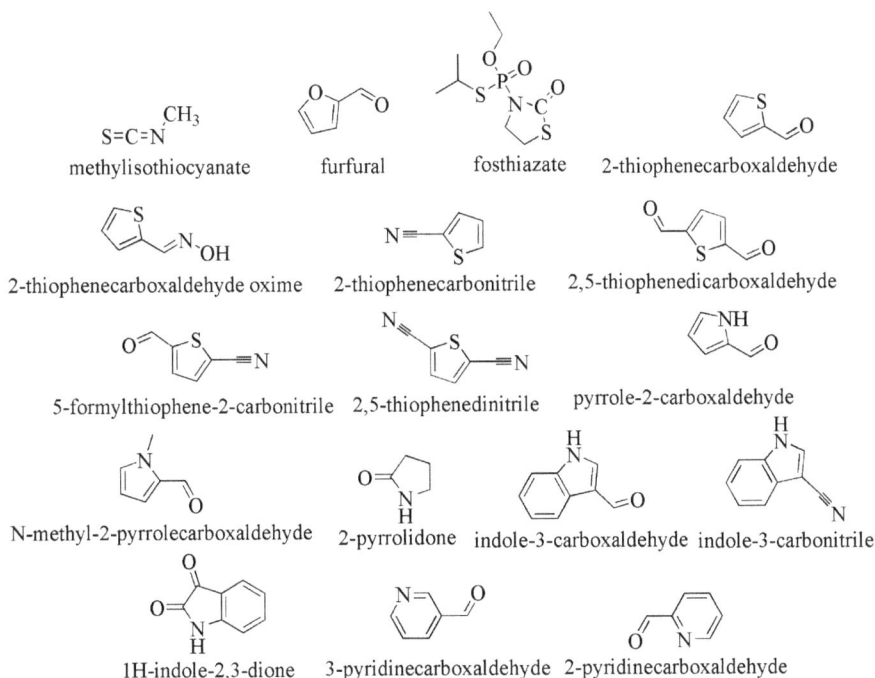

Figure 4. Caper methanol extracts as nematicidal agents.

Plant volatiles, such as aldehydes, and alcohols with short-chain are secreted by plants in response to wounds and insect attacks. These effectively trigger antimicrobial and insecticidal properties. The *trans*-2-Hexenal is an important volatile component found in fruits and vegetable tissues (Wang et al. 2015). It is reported with a 100% mortality rate in *M. incognita* and *Heterodera avenae* (Miao et al. 2012). The *in vitro* study of volatile by fumigation and aqueous method revealed that in the aqueous phase nematicidal potential of volatile component decreased as the rate with which it enters in nematode becomes equal to the diffusion rate of oxygen. Therefore, it acts as a potential soil fumigant for *M. incognita* treatment as its efficacy is compared to dazomet and abamectin. Apart from this at the dose of 250–500 l/ha, it significantly enhanced plant height, stem diameter, inhibit mycelia growth, induces resistance-related enzymes, enhances methyl jasmonate response, accumulates antitoxin and lignin in tomato and *Eruca sativa* (Aissani et al. 2015). It can penetrate strongly in soil and nematode and can be used as a substitute for traditional fumigants chloropicrin and sulfuryl fluoride (Lu et al. 2017).

The aldehydes and ketones derived from the plant materials were found to be effective fumigants with potential nematicidal capability as they easily identify the target in the soil. Among these plant materials, species belonging to the Ruta family such as *Ruta graveolens* L. and *Ruta chalepensis* are perennial herbs found in the Mediterranean area. The herb is embedded with volatile sulfur components, ketones, flavonoids, amino acids, saponins, alkaloids, phenols, and furocoumarins (Mejrib et al. 2010, Escher et al. 2006). These are reported to exhibit insecticidal, bactericidal, fungicidal, anthelmintic activity with no detrimental effects, and these were found to affect the central nervous system. The nematicidal activity of the essential oil and methanolic extract of *R. chalepensis* were evaluated against *M. incognita*, *M. javanica* (Ntalli et al. 2011), *Panagrellus redivivus*, and *B. xylophilus* (Gu et al. 2007). The study revealed that the essential oil of tested Ruta species induce paralysis in both tested nematodal species (EC_{50} = 77.5 and 107.3 mg/l), but the methanol extract was only selective against *M. incognita* species (EC_{50} = 1001 mg/l). The essential oils being highly concentrated in nature were found to be composed of ketones-2-nonanone, 2-nonanol, 2-decanone, octyl acetate, 2-undecanone, 2-dodecanone, 2-tridecanone, which exhibited potential nematode paralysis activity. The 2-Undecanone was found to be a highly active ketone derivative that exhibited potential nematicidal activity. Apart from this, it is an active constituent of plant growth promoting bacteria such as *Bacillus megaterium* (Huang et al. 2010) and acts as a repellant (Antonious and Snyder 2006).

El-Nagdi and Youseff (2013) conducted the nematicidal activity of the garlic clove's aqueous extract against *M. incognita*, infecting the crop of tomato. The aqueous extract decreased the galling and the number of egg masses on the roots of tomato, and root and soil nematode populations as compared to plants populated with nematode species. Abd Elgawad et al. (2009) revealed that treatment of soil with the commercial products that contain the aqueous garlic extract declined the root galling index of nematode and enhanced the activity of β-1,3-glucanase, catalase, and enzyme of leaves of tomato as compared to nematode-infected plants. Osman et al. (2005) studied that decrease in nematode population with cloves of garlic is due to the presence and effect of pyruvic acid, ammonia as well as diallyl disulfide. In addition to this, Nigh 1985 reported the properties of garlic that it possesses several biochemical substances as well as the allelopathic structures, which are highly toxic to nematode species. Ameen (1996) in a study reported that the soil population of *Rotylenchulus reniformis* and *M. incognita* were significantly reduced in monoculture of garlic or when intercropping of garlic was done with cowpea or tomato. It was revealed that the reduction in nematode population is due to the allelopathic effect of garlic or that garlic did not provide essential elements for nematode development.

Qamar et al. (2005) reported the presence of several acids viz-a-viz camaric acid, lantanilic acid, and oleanolic acid obtained from the methanolic extract of aerial parts of *Lantana camara* and caused J_2 mortality in *M. incognita*. Begum et al. (2015) reported a nematicidal assay of several triterpenes isolated from *Lantana camara*, such as lancamarolide, oleanonic acid, lantadene A, 11-α-hydroxy-3-oxours-12-en-28-oic acid, betulinic acid, lantadene B, and lantaninilic acid. These triterpenes were found in aerial parts and were assessed against *M. incognita*. The most effective compound was found to be oleanonic acid which exhibited mortality of around 80% after 72 hours at the concentration of 0.0625%, which was found to be comparable to a standard used Furadan. The leaf extracts of *L. camara* acts as nemostatic in nature as they killed 96% and 75% of J_2 at S and S/2 concentrations, respectively (Ahmad et al. 2010).

Several natural compounds such as glycoside (asparagusic acid) have been reported with suppressive against *Meloidogyne* spp. Glycoside is obtained from

Asparagus officinalis (Chitwood 2002). Another includes nonacosane-10-ol and 23a-homostigmast-5-en-3b-ol extracted from the roots of *F. parviflora* (Claudius-Cole 2010). Glycoside is also identified from *Ageratum conyzoides* (billygoat-weed, chick weed, goatweed, and whiteweed) which is detrimental to *M. incognita* (Khan et al. 2017). One of the essential components of plant secondary metabolites exhibiting biocidal activity are cyanogenic glucoside (Vetter 2000). In *Sorghum bicolor*, dhurrin is the most important cyanogenic glucoside which is found in shoot tips of seedlings. This compound is made from the L-tyrosine (amino acid) catalyzed by enzymes cytochrome P450 present in the cell membrane and soluble UDP (uridine diphosphate)-glucosyltransferase (Laursen et al. 2016). Like the GSL-MYR system in the case of cruciferous plants intact tissues, substrate and enzymes are located in separate cells, as in dhurrin in the vacuole of the epidermal cell and catabolic enzymes in mesophyll cells. Due to biotic invasion, there is disruption of cellular integrity and then hydrolysis of dhurrin is carried out by the endogenous β–D-glucoside glucohydrolase (dhurrinase) for the liberation of glucose and then there is a production of *p*-hydroxymandelonitrile, which is highly unstable and immediately converted to toxin hydrogen cyanide (HCN) and *p*-hydroxybenzaldehyde by the action of an enzyme α-hydroxynitrile lyase or at basic pH (De Nicola et al. 2011). With an increase in plant age, the concentration of dhurrin decreases (Bolarinwa et al. 2016). Various crops like clover, flax and sudangrass (*Sorghum sudanense*) produce nematotoxic amino acid cyanogenic glycosides (Widmer and Abawi 2000). Cyanogenic glycosides release toxic hydrogen cyanide, which protects plants from herbivores. Cassava roots, which produce linamarin as cyanogenic glycoside, were used typically as antinemic agents in Brazil by the release of Manipueira (Chitwood 2002).

The sustainable eco-friendly and toxic potential of neem (*Azadirachta indica*) has been widely studied as a nematicide, insecticide, germicidal, and pharmacological agent. The different products of neem, like crude extracts, oils, cakes, and dry powders, derived from leaves, roots, barks, fruits, and seeds contain various bioactive compounds and are reported with a distinct pharmacological profile. The different chemical constituents of neem are nimbin, nimbidin, azadirachtin, salannin, thionemon, meliantroil, nimbanene, 6-desacetylnimbinene, nimbandoil, nimbolide, ascorbic acid, *n*-hexacosanol, 7-desacetyl-7-benzoylazadiradione, 7-desacetyl-7-benzoylgedunin, 17-hydroxyazadiradione, and nimbiol. The study conducted on aerial parts of neem revealed that the standard and undiluted neem extracts can exhibit a lethal effect on the J_2 and egg hatching (more than 80%) as they efficiently reduced gall production. But the alcoholic and aqueous extracts of neem are very less effective. Apart from nematicidal character, these extracts significantly increase root length and enhance the activity of enzymes viz. Phenylalanine ammonia lyase (PAL), Polyphenol oxidase (PPO), and Peroxidase (POX) in vegetable crops (Nile et al. 2018). It was reported that extracts of neem leaf completely inhibited hatching of egg masses of *M. incognita* and were lethal to larvae (Agbenin et al. 2005). Haroon et al. (2018) reported the active chemical compounds found in neem extract and identified it as the best effective extract in inhibition of egg hatching are alkaloids, flavonoids, saponins, and amides including benzamide and ketones.

Bhattacharyya (2017) reported marigold (*Tagetes* spp.) as an exceptional plant for the management of root-knot nematode due to the presence of polyacetylenes and polythienyls (Wat et al. 1981). Marigold releases bioactive chemicals from root exudates that are effective in managing the nematode population. Hence, marigold usage in nematode control is an eco-friendly and cost-effective method. It has been reported to contain certain compounds in a synthetic form viz. 5-(3-buten-1-ynyl)-2,2-bithienyl, and α-terthienyl which act as nematode suppressant. The presence of flavonoids, flavones, di-hydroflavonoid, and flavones lacking a free OH group has been reported in the roots of marigold plants. In addition, roots also have been mentioned to restrain various chemicals viz. phenols, amines, amides, and ketones. When marigold is sown as an intercrop with susceptible hosts of nematodes, it acts as nematode suppressants. In the case of pot trials, marigold plants and test host plants remain near each other and thus allow the leachates of root from marigold to reach the root zone of the nematode host plant. The marigold's nematicidal activity has been reported in growing plants' roots but not in extracts of leaves or roots. In marigold roots, there is a sequence of several events which are triggered by the penetration and movement of PPNs through root vascular tissue and the final products of these reactions have a nematicidal effect. The nematicidal compounds immediately permeate from the root tissues of the marigold into roots attached to the nematode. However, they are also supposed to suppress oviposition and nematode egg hatching found in the rhizosphere. It has been studied that α-terthienyl was a major bioactive compound released by roots of marigold. In addition to this compound, another compound, i.e., bithienyl compounds present in aerial parts of *Tagetes* sp. are extremely bioactive in nature. El-Gengaihi et al. (2001) extracted three nematicidal compounds with the help of chloroform from *Tagetes patula, Tagetes erecta*, and *Tagetes minuta*. These compounds contained sigma-4, 22-dien-3-b-ol, 5(-ent-1-ol)-2,2-bithienyl and 5-(4-acetoxy-1-butenyl)-2,2-bithienyl. The α-terthienyl is a sulfur-containing heterocyclic compound that is plentiful in *Tagetes* tissue. The biological activity is triggered by photoactivation with near UV light, leading to the production of singlet active oxygen molecules which is toxic in nature.

Hamaguchi et al. (2020) reported nematicidal activities of the marigold exudate α-terthienyl oxidative stress-inducing compound against nematodes. Related to the activity of α-terthienyl, various mechanisms have been reported. The oxygen-dependent phototoxicity of α-terthienyl is reported to generate single oxygen and superoxide anion radicals (Nivsarkar et al. 2001). These reactive oxygen species (ROS) *in vivo* behave as substrates for the detoxification enzymes catalases (CTLs) and superoxide dismutases (SODs). In addition to this, photoactivated α-terthienyl (PAT) exhibited significant concentration-dependent ROS-induction activity in lepidopteran ovarian Tn5B1-4 and Sf-21 cells, which declined the activity of peroxidase (POD), SOD, and CTL (Huang et al. 2017). Biomolecules that are oxidized by ROS become harmful substances, Phase II metabolism that employs these enzymes as glutathione S-transferases (GSTs), and UDP-glucuronosyl transferases (UGTs) (Lindblom and Dodd 2006) is believed to be of utmost importance for resistance to α-terthienyl. In the free-living model nematode, *Caenorhabditis elegans*, there are oxidative stress-related enzymes that are majorly controlled by the cap 'n' collar transcription factor SKN-1, which is a functional and structural homolog of the mammalian Nef2. SKN-1 activity in such stress responses is negatively regulated by the WD40 repeat protein and WDR-23 (Choe et al. 2009).

Liu et al. (2016) performed a screening experiment (Program) for new agrochemicals, which were taken from Chinese medicinal herbs. The medicinal herb was ethanol extract of rhizomes of *Notopterygium incisum* which was reported to exhibit nematicidal activity against the two nematode species, *M. incognita*, and *Bursaphelenchus xylophilus*. Based on bioactivity-guided fractionation of ethanol extract, isolation of the four constituents was done which were identified as falcarinol, columbianetin, falcarindiol, and isoimperatorin. Amongst the four isolated components, two acetylenic compounds, falcarinol, and falcarindiol (2.20–12.60 µg/ml and 1.06–4.96 µg/ml, respectively) showed stronger nematicidal activity than two columbianetin, furanocoumarins, and isoimperatorin (21.83–103.44 µg/ml and 17.21–30.91 µg/ml, respectively) against the two nematode species, *B. xylophilus*, and *M. incognita*. The four constituents which were isolated also exhibited phototoxic activity against the nematodes. The results showed that *N. incisum's* ethanol extract and its isolated constituents (four) can be developed as natural nematicides for plant-parasitic nematode's control.

Naturally occurring polyacetylenes (derived from *Carthamus tinctorius*) have also been reported as antinemic agents against rice white tip nematode (*Aphelencoides besseyi*) (Kogiso et al. 1976). In addition to this, 14 polyacetylenes or their thiophene derivatives, which were isolated from Asteraceae spp. were reported to exhibit toxic effects against adult nematodes (*C. elegans*) and toxicity of these compounds was reported to be increased more by irradiation with near-UV radiation or natural sunlight (Wat et al. 1981). However, their studies with falcarindiol exhibited no photoactivation with UV_A radiation (Eckenbach et al. 1993). Moreover, both these constituents, i.e., falcarinol and falcarindiol have been reported as antifungal compounds in several Apiaceae plant species, decreasing germination of spores of several fungi. Several activities like antibacterial, anti-*Candida* and antimycobacterial have been shown by Falcarinol and falcarindiol with an ability to kill *Mycobacterium tuberculosis* and isoniazid-resistant *Mycobacterium avium* at 10 µg/disc in a disc diffusion assay. However, this was the first most important report of nematicidal activity and photoactivation with UV treatment of falcarinol and falcarindiol against the two nematode species. The mode of action of falcarinol and falcarindiol's against nematodes was not investigated; however, the process for the antifungal mechanism of falcarinol and falcarindiol is thought to involve disruption of cell membranes (Garrod et al. 1979). Thus, the lipophilic falcarinol compound was more toxic than falcarindiol against the two nematode species, i.e., the more polar acetylene. It was also supposed that the insecticidal action of falcarinol may be associated with the GABAergic block which is related to the higher intake expected in herbivorous insects because $GABA_A$ receptors are important targets of neuroactive pesticides (Czyzewska et al. 2014, Kuriyama et al. 2005).

The *Artemisia* spp. widely known as Mugwort from family Asteraceae comprised of a wide number of biodynamic compounds and secondary metabolites (Saadali et al. 2001). Species of *Artemisia* such as *Artemisia vulgaris, Artemisia elegantissima*, and *Artemisia incisa* are reported with nematicidal activity against *M. incognita* as used under integrated pest management programme. It was reported that species reduced the galling index when applied in a proper dose-dependent way (Costa et al. 2003). These are highly toxic to second-stage juveniles as they caused mortality within 24 hours of application (Dias et al. 2000, Shakil et al. 2004). The phytochemicals of this family such as isoscopletin and apigenin showed more than 90% mortality of juveniles at 0.3 mg/ml of concentration. There was a significant reduction in the number of galls, galling index, and egg masses formed but this positively provided boost to the various plant growth parameters (Khan et al. 2019). The -OH functional group present at 6th and 7th position and carbonyl group in the structure of isoscopletin induce the highest mortality. The hydroxyl group is reported with a varied number of biological activities, such as termiticidal, cytotoxic, and bactericidal activities (Takaishi et al. 2008, Adfa et al. 2012).

El-Ansary and Al-Saman (2018) studied that crude proteins present in the seeds of *Moringa oleifera* could be used as a safe and effective nematicide which are eco-friendly and safer to use for farmers and other users. At room temperature, both 50% and 60% concentrations of crude protein of ammonium sulfate were selected for detection of activity of lectin in the parts with the help of the agglutination technique. The trypsinized cattle erythrocytes were placed on a glass slide of a microscope after dialysis against PBS instead of a lectin. Lectins belong to the class of glycoproteins that are abundantly found in the seeds of the plant, agglutinate blood erythrocytes, and these exhibit a significant role in the defense system of plants prone to various pathogenic microorganisms, including insects and pests. Phytolectins are reported with their antinemic activity against the control root-knot nematodes (Marban-Mendoza et al. 1987, Al-Saman et al. 2015). Moreover, Santos et al. (2009) reported the presence of lectins in *M. oleifera* seed extract. Lectins were classified into two categories based on affinity support used and extraction solvent, i.e., water-soluble *M. oleifera* lectin (WSMoL) and coagulant *M. oleifera* lectin (cMoL). WASMoL is a carbohydrate binder recognized as D (+)-fructose and N acetylglucosamine (Coelho et al. 2009) and cMoL from saline solvent is recognized except for D (+)-fructose. Salles et al. (2014) reported that the antinemic activity of seeds of *M. oleifera* includes different biomolecules, particularly low molecular weight molecules.

The increase in growth parameters (plant length, leaf numbers, and number and weight of fruits), higher accumulation of dry matter, less number of galls, and decreased nematode population have been reported when the plants were treated with powder of cocoa bean testa. The application of powder of cocoa bean testa as a mulch has been reported toxic to nematodes juveniles. The applicated powder had shown its direct effect on the second stage juveniles residing in the soil, thus diminishing the motile juvenile's number which can penetrate the roots of tomato plants (Ojo et al. 2013). Similarly, Agbenin et al. (2005) conducted a study on tomatoes and used botanicals for the management of *M. incognita*. The presence of a large number of phytochemicals, like flavonoids, saponins and glycosides, may be responsible for the nematicidal property of this extract or oxygenated compounds which have lipophilic properties due to which they can solubilize the cells cytoplasmic membrane and their functional groups obstructing with enzyme-protein structures of the target nematodes (Knobloch et al. 1989). The result obtained exhibited that the Cocoa bean testa contained these phytochemicals.

Seed and leaf extracts of *Ricinus communis* (Castor) has an impact on root-knot nematode population, root galling index, and nematode reproduction factor as well as increased growth parameters and crop yield of root-knot nematode infected tomato plants in the screen house (Oluwatayo et al. 2019). Biodegradation of castor's oil cakes has been reported as an effective measure to control of *M. incognita* population (Tiyagi et al. 2002). Seeds of several species of plants had been reported to exhibit antinemic potential (Khurma and Kumari 1996, Khurma and Singh 1997, Khurma and Chaudhry 1999). Another study revealed that the seeds of Amaltas and Castor were found to be effective and has been reported to inhibit the hatching of eggs of root-knot nematodes (*M. incognita*). The aqueous extract has reduced egg hatching significantly and larval mortality of root-knot nematode at higher concentrations of aqueous extract had increased its potency. The phytochemical screening of Castor and Amaltas exhibited the occurrence of steroids, terpenoids, protein, tannin, coumarins, phenols, saponins, carbohydrates, and flavonoids in the methanol residue of these seed extracts, and phlobatannins were missing in both the seed extracts. Alkaloids, carbohydrates, tannins, and flavonoids constituents display remedial activity against various diseases and rationalize their use as conventional medicine.

Mainoo and Banful (2019) studied the population of *Scutellonema* spp. under mulches of *Chromolaena odorata* (Siam weed) and *M. oleifera* (Moringa) did not change over the same period of cropping and application. They concluded that despite cropping of yam, *C. odorata* and *M. oleifera* mulches were able to prevent the *Scutellonema* spp. multiplication, which were the most damaging species that causes huge loss to tubers of yam in terms of quality and quantity. These mulches can act as an efficient good biocontrol agent for the management of *Scutellonema* spp. of nematode. This was the first study reported of nematicidal properties of mulch of *M. oleifera* for the elimination of *Scutellonema* spp. In the case of *C. odorata*, by corroborating the findings of previous studies, it was concluded that *C. odorata's* direct application either as a mulch or as in natural fallows, decreased the *Scutellonema* spp. population (Ajith and Sheela 1996, Adekunle and Fawole 2003, Adediran et al. 2005). The mechanism behind the control of nematodes was the combined activity of amides, flavonoids, alkaloids, saponins, amides, and ketones which are released during the mulch's material decomposition (Thoden et al. 2007, Akpheokhai et al. 2012). Many researchers Adekunle and Fawole (2003) and Adegbite (2005) in their separate reports concluded that *C. odorata* contain flavonoids and alkaloids and other phytochemicals, which exhibited nematicidal properties on it. In addition, Umar and Mamman (2014) had reported that the tannins and saponins in *C. odorata* were responsible for egg hatch inhibition of such nematodes. Adding to this, when the C:N ratio of the amendment is < 20:1, there is more effect on nematodes (Stirling 1991). In a study by Mainoo and Banful (2019), *M. oleifera* and *C. odorata* mulches have C/N ratios < 20:1. Higher yields obtained under *M. oleifera* and *C. odorata* was due to the high nutrient content of leaves that were produced during decomposition in synchrony with yam tuberization.

Ogwudire et al. (2019) studied the effect of root extracts of *Jatropha curcas* L. for controlling root-knot nematode (*M. incognita*). Phytochemical root extracts of *J. curcas* caused *M. incognita* J_2 mortality as compared to control. When saponins, tannins, alkaloids, and flavonoids were applied, the highest rate of mortality was reported (P > 0.05). The mechanism of action of the phytochemicals could be due to their cytotoxicity (Matsuhashi et al. 2002, Kuljanabhagavad and Wink 2009). There was a significant (P < 0.05) increase in juvenile mortality with increased rates of phytochemical treatment applied. The highest mortality rate with the application of alkaloids (98.9%), saponins (96.1%), flavonoids (94.73%), and tannin (94.17%) occurred when applied at 10 ml after an incubation period of 72 hours. This is in confirmation with similar studies revealed by other scientists that there is an increase in juvenile mortality with an increase in doses or rates of phytochemical extracts (Sharma and Prasad 1995, Hussein et al. 2016, Khan et al. 2017). Gamal et al. (2008) also revealed mortality of around 80–94% juveniles of *M. incognita* after 72 hours of exposure when methanol extracts of plants were applied. Mortality of J_2 increased with increased exposure hours of phytochemicals.

Ismail (2014) screened Jatropha species and reported suppressiveness of development and reproduction in *M. javanica*. Several jatropha plants significantly decreased nematode density on roots and nematode population recorded from soil and roots of the host. Moreover, a positive correlation was reported between the number of both species of jatropha and the percentage decline in the final nematode population as well as the present reduction in RGI due to the root-knot nematodes. Therefore, the application of four plants of jatropha resulted in the highest reduction in root galling index. *J. curcas* was reported with 83% and *Jatropha gossypiifolia* with 72%, respectively. The study revealed that the population of *M. javanica* increased around sunflower alone, but there was a significant decrease in nematode population around jatropha seedlings grown in an intercultural operation with sunflower (1–4 *Jatropha curcas* or *J. gossypiifolia*/pot). The build-up of nematode also decreased. The toxic character of the root contributes to the reduction in the final nematode population and build-up rate by growing *J. gossypiifolia* or *J. curcas*. These results are in conformity, as reported by Korayem and Osman (1992), Claudius et al. (2010), Umeh and Ndana (2010), Onyeke and Akueshi (2012), and Ismail (2013). The studies revealed that the plants, such as *J. curcas* or *J. gossypiifolia*, decrease the population of plant-parasitic nematodes through the production of nematotoxins in the soil environment, which are not found to be phytotoxic to the plants and contributes to the growth parameters of the plant. Fassuliotis 1979 reported two categories of resistance. The first one is pre-infectional resistance, which operates before the entry of nematodes into the root surface. The second was the post-infectional resistance which operated after penetration of nematode in the tissues of the infected plant. The results revealed that the jatropha plants exhibited have two different types of resistance; the first is to the larvae of *M. javanica* larva, which due to resistance possessed by the plant in combination with sunflower roots, and the second is larvae become unable to penetrate the host plant. Another study carried out by Umeh and Ndana (2010) revealed that the *M. incognita* is not able to reproduce or multiply its generations in the presence of both *J. gossypiifolia/J. curcas*. In general, these nematicidal phytochemicals are eco-friendly and safe. These substances constitute attractants, repellents, egg hatch inhibitors/stimulants, nematotoxicants, etc., formed in opposite and equal action of nematode presence (Chitwood 2002).

Hinmikaiye et al. (2016) studied the effect of *J. curcas* for management of root-knot nematode, *Meloidogyne* spp. They studied that bioactive compounds present in the test plant, saponin, tannin, alkaloids, flavonoids, and anthraquinone could have been the reason for *J. curcas* efficacy in managing pepper affected by root-knot nematode; this is in confirmation with the results of Deverall (1972) on astringency of pesticides and Phyto-repellant nature of flavonoids and other bioactive compounds present in plants.

Khan et al. (2017) studied all the aqueous extracts of weeds viz., *Ageratum conyzoides, Ipomoea carnea, Pontederia crassipes, Nicotiana plumbaginifolia,* and *Trianthema portulacastrum*, which efficiently inhibited egg hatching and elevated the mortality of second-stage juveniles *M. incognita*. But the aqueous extract of *A. conyzoides* (billygoat-weed, chick weed, goatweed, and whiteweed) was found to be most active against *M. incognita*. The aqueous extract of weed was found to be embedded with antinemic potential as studied against *M. incognita* due to the presence of phytochemicals. These include saponins, alkaloids, tannins, flavonoids, and terpenoids. *A. conyzoides* also showed antibacterial as well nematicidal activity due to the huge presence of various phytocompounds (Phadungkit et al. 2012), including tannins, saponins, alkaloids, flavonoids, and phenol.

Among phytochemicals identified in water hyacinth as *n*-hexadecanoic, palmitic acid, and ethyl ester have pesticide, nematicidal, and antifungal properties (Upgade and Anusha 2013). Phytochemical analysis showed that plants are rich in terpenoids, alkaloids, flavonoids, and phenols which have the highest nematicidal activity (Pavela 2004). Umar and Mohammed (2013) studied the efficacy of water hyacinth leaf extract of Solms on the J_2 mortality of *M. incognita*. Results revealed that there is 100% juvenile mortality by crude extracts. Lata and Dubey (2010) reported that secondary metabolites present in shoots and rhizomes of water hyacinth like quinones, terpenoid, and flavonoids have curative activity against various pathogens. Anuja and Satyawati (2007) also studied extracts of plants that contained alkaloids, phenolic compounds, flavonoids, and glucosides were found to be effective against *M. incognita*.

Shoots of *Datura stramonium* were evaluated against root-knot nematodes by Oplos et al. (2018) with two types of assays, i.e., Ovicidal and Larvicidal assay. Efficacy was also studied in host roots for the inhibition of nematode development in roots. The soil of the host was amended with *Solanum nigrum* seeds. *D. stramonium* was able to retard the motility of *M. javanica* and *M. incognita* at $EC_{50} = 427\,\mu g/ml$ after 72 hours of duration. Datura species are a rich sources of alkaloids, like meteloidine, atropine, nicotine, hyoscyamine, scopolamine, flavonoids, and terpenoids which have the highest rate of nematicidal activity (Pavela 2004).

The leaves of *Datura innoxia, Datura metel*, and *Brugmansia suaveolens* and their phytochemical constituents were screened for nematicidal activity by Nandakumar et al. 2017. The leaves were found to be positive for the secondary metabolites, like steroids, alkaloids, saponins, terpenoids, triterpenes, tannins, total phenolics, flavonoids, but it showed negative value for the anthraquinone glycosides and cardiac glycosides. Due to the presence of phenol functional groups as part of a constituent in leaves, these were characterized as phenolic compounds. These phenol groups have a major role to play as they behave as chemical barriers and protect plants from pathogens (Petti and Scully 2009). Alkaloids have a protective function and have a usage in medicine, especially steroidal alkaloids (Lata and Dubey 2010). These plants are good sources of histamine, serotonin, and prostaglandins, which have significance in the pharmaceutical industry as anti-inflammatory and anti-analgesic polyherbal formulations tropane alkaloids that are reported with spasmolytic properties (Capasso et al. 1997, Sharma and Sharma 2010).

The secondary metabolites produced by plants are essential for the plant defense system which is comprised of alkaloids, terpenoids, and phenols. Phenols consist of simple phenols, phenolic acids, phenylpropanoids, flavonoids, tannins, and quinone. Among the simple phenols and phenolic acids, trans-cinnamic acid, pyragallol, o-hydroxy naphthoic acid, and ethyl gallate are commendable nematicides with a 95% mortality rate. The o-Hydroxy naphthoic acid and trans-cinnamic acid are highly effective to combat egg hatching. It is reported that the phenolic compounds substituted with electron-donating groups are more active than the electron-withdrawing groups. Salicylic acid (SA) is another major phenolic compound that apart from nematicidal action contributes toward the shoot length, shoot weight, and root length. The SA significantly interferes in the gall formation process, impair egg mass production, and reduces the nematode population. Phenylpropanoid are the type of signal molecules produced during the interaction of plants and microbes which induce resistance against the nematode (Ohri and Pannu 2010). Phenolic acids with the substitution of electron-donating groups such as chloro, and phenoxyacetic esters act as potent nematicidal agents. The addition of phenols along with other oilcake in soil reduces the *M. incognita* infestation. The nematicidal potential of *Haplophyllum tuberculatum*, *Plectranthus cylindraceus* oil, *Viola betonicifolia*, and *Eucalyptus exserta* bark are all due to phenolic compounds present in them and are widely used to control *M. incognita* and *M. javanica* (Li and Xu 2012, Sahebani et al. 2011).

Semiochemicals or secondary metabolites produced by *Melia azedarach* L. or Paraiso, Chinaberry, of Meliacaceae family showed potent nematicidal activity against *M. incognita* (Aoudia et al. 2012). *M. azedarach* is reported with a higher amount of phenolic compounds, such as caffeic acid, (+)-epicatechin, and kaempferol. Fruits of *M. azedarach* showed second-stage juvenile mortality of *M. incognita* at EC_{50} value 3,400 mg/kg and in the soil these exhibited activities at 0.02 g a.i./kg comparable to fenamiphos. The botanical is reported to enhance defense mechanism of the plant by retarding the activity of catalase and peroxidase. The compounds responsible for the activity of *M. azedarach* are acetic acid, hexanoic acid, hexadecenoic acid, furfural, 5-hydroxymethylfurfural, 5-methylfurfural, and furfurol. Among these active components, furfural emerged as the most active bionematicidal compound with an EC_{50} value of 24 µg/ml. The stem bark of botanical is comprised of hexasosylferulate, tetracosyferulate, pentacosylferulate, heptacosylferulate, and octacosylferulate due to which it is reported to exhibit insecticidal and genotoxic effects (Orhan et al.

2011). The organic acids, such as acetic acid, butyric acid and hexanoic acid, were reported with significant nematicidal activity against *M. incognita* with EC_{50} values 38.3, 40.7, and 41.1 mg/l. The organic acids such as chlorogenic acid, rosmarinic acid, 3,4-dihydroxybenzoic acid, salicylic acid, β-aminobutyric acid deform the egg, induce plant defense mechanism and destroy the inner layer of the egg shell (Caboni and Ntalli 2014).

Triumfetta grandidens belongs to the Tiliaceae family is reported with several biological properties due to the presence of triterpenoids, alkaloids, triterpenes, polyols, steroids, lupeol, oleanolic acids, tormentic, heptadecanoic acid, glycosides, and β-carotene. *T. grandidens* was found to be highly active against *M. incognita* as it caused 100% mortality of second-stage juveniles even after 48 hours duration with 500 µg/ml concentrations. The compounds from waltherione E and weltherione A isolated from *T. grandidens* had shown the highest larvicidal potential with EC_{50} value 0.09 and 0.27 µg/ml and shown ovicidal potential at 1.25 µg/ml comparable to that of abamectin. The nematicidal potential of *T. grandidens* is supported with the presence of 4-quinolone alkaloids, waltherione E, and weltherione A (Figure 5), which cause acetylcholinesterase inhibition retarding nematode locomotion (Jang et al. 2014).

Some species of the genus *Bacillus*, *Pseudomonas*, *Azotobacter*, and *Paenibacillus* belongs to plant growth regulating bacteria such as rhizobacteria and helps in plant growth regulatory activity. These are found to be very toxic for second-stage juveniles (J_2). There are certain bioactive and antibiotics secreted by these species, such as lipases, phospholipases, proteases, polymyxins, fusaricidins, phytohormones, proteins, and the broad spectrum of antibiotics exhibiting a wide range of biological activities, such as antibacterial and antifungal. The plant growth regulating character of these species helps the nutrient-deficient soils in N, P, and K uptake for better yields. As reported the appropriate ratio of carbon and nitrogen effectively control nematode as organic materials with a ratio of 20:1 has a high degradation rate and antinemic character. Another set of bacteria, such as *Pseudomonas fluorescens*, *Trichoderma viride*, are widely used bio-pesticides against a wide range of crops as *P. fluorescens* bacteria produce an antagonistic effect by the release of phenazine and hydrogen cyanide. Whereas the fungus *T. viride* is nematophagous in nature as it infects the eggs and larvae of the nematode

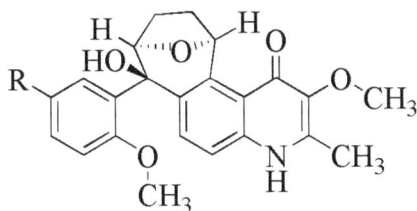

R = -OCH_3 , waltherione E
R= -H, waltherione A

Figure 5. Structure of *T. grandidens* active components.

by enhancing the chitinase and protease activity. These also protect roots from wilt diseases and enhance plant growth and absorption of nutrients (Terefe 2015).

Strains of fungus *Pochonia chlamydosporia* consist of an important metabolite called aurovertin which damage the internal structure of *Panagrellus redivivus*, inhibit Adenosine triphosphate (ATP) synthesis, and ATP-hydrolysis uncoupling oxidative phosphorylation (Niu et al. 2010). *Pseudomonas aeruginosa* also contain phenazine-pyocyanin metabolite which effectively inhibit Vacuolar ATPase (V-ATPase) (Ran et al. 2003). V-ATPase is essential for nematode nutrition, cuticle synthesis, reproduction, osmoregulation, and neurobiology (Knight and Behm 2011). The compounds isolated from ascomycetes-cochlioquinones and mycorrhizins were reported to be strongly active against saprophytic *Caenorhabditis elegans* but not towards *M. incognita*. From basidiomycetes, the cyclic dodecapeptide omphalotin was found to be selective against *M. incognita*. The compounds macrodiolides clonostachydiol and helminidiol were found to be very effective.

Basidiomycetes, such as *Pleurotus ostreatus*, *Pleurotus eryngii*, *Coprinus comatus*, *Pleurotus cornucopiae*, *Lentinula edodes*, *Neonothopanus nambi*, are another class of nematicidal agents that inhibit nematodal movement by hyphal penetration which digest the body by enzymatic action, especially myceliophagous nematodes and production of nematode-toxin. These are reported to exhibit more than 95% mortality in *Panagrellus redivivus*, *Bursaphelenchus xylophilus*, *Heterodera schantii*, and *M. arenaria*. The strong nematicidal activity is contributed by basidiomycetes is due to the presence of linoleic acid (LD_{50} = 10 and 5 µg/ml), 1-hydroxypyrene, 5-pentyl-2-furaldehyde, 5(4-pentenyl)-2-furaldehyde, omphalotin, cheimonophyllons, cheimonophyllal, 1,2-dihydroxymintlactone, whereas *p*-anisaldehyde reduces nematicidal character (Figure 6) (Subramaniyan et al. 2017).

Another important biocontrol agent is fungi such as *Chaetomium globosum* which is found on decay plants and animals and secretes secondary metabolites chaetoglobosins. Cheatoglobosins actively binds actin filaments resulting in movement inhibition and proliferation of mammalian cells. These are reported to

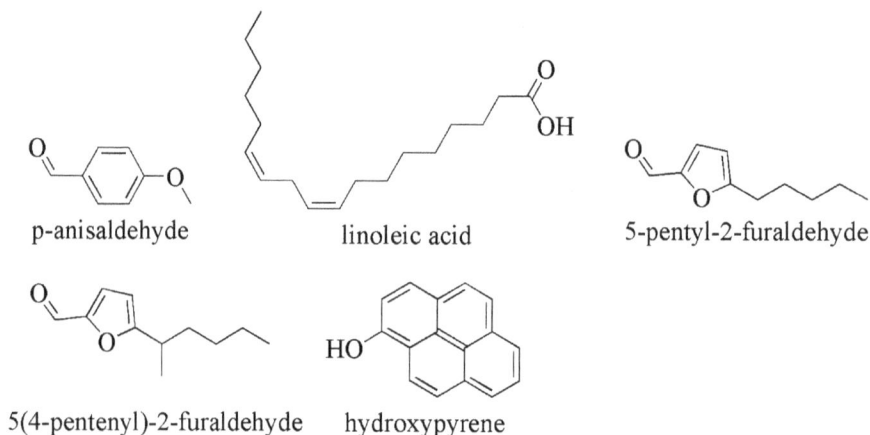

Figure 6. Active components of Basidiomycetes.

have a cytotoxic effect and are widely used as a biocontrol agent against a wide variety of plant pathogens. The culture broth of *C. globosum* is reported for egg hatch inhibition and second stage juveniles' mortality of *M. incognita* and *Heterodera glycines*. Flavipin nematicidal compounds from fungus are reported to exhibit several biological activities (Figure 7). For the integrated pest management, the nematicidal activity of *C. globosum* NK102, culture filtrates, and chaetoglobosin A (ChA) was evaluated against *M. incognita*. *C. globosum* NK102 repel second-stage juveniles, which was followed by filtrates and ChA with 99.8% mortality at 300 µg/ ml at 72 hours; *C. globosum* showed repellant activity as chemicals excreted by these contribute to activity analyzed by chemotaxis assay. But ChA and filtrates were not able to affect the egg hatch until 72 hours of exposure. The egg per plant was also found to be reduced. These NK102 can be widely used for control of ectoparasite nematode *Tylenchorhynchus ventralis*, *Filenchus misellus*, and *Filenchus discrepans* by local interaction with soil microorganisms (Hu et al. 2013).

The synthetic nematicides are not able to easily penetrate the eggshell of nematodes comprised of Vitelline, chitinous, and lipid layers. But the bioagents used secretes some toxins which directly damage the eggshell, such as *Trichoderma harzianum*, *Bacillus firmus*, *Bacillus sphaericus*, *Bacillus laterosporus*, *Xenorhabdus* spp., *Photorhabdus* spp. and *Verticillium chlamydosporium*.

The novel 'honey-trap' mechanism of one of its species *Paenibacillus polymyxa* crowned it as next-generation nematicide because it has the potential to suppress disease caused by nematode and its vector fungus (Cheng et al. 2017). It is comprised of volatile organic compounds (VOCs) identified as acetone, 2-heptanone, 2-nonanone, 2-nonanol, 2-decanone, 2-decanol, 4-acetylbenzoic acid, furfural acetone, 2-undecanone, and 2-undecanol (Figure 8). These act as neuropeptides as they affect the nervous system, surface coat intestine, pharynx, and other tissues to effectively inhibit target pathogen growth and reside in the target for a longer time. Out of studied VOCs, 2-decanol, 4-actylbenzoic acid, furfural acetone, and 2-undecanone were found to be most active as LC_{50} ranges from 4.44–23.12 mg/l since they kill nematode by contact or fumigation and act as repellants. But the acetone and 2-heptanone were found least effective as the calculated mortality rate was below 10% but can act as attractants and be used in combination with other nematicides to improve the efficacy of the compound. The dual characteristic of attractant and repellant was exhibited by 4-acetylbenzoic acid as it is classified as attractant at low concentration and repellant at high concentration (ALRH). The

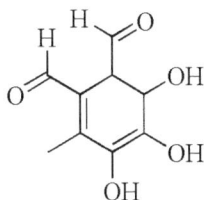

Flavipin

Figure 7. Structure of Flavipin.

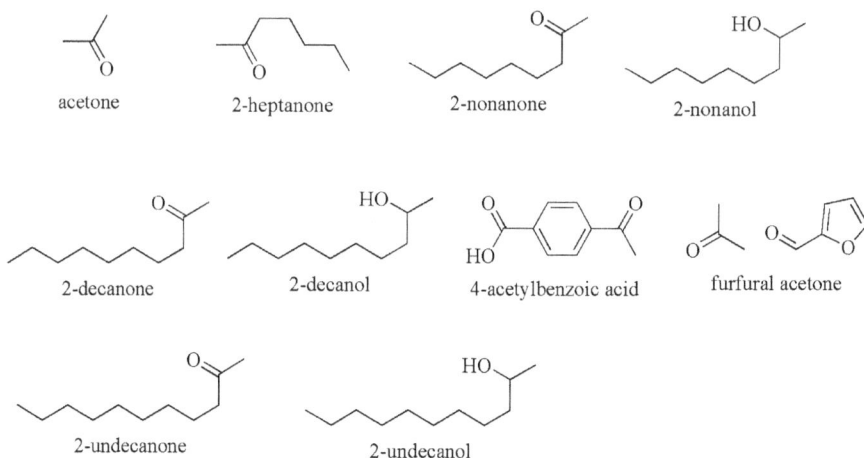

Figure 8. Volatile organic compounds of *Paenibacillus polymyxa.*

mechanism of action showed that the disrupted pharynx and intestine were a major cause for the mortality rate (80 to 87%) by culture filtrates of *P. polymyxa* and longer carbon chain length was reported to contribute to the better nematicidal activity of studied compounds.

Natural products consist of various primary and secondary metabolites which exhibit low toxicity, high potency and are environmentally acceptable substitutes for their use as commercial nematicides. With the legacy of certainly known molecules used traditionally, efforts need to be made to take them to the practical levels for their vast use. Besides, there is also a need to identify active molecules and development of strategies for isolating, identifying, and bioassaying these compounds so that they prove to be effective methods to combat pathogens in the agricultural industry. Conclusively, the natural products have shown putative antinematodal activity. Therefore, there should be empirical screening or rational designing of compounds for the development of prospective compounds and successful transfer of rationally designed molecules from the laboratory to the field.

References

Abd-Elgawad, M.M., Kabeil, S.S. and Abd-El-Wahab, A.E. 2009. Changes in protein content and enzymatic activity of tomato plants in response to nematode infection. Egypt J. Agronematol. 7(10): 49–61

Abdel-Rahman, F.H., Alaniz, N.M. and Saleh, M.A. 2013. Nematicidal activity of terpenoids. J. Env. Sci. Health 48: 16–22.

Adediran, J.A., Adegbite, A.A., Akinlosotu, T., Agbaje, G.O., Taiwo, L.B., Owolade, O.F. and Oluwatosin, G.A. 2005. Evaluation of fallow and cover crops for nematode suppression in three agroecologies of south western Nigeria. African J. Biotechnol. 4(10): 1034–1039.

Adegbite, A.A. and Adesiyan, S.O. 2005. Root extracts of plants to control root-knot nematodes on edible soybean. World J. Agri. Sci. 1(1): 18–21.

Adekunle, O.K. and Fawole, B. 2003. Chemical and non-chemical control of *Meloidogyne incognita* infecting cowpea under field conditions. Moor J. Agri. Res. 4(1): 94–99.

Adfa, M., Hattori, Y., Yoshimura, T. and Koketsu, M. 2012. Antinemic activity of 7-alkoxycoumarins and related analogs against *Copototermes formosanus* Shikari. Int. Biodeterior. Biodegradation 74: 129–135.

Agbenin, N.O., Emechebe, A.M., Marley, P.S. and Akpa, A.D. 2005. Evaluation of nematicidal action of some botanicals on *Meloidogyne incognita in vivo* and *vitro*. J. Agric. Rural Dev. Tropics Subtropics 106(1): 29–39.

Ahmad, F., Rather, M.A. and Siddiqui, A.M. 2010. Nematicidal activity of leaf extracts from *Lantana camara* L. against *Meloidogyne incognita* (kofoid and white) chitwood and its use to manage roots infection of *Solanum melongena* L. Brazilian Archives Bio Technol. 53(3). http://dx.doi.org/10.1590/S1516-89132010000300006.

Aissani, N., Urgeghe, P.P., Oplos, C., Saba, M., Tocco, G., Petretto, G.L. and Caboni, P. 2015. Nematicidal activity of the volatilome of *Eruca sativa* on *Meloidogyne incognita*. J. Agric. Food Chem. 63: 6120–25.

Ajith, K. and Sheela, S. 1996. Utilization of green leaves of neem and upatorium for the management of soil organisms in bhindi and cowpea. Indian J. Nematol. 26: 139–143.

Akpheokhai, I.L., Cole, A.O.C. and Fawole, B. 2012. Evaluation of some plant extracts for the management of *Meloidogyne incognita* on soybeans (*Glycine max*). World J. Agric. Sci. 8(4): 429–435.

Al-Saman, M.A., Farfour, S.A., Tayel, A.A. and Rizk, N.M. 2015. Bioactivity of lectin from Egyptian *Jatropha curcas* seeds and its potentiality as antifungal agent. Glob Adv. Res. J. Microbiol. 4(7): 87–97.

Ameen, H.H. 1996. Influence of garlic *Allium sativum* on populations of *Rotylenchulus reniformis* and *Meloidogyne incognita* infecting cowpea and tomato. Al-Azhar J. Agric. Res. 23: 77–85.

Anamika, Utreja, D., Kaur, J. and Sharma, S. 2019. Synthesis of Schiff bases of coumarin and their antifungal activity. Indian J. Heterocycl. Chem. 28(4): 433–39.

Antonious, G.F. and Snyder, J.C. 2006. Natural products: repellency and toxicity of wild tomato leaf extracts to the two-spotted spider mite, Tetranychusurticae Koch. J. Environ. Sci. Health 41: 43–55.

Anuja, B. and Satyawati, S. 2007. Effect of some plant extracts on the hatch of *M. incognita* eggs. J. Bot. 3: 312–316.

Anulika, N.P., Ignatius, E.O., Raymond, E.S., Osasere, O.I. and Abiola, A.H. 2016. The chemistry of Natural product: Plant secondary metabolites. Int. J. Technol. Enhncmnt. Em. Engg. Res. 4(8): 1–9.

Aoudia, H., Ntalli, N., Aissani, N., Yahiaoui-Zaidi, R. and Caboni, P. 2012. Nematoxic phenolic compounds from *Melia azedarch* against *Meloidogyne incognita*. J. Agric. Food Chem. dx.doi.org/10.1021/jf3038874.

Atkinson, N.J., Lilley, C.J. and Urwin, P.E. 2013. Identification of genes involved in the response to simultaneous biotic and abiotic stress. Plant Physiol. 162: 2028–2041.

Avato, P., D'Addabbo, T., Leonetti, P. and Argentieri, M.P. 2013. Nematicidal potential of Brassicaceae. Phytochem. Rev. 12: 791–802.

Baidoo, R., Mengistu, T., McSorley, R., Stamps, R.H., Brito, J. and Crow, W.T. 2017 Management of root-knot nematode (*Meloidogyn eincognita*) on *Pittosporum tobira* under greenhouse, field, and on-farm conditions in Florida. J. Nematol. 49: 133–139. https://doi.org/10.21307/jofnem-2017-057.

Bakkali, F., Averbeck, S., Averbeck, D. and Idaomar, M. 2008. Biological effects of essential oil—A review. Food Chem. Toxicol. 46: 446–475.

Begum, S., Ayub, A., Siddiqui, B.S., Fayyaz, S. and Kazi, F. 2015 Nematicidal triterpenoids from *Lantana camara*. Chem. Biodiv. 2(9): 1435–42.

Bernard, G.C., Egnin, M. and Bonsi, C. 2017. The impact of plant-parasitic nematodes on agriculture and methods of control. Nematology-Concepts, Diagnosis and Control, Chapter 7: 121–51.

Bhattacharyya, M. 2017. Use of marigold (*Tagetes* sp.) for the successful control of nematodes in agriculture. Pharma Innov. J. 6(11): 01–03.

Bolarinwa, I.F., Oke, M.O., Olaniyan, S.A. and Ajala, A.S. 2016. A review of cyanogenic glycosides in edible plants. pp. 179–191. *In*: Larramendy, M.L. (ed.). Toxicology-New Aspects to This Scientific Conundrum. InTech Open publishers, London, UK.

Caboni, P., Sarias, G., Aissani, N., Tocco, G., Sasanelli, N., Liori, B., Carta, A. and Angioni, A. 2012. Nematicidal activity of 2-Thiophenecarboxaldehyde and Methylisothiocyanate from Caper (*Capparis spinosa*) against *Meloidogyne incognita*. J. Agric. Food Chem. 60: 7345–7351.

Caboni, P. and Ntalli, N.G. 2014. Botanical nematicides, recent findings. In Biopesticides: State of the Art and Future Opportunities. ACS Symposium Series, Washington, DC, 2014.

Cal, de A., Larena, I., Guijarro, B. and Melgarejo, P. 2012. Use of biofungicides for controlling plant diseases to improve food availability. Agri 2(2): 109–124. https://doi.org/10.3390/agriculture20200109.

Capasso, A., De Feo, V., De Simone, F. and Sorrentino, L. 1997. Activity directed isolation of Spasmolytic (anticholinergic) alkaloids from *Brugmansia arborea* (L.) Lagerheim. Int. J. Pharmacognosy 35(1): 43–48.

Charles, K., Agathar, K., Ronald, M., Cosmas, P., Ignitius, M. and Blessing, M. 2015. Nematicidal effects of Brassica formulations against root knot nematodes (*Meloidogyne javanica*) in tomatoes (*Solanum lycopersicum* L.). Pak. J. Phytopathol. 27(02): 109–114.

Cheng, W., Yang, J., Nie, Q., Huang, D., Yu, C., Zheng, L., Cai, M., Thomashow, L.S., Weller, D.M., Yu, Z. and Zhang, J. 2017. Volatile organic compounds from *Paenibacillus polymyxa* KM2501-1 control *Meloidogyne incognita* by multiple strategies. Sci. Rep. 7: 16213.

Chitwood, D.J. 2002. Phytochemicals based strategies for nematode control. Ann. Rev. Phytopathol. 40(1): 221–249.

Choe, K.P., Przybysz, A.J. and Strange, K. 2009. The WD40 repeat protein WDR-23 functions with the CUL4/DDB1 ubiquitin ligase to regulate nuclear abundance and activity of SKN-1 in *Caenorhabditis elegans*. Mol. Cellular Biol. 29: 2704–2715.

Claudius, C.A.O., Aminu, A.E. and Fawole, B. 2010. Evaluation of plant extracts in the management of rootknot nematode *Meloidogyne incognita* on cowpea. *Vignaunguiculata* L. (Walp). Mycopath 8(1): 53–60.

Coelho, J.S., Santos, N.D., Napoleão, T.H., Gomes, F.S., Ferreira, R.S., Zingali, R.B., Coelho, L.C., Leite, S.P., Navarro, D.M. and Paiva, P.M. 2009. Effect of *Moringa oleifera* lectin on development and mortality of *Aedes aegypti* larvae. Chemosphere 77(7): 934–938.

Costa, S.D.S.D.R., Santos, M.D.A. and Ryan, M.F. 2003. Effect of *Artemisia vulgaris* rhizome extracts on hatching, mortality and plant infectivity of *Meloidogyne megadora*. J. Nematol. 35(4): 437–442.

Curto, G., Lazzeri, L., Dallavale, E., Santi, R. and Malaguti, L. 2006. Effectiveness of crop rotation with Brassicaceae species for the management of the southern root knot nematode *Meloidogyne incognita*, Second International Biofumigation Symposium p 51, June 25–29, Moscow.

Czyzewska, M.M., Chrobok, L., Kania, A., Jatczak, M., Pollastro, F., Appendino, G. and Mozrzymas, J.W. 2014. Dietary acetylenic oxylipin falcarinol differentially modulates $GABA_A$ receptors. J. Nat. Prod. 77: 2671–2677. doi: 10.1021/np500615j.

De Nicola, G.R., Leoni, O., Malaguti, L., Bernardi, R. and Lazzeri, L. 2011. A simple analytical method for dhurrin content evaluation in cyanogenic plants for their utilization in fodder and biofumigation. J. Agric. Food Chem. 59: 8065–8069.

Deverall, B.J. 1972. In Phytochemical Ecology, ed. Harnborne, J.B. Academic Press. London. pp. 217–234.

Dias, C.R., Schwan, A.V., Ezequiel, D.P., Sarmento, M.C. and Ferraz, S. 2000. Effect of aqueous extracts of medicinal plant on the survival of juveniles of *Meloidogyne incognita*. Nematologia Brasileira 24: 203–210.

Dresselhaus, T. and Huckelhoven, R. 2018. Biotic and abiotic stress response in crop plants. Agronomy 8: 267–272. doi:10.3390/agromnomy8110267.

Echeverrigaray, S., Zacaria, J. and Beltrao, R. 2010. Nematicidal activity of monoterpenoids against the root knot nematode *Meloidogyne incognita*. Am. Phytopathol. Soc. 100(2): 199–203.

Eckenbach, U., Lampman, R.L., Seigler, D.S., Ebinger, J. and Novak, R.J. 1999. Mosquitocidal activity of acetylenic compounds from *Cryptotaenia canadensis*. J. Chem. Eco. 25: 1885–1893. doi: 10.1023/A:1020938001272.

El-Ansary, M.S.M. and Al-Saman, M.A. 2018. Appraisal of *Moringa oleifera* crude proteins for the control of root-knot nematode, *Meloidogyne incognita* in banana. RendicontiLincei. ScienzeFisiche e Naturalihttps://doi.org/10.1007/s12210-018-0692-9.

El-Gengaihi, S.E., Osman, H.A., Youssef, M.M.A. and Mohamed, S.M. 2001. Efficacy of *Tagetes* species extracts on the mortality of the reniform nematode, *Rotylenchusreniformis*. Bull NRC, Egypt 26: 441–450.

Elling, A.A. 2013. Major emerging problems with minor *Meloidogyne* species. Phytopathol. 103: 1092–1102.

El-Nagdi, W.M.A. and Yousef, M.M.A. 2013. Comparative efficacy of garlic clove and castor seed aqueous extracts against the root-knot nematode, *Meloidogyne incognita* infecting tomato plant. J. Plant Prot. Res. 53(3): 285–288.

Enan, E. 2001. Insecticidal activity of essential oils: Octopaminergic sites of action. Comparative Biochemistry and Physiology Part C: Toxicol. Pharmacol. 130: 325–337.

Escher, S., Niclassa, Y., van de Waalb, M. and Starkenmann, C. 2006. Combinatorial synthesis by nature: volatile organic sulfur containing constutients of *Ruta chalepensis* L. Chem. Biodiversity 3: 943–957.

Fassuliotis, G. 1979. Plant breeding for root knot nematode resistance. pp. 425–453. *In*: Lamberti, F. and Taylor, C.E. (eds.). Root Knot Nematodes (Meloidogyne species). Systematics, Biology and Control. Academic Press, New York, USA.

Gad, S.C. 2014. Encyclopedia of Toxicology, Reference Module in Biomedical Sciences, 3rd ed., 473–474. https://doi.org/10.1016/B978-0-12-386454-3.00888-5.

Gamal, A.E., Dong, W.L., Jung, C.P., Hwang, B.Y. and Ho, Y.C. 2008. Evaluation of various plant extracts for their nematicidal efficacies against juveniles of *Meloidogyne incognita*. J. Asia Pacific Entomol. 11(2): 99–102.

Garrod, B., Lea, E.J.A. and Lewis, G. 1979. Studies on the mechanism of action of the antifungal compound falcarindiol. New Phytol. 83: 463–471. doi: 10.1111/j.1469-8137.1979.tb07471.x.

Gu, Y.Q., Mo, M.H., Zhou, J.P., Zou, C.S. and Zhang, K.Q. 2007. Evaluation and identification of potential organic nematicidal volatiles from soil bacteria. Soil Biol. Biochem. 39: 2567–2565.

Gull, A., Lone, A.A. and Wani, N.I. 2019. Biotic and abiotic stresses in plants. Abiotic and Biotic stress in Plants, Alexandre BO, IntechOpen. Doi:10.5772/intechopen.85832.

Hamaguchi, T., Sato, K., Vicente, C.S.L. and Hasegawa, K. 2020. Nematicidal Actions of the Marigold Exudate α-terthienyl: Oxidative Stress-Inducing Compound Penetrates Nematode Hypodermis. https://bio.biologists.org/content/8/4/bio038646.

Haroon, S.A., Hassan, B.A.A., Hamad, F.M. and Rady, M.M. 2018. The efficiency of some natural alternatives in root knot nematode control. Adv. Plants Agric. Res. (4): 355–362.

Hinmikaiye, A.S., Abolusoro, S.A., Oloniruha, J.A., Ogundare, S.K., Babalola, T.S., Kadiri, W.O.J. and Ayodele, F.G. 2016. Studies on the comparative toxicity of Jatrophacurcas and synthetic nematicide on the root-knot nematode infected sweet pepper (*Capsicum Annuum*). Am. Res. J. Agric. 2378–9018.

Hu, Y., Zhang, W., Zhang, P., Ruan, W. and Zhu, X. 2013. Nematicidal activity of Chaetoglobosin A produced by *Chaetomium globosum* NK102 against *Meloidogyne incognita*. J. Agric. Food Chem. 61: 41–46.

Huang, Q., Yun, X., Rao, W. and Xiao, C. 2017. Antioxidative cellular response of lepidopteran ovarian cells to photoactivated alpha-terthienyl. Pest Biochem. Physiol. 137: 1–7. doi: 10.1016/j.pestbp.2016.09.006.

Huang, Y., Xu, C., Ma, L., Zhang, K., Duan, C. and Mo, M. 2010. Characterisation of volatiles produced from *Bacillus megaterium* YFM3.25 and their nematicidal activity against *M. incognita*. Eur. J. Plant Pathol. 126: 417–422.

Hussein, A., Salim, Iman, S., Salman, Ishtar, I.M. and Hatam, H.H. 2016. Evaluation of some plant extracts for their nematicidal properties against root-knot nematode, *Meloidogyne* sp. J. Genetic Env. Res. Conser. 4(3): 241–244.

Ismail, A.E. 2013. Feasibility of growing moringa oleifera as a mix-crop along with tomato for control of *Meloidogyne incognita* and *Rotylenchulus reniformis* in Egypt. Archiv. Phytopathol. Plant Prot. 46(12): 1403–1407.

Ismail, A.E. 2014. Growing *Jatropha curcas* and *Jatropha gossyliifolia* as a interculture with sunflower for control of *Meloiodogyne javanica* in Egypt. Int. J. Sus. Agric. Res. 1(2): 39–44.

Jain, N., Utreja, D. and Dhillon, N.K. 2019. A convenient one-pot synthesis and nematicidal activity of nicotinic acid amides. Russ J. Org. Chem. 55(6): 845–851.

Jain, R.K., Mathur, K.N. and Singh, R.V. 2007. Estimation of lossess due to plant parasitic nematodes on different crops in India. Indian J. Nematol. 37: 219–221.

Jang, J.Y., Dang, Q.L., Choi, Y.H., Choi, G.J., Jang, K.S., Cha, B., Luu, N.H. and Kim, J.C. 2014. Nematicidal activities of 4-quinolone alkaloids isolated from the aerial part of the *Triumfetta grandidens* against *Meloidogyne incognita*. J. Agric. Food Chem. dx.doi.org/10.1021/jf504572h.

Kaur, G., Utreja, D., Jain, N. and Dhillon, N.K. 2020. Synthesis and evaluation of pyrazole derivatives as potent antinemic agents. Russ. J. Org. Chem. 56(1): 113–118.

Kaur, J., Utreja, D., Dhillon, N.K. and Sharma, S. 2018. Synthesis of series of triazine derivatives and their evaluation against root knot nematode *Meloidogyne incognita*. Lett. Org. Chem. 15: 870–877.

Kaur, J., Utreja, D., Dhillon, N.K. and Sharma, S. 2019. Synthesis of indole derivatives and their evaluation against root knot nematode *Meloidogyne incognita*. Lett. Org. Chem. 16: 759–767.

Kaur, K., Utreja, D., Dhillon, N.K., Pathak, R.K. and Singh, K. 2021. N-alkylisatin derivatives: Synthesis, nematicidal evaluation and protein target identifications for their mode of action. Pest Biochem. Physiol. 171: 104736.

Khalil, M.S. 2014. Bright future with nematicidal phytochemicals. Bio Med. 6: 2.

Khan, A., Asif, M., Tariq, M., Rehman, B., Parihar, K. and Siddiqui, M.A. 2017. Phytochemical investigation, nematostatic and nematicidal potential of weeds extract against the root-knot nematode, *Meloidogyne incognita in vitro*. Asian J. Biol. Sci. 10: 38–46.

Khan, A., Tariq, M., Asif, M. and Siddiqui, M.A. 2017. Evaluation of botanicals toxicants against root knot nematode, *Meloidogyne incognita in vitro*. Asian J. Biol. 4(3): 1–7.

Khan, R., Naz, I., Hussain, S., Khan, R.A.A., Ullah, S., Rashid, M.U. and Siddique, I. 2019. Phytochemical management of root knot nematode (*Meloidogyne incognita*) kofoid and white chitwood by *Artemisia* spp. in tomato (*Lycopersicon esculentum* L.). Braz J. Biol. https://doi.org/10.1590/1519-6984.222040.

Khurma, U.R. and Kumari, S. 1996. Effect of four seed extracts on juvenile mortality of *Meloidogyne javanica*. Indian J. Nematol. 26: 214–215.

Khurma, U.R. and Singh, A. 1997. Nematicidal potential of seed extracts: *In vitro* effects on juvenile mortality and egg hatch of *Meloidogyne incognita* and *M. javanica*. Nematol. Medit. 25: 49–54.

Khurma, U.R. and Chaudhary, P. 1999. Comparative effects of extracts of different parts of *Calotropis procera*, *Cassia fistula*, *Ricinus communis* and *Sesbania sesban* on *Meloidogyne javanica* juveniles. J. Env. Biol. 20(4): 287–288.

Knight, A.J. and Behm, C.A. 2011. Minireview: The role of vaculor ATPase in nematodes. Exp. Parasitol. in press.

Knobloch, K., Pauli, A., Iberi, N., Weigand, N. and Weis, H.M. 1989. Antibacterial and antifungal properties of essential oil components. J. Essential Oil Res. 1: 119–128.

Kogiso, S., Wada, K. and Munakata, K. 1976. Isolation of nematicidal polyacetylenes from *Carthamus tinctorius* L. Agric Biol. Chem. 40: 2085–2089. doi: 10.1080/00021369.1976.10862338.

Kong, J.O., Lee, S.M., Moon, Y.S., Lee, S.G. and Ahn, Y.J. 2006. Nematicidal activity of plant essential oils against *Bursaphelenchus xylophilus* (Nematoda: *Alphelenchoididae*). J. Asia Pac. Entomol. 9: 173–178.

Kong, J.O., Park, I.K., Choi, K.S., Shin, S.C. and Ahn, Y.J. 2007. Nematicidal activities of thyme red and white oil compounds toward *Bursaphelenchus xylophilus* (Nematoda: Parasitaphelenchidae). J. Nematol. 39: 237–242.

Korayem, A.M. and Osman, H.A. 1992. Übernematizidewirkungen der henna-pflanzelawsoniainermisgegen den wurzelnematoden *Meloidogyne incognita*. Anzeiger Fur SchädlingskundePflanzenschutzUmweltschutz 65(4): 14–16.

Kostyukovsky, M., Rafaeli, A., Gileadi, C., Demchenko, N. and Shaaya, E. 2002. Activation of octapaminergic receptors by essential oil constitutents isolated from aromatic plants: Possible mode of action against insect pests. Pest Manag. Sci. 58: 1101–1106.

Kuljanabhagavad, T. and Wink, M. 2009. Biological activities and chemistry of saponins from *Chenopodium quinoa* Willd. Phytochem. Rev. 8: 473–490.

Kuriyama, T., Ju, X.L., Fusazaki, S., Hishinuma, H., Satou, T., Koike, K., Nikaido, T. and Ozoe, Y. 2005. Nematocidal quassinoids and bicyclophosphorothionates: A possible common mode of action on the GABA receptor. Pest Biochem. Physiol. 81: 176–187. doi: 10.1016/j.pestbp.2004.11.008.

Lata, N. and Dubey, V. 2010. Preliminary phytochemical screening of Eichhorniacrassipes: the world's worst aquatic Weed. J. Pharm. Res. 3: 1240–1242.

Laursen, T., Borch, J., Knudsen, C., Bavishi, K., Torta, F., Martens, H.J., Silvestro, D., Hatzakis, N.S., Wenk, M.R., Daffron, T.R., Olsen, C.E., Motawia, M.S., Hamberger, B., Moller, B.L. and Bassard, J. 2016. Characterization of a dynamic metabolon producing the defense compound dhurrin in sorghum. Sci. 354: 890–893.

Lindblom, T.H. and Dodd, A.K. 2006. Xenobiotic detoxification in the nematode *Caenorhabditis elegans*. J. Exp. Zoo 305: 720–730. doi: 10.1002/jez.a.324.

Liu, G., Lai, D., Liu, Q.Z., Zhou, L. and Liu, Z.L. 2016. Identification of nematicidal constituents of *Notopterygium incisum* rhizomes against *Bursaphelenchus xylophilus* and *Meloidogyne incognita*. Mol. 21(10): 1276.

Lord, J.S., Lazzeri, L., Atkinson, H.J. and Urwin, P.E. 2011. Biofumigation for control of pale potato cyst nematodes: activity of Brassica leaf extracts and green manures on *Globodera pallida in vitro* and in soil. J. Agric. Food Chem. 59: 7882–7890.

Lu, H., Xu, S., Zhang, W., Xu, C., Li, B., Zhang, D., Mu, W. and Liu, F. 2017. Nematicidal activity of trans-2-hexanal against southern root knot nematode (*Meloidogyne incognita*) on tomato plants. J. Agric. Food Chem. doi: 10.1021/acs.jafc.6b04091.

Madani, N., Kimball, J.S., Ballantyne, A.P., Affleck, D.L.R., Bodegom, P.M., Reich, P.B., Kattge, J., Sala, A., Nazeri, M., Jones, M.O., Zhao, M. and Steven, W. 2018. Future global productivity will be affected by plant trait response to climate. Sci. Rep. 8: 2870.

Mainoo, A.A. and Banful, B.K. 2019. Nematicidal properties of *Moringa oleifera*, *Chromolaena odorata* and Panicum maximum and their control effects on pathogenic nematodes of Yam. J. Exp. Agric. Int. 31(2): 1–7.

Maqbool, A., Hayat, C.S. and Tanveer, A. 2004. Comparative efficacy of various indigenous and allopathic drugs against fascioliasis in buffaloes. Veterinary Archive 74: 107–14.

Marban-Mendoza, N., Jeyaprakash, A., Jansson, H.B., Damon, J.R.R.A. and Zuckerman, B.M. 1987. Control of root-knot nematodes on tomato by lectins. J. Nematol. 19(3): 331–335.

Matsuhashi, R., Satou, T., Koike, K., Yokosuka, A., Mimaki, Y., Sashida, Y. and Nikaido, T. 2002. Nematicidal activity of isoquinoline alkaloids against a species of *Diplogastridae*. Planta Med. 68: 169–171.

Matthiessen, J.N. and Kirkegaard, J.A. 2006. Biofumigation and biodegradation: opportunity and challenge in soil-borne pest and disease management. Crit. Rev. Plant Sci. 25: 235–265.

Mejrib, J., Abderrabbab, M. and Mejria, M. 2010. Chemical composition of the essential oil of *Ruta chalepensis* L.: influence of drying, hydrodistillation duration and plant parts. Ind. Crop Prod. 32: 671–73.

Miao, J.Q., Wang, M., Li, X.H., Yang, F.M. and Liu, F. 2012. Antifungal and nematicidal activities of five volatile organic compounds against soil borne pathogenic fungi and nematode. Acta Phytophy Sin 6: 017.

Mostafanezhad, H., Sahebani, N. and Nourinejhad, Z.S. 2014. Control of root-knot nematode (*Meloidogyne javanica*) with combination of *Arthrobotrys oligospora* and salicyclic acid and study of some plant defense responses. Biocontrol Sci. Technol. 24(2): 203–215.

Motisi, N., Doré, T., Lucas, P. and Montfort, F. 2010. Dealing with the variability in biofumigation efficacy through an epidemiological framework. Soil Biol. Biochem. 42: 2044–2057

Nandakumar, A., Vaganan, M.M., Sundararaju, P. and Udayakumar, R. 2017. Nematicidal activity of aqueous leaf extracts of *Datura metel*, *Datura innoxia* and *Brugmansiasuaveolens*. Am. J. Entomol. 1(2): 39–45.

Narsai, R., Wang, C., Chen, J., Wu, J., Shou, H. and Whelan, J. 2013. Antagonistic, overlapping and distinct responses to biotic stress in rice (*Oryza sativa*) and their interactions with abiotic stress. BMC Genomics 14: 93.

Naz, I., Palomares-Rius, J.E., Saifullah, Blok, V., Khan, M.R., Ali, S. and Ali, S. 2012. *In vitro* and in planta nematicidal activity of *Fumaria parviflora* (Fumariaceae) against southern root-knot nematode *Meloidogyne incognita*. Plant Pathol. doi:10.1111/j.1365-3059.2012.02682.x.

Nigh, E.L. 1985. Allelopathic activity of plants to nematode. J. Nematol. 17(4): 518.

Nile, A.S., Nile, S.H., Keum, Y.S., Kim, D.H., Venkidasamy, B. and Ramalingam, S. 2018. Nematicidal potential and specific enzyme activity enhancement potential of neem (*Azadirachta indica* A. Juss.) aerial parts. Env. Sci. Pol. Res. 25: 4204–4213.

Niu, X.M., Wang, Y.L., Chu, Y.S., Xue, H.X., Li, N., Wei, L.X., Mo, M.H. and Zhang, K.Q. 2010. Nematodetoxic aurovertin-type metabolites from a root knot nematode parasitic fungus *Pochoniachamydosporia*. J. Agric. Food Chem. 58(2): 828–34.

Nivsarkar, M., Cherian, B. and Padh, H. 2001. Alpha-terthienyl: A plant-derived new generation insecticide. Curr. Sci. 81: 667–672.

Ntalli, N.G., Manconi, F., Leonti, M., Maxia, A. and Caboni, P. 2011. Aliphatic ketones from *Ruta chalepensis* (Rutaceae) induce paralysis on root knot nematodes. J. Agric. Food Chem. 59: 7098–7103.

Ntalli, N.G. and Caboni, P. 2012. Botanical nematicides: A review. J. Agric. Food Chem. 60: 9929–9940.

Ogwudire, V.E., Agu, C.M., Ojiako, F.O., Cookey, C.O. and Nwokeji, C.M. 2019. Nematicidal effects of *Jatropha curcus* L. root extracts for the control of root knot nematode *Meloidgyne incognita*. Int. J. Agric. Rural Dev. 22(2): 4468–4473.

Ohri, P. and Pannu, S.K. 2009. Effect of terpenoids on nematodes: A review. J. Env. Res. Dev. 4(1): 171–178.

Ohri, P. and Pannu, S.K. 2010. Effect of phenolic compounds on nematodes—A review. J. Appl. Nat. Sci. 2(2): 344–350.

Ojo, G.T. and Umar, I. 2013. Evaluation of some botanicals on root–knot nematode (*Meloidogyne javanica*) in Tomato (*Lycopersicon esculentum*, Mill) in Yola Adamawa State, Nigeria. Biological Forum – Am. Int. J. 5(2): 31–36.

Oluwatayo, J.I., Jidere, C.I. and Nwankiti, A. 2019. Nematicidal effect of some botanical extracts for the management of *Meloidogyne incognita* and on growth of tomato. Asian J. Agric. Horti. Res. 4(2): 1–8.

Onyeke, C.C. and Akueshi, C.O. 2012. Infectivity and reproduction of *Meloidogyne incognita* (Kofoid and White) chitwood on Africa yam bean, sphenostylisstenocarpa (Hochst Ex. A. Rich) harms accessions as influenced by botanical soil amendments. African J. Biotechnol. 11(18): 13095–13103.

Osman, H.A., El–Gindi, A.Y., Ameen, H.H., Youssef, M.M.A. and Lashein, A.M. 2005. Evaluation of the nematicidal effects of smashed garlic, sincocin and nemaless on the root-knot nematode, *Meloidogyne incognita* infecting cowpea plants. Bull Nat. Res. Centre Egypt 30(3): 297–305.

Pavela, R. 2004. Insecticidal activity of certain medicinal plants. Fitoterapia 75: 745–749.

Petti, S. and Scully, C. 2009. Polyphenols, oral health and disease: A review. J. Dent. 7(6): 413–423.

Phadungkit, M., Somdee, T. and Kangsadalampai, K. 2012. Phytochemical screening, antioxidant and antimutagenic activities of selected Thai edible plant extracts. J. Med. Plants Res. 6: 662–666.

Priestley, C.M., Williamson, E.M., Wafford, K.A. and Sattelle, D.B. 2003. Thymol, a constituent of thyme essential oil, is a positive allosteric modulator of human GABA receptors and a homo-oligomeric GABA receptor from *Drosophilla melanogaster*. Brit. J. Pharmacol. 140: 1363–1372.

Qamar, F., Begum, S., Raza, S.M., Wahab, A. and Siddiqui, B.S. 2005. Nematicidal natural products from the aerial parts of *Lantana camara* Linn. Nat. Prod. Res. 9(6): 609–13.

Qiao, K., Liu, X., Wang, H., Xia, X., Ji, X. and Wang, K. 2012. Effect of abamectin on root knot nematodes and tomato yield. Pest Manag. Sci. 68: 853–857. https://doi.org/10.1002/ps.2338.

Ralmi, N.H.A.A., Khandaker, M.M. and Mat, N. 2016. Occurrence and control of root nematode in crops: A review. Aus J. Crop Sci. 10(12): 1649–1654. doi:10.21475/ajcs.2016.10.12. p7444.

Ran, H., Hassett, D.J. and Lau, G.W. 2003. Human targets of *Pseudomonas aeruginosa* pyocyanin. Proc. Natl. Acad. Sci. 100(24): 14315–20.

Riga, E., Mojtahedi, H., Ingham, R. and McGuire, A.M. 2004. Green manure amendments and management of root knot nematodes on potato in the Pacific Northwest of USA. pp 151–158. *In*: Cook, R.C. and Hunt, D.J. (eds.). Nematology Monographs and Perspectives. Proceedings of the Fourth International Congress of Nematology. 2nd edition, Leiden, The Netherlands. Brill Academic Publishers, Inc.

Riga, E. 2011. The effects of Brassica green manures on plant parasitic and free living nematodes used in combination with reduced rate of synthetic nematicides. J. Nematol. 43(2): 119–121.

Saadali, B., Boriky, D., Blaghen, M., Vanhaelen, M. and Talbi, M. 2001. Alkamides from *Artemisia dracunculus*. Phytochem. 58(7): 1083–1086.

Salles, H.O., Braga, A.C.L., Nascimento, M.T.C., Sousa, A.M.P., Lima, A.R., Vieira, L.S., Cavalcante, A.C.R., Egito, A.S. and Andrade, L.B.S. 2014. Lectin, hemolysin and protease inhibitors in seed fractions with ovicidal activity against Haemonchuscontortus. Revista Brasileira de ParasitologiaVeterinária 23(2): 136–143.

Santos, A.F.S., Luz, L.A., Argolo, A.C.C., Teixeira, J.A., Paiva, P.M.G. and Coelho, L.C.B.B. 2009. Isolation of a seed coagulant *Moringa oleifera* lectin. Process Biochem. 44(4): 504–508.

Savary, S., Willicoquet, L., Pethybridge, S.J., Esker, P., McRoberts, N. and Nelson, A. 2019. The global burden of pathogens and pests on major food crops. Nat. Eco Evo. 3: 430–439. DOI: 10.1038/s41559-018-0793-y.

Serge, S., Willocquet, L., Pethybridge, S.J. and Esker, P. 2019. The global burden of pathogens and pests on major food crops. Nat. Eco Evo. 3(3): 1.

Shakil, N.A., Prasad, D., Saxena, D.B. and Gupta, A.K. 2004. Nematicidal activity of essential oil of *Artemisia annua* against root-knot and reniform nematodes. Annals Plant Prot. Sci. 12: 397–402.

Sharma, A., Singh, S. and Utreja, D. 2016. Recent advances in synthesis and antifungal activity of 1,3,5-triazines. Curr. Org. Synth. 13: 484–503.

Sharma, H.K. and Prasad, D. 1995. Potentialities of a few botanical in Management of rootknot nematode *M. incognita* on soybean. Annals Agric. Res. 16: 73–476.

Sharma, M.C. and Sharma, S. 2010. Phytochemical, preliminary pharmacognostical and antimicrobial evaluation of combined crude aqueous extract. Int. J. Microbiol. Res. 1(3): 166–170.

Singh, A.U. 2015. Yield losses in crops due to phytonematodes. Project coordinating cell, AICRP on nematodes ICAR-IARI, New Delhi, p 36.

Singla, J. and Krattinger, S.G. 2016. Biotic stress resistance genes in wheat. Encyclopedia of Food Grains (Second Edition) 4: 388–392. https://doi.org/10.1016/B978-0-12-394437-5.00229-1.

Stirling, G.R. 1991. Biological Control of Plant Parasitic Nematodes. Progress, Problems and Prospects. Wallingford, Oxon, UK: CAB International, 275.

Subramaniyan, S., Paulraj, M.G., Khusro, A. and Al-Dhabi, N.A. 2017. Biocontrol properties of Basidiomycetes: An overview. J. Fungi 3(2): 1–14.

Takaishi, K., Izumi, M., Bba, N., Kawazu, K. and Nakajima, S. 2008. Synthesis and biological evaluation of alkoxycoumarin as novel nematicidal constitutents. Bio Med. Chem. Lett. 18(20): 5614–5617.

Terefe, M. 2015. Evaluation of nematicidal action of some bio-agents and botanicals for eco-friendly management of root-knot nematodes, *Meloidogyne incognita* in Tomato. Sci. Tech Arts Res. J. 4(3): 71–78.

Thoden, T.C., Oppre, M.B. and Allmann, J.H. 2007. Pyrrolizidine alkaloids of *Chromolaena odorata* act as nematicidal agents and reduced infections of lettuce roots by *Meloidogyne incognita*. Nematol. 9(3): 343–349.

Tiyagi, S.A., Khan, A.V. and Alam, M.M. 2002. Biodegradable effect of oil-seed cakes on plant-parasitic nematodes and soil inhabiting fungi infesting *Trigonellafoenum-greacum* and *Phaseolus aureus*. Indian J. Nematol. 32: 47–57.

Tsao, R. and Yu, Q. 2011. Nematicidal activity of monoterpenoid compounds against economically important nematodes in agriculture. Tandfonline, 350–354, https://dx.doi.org/10.1080/10412905.2000.9699533.

Tsygankova, V., Andrusevich, Y., Shtompel, O., Pilyo, S., Prokopenko, V., Kornienko, A. and Brovarets, V. 2016. Study of growth regulating activity derivatives of [1,3] Oxazolo [5,4-d] pyrimidine and N-sulfonyl substituted of 1,3-Oxazole on soybean, wheat, flax and pumpkin plants. Int. J. Chem. Studies 4(5): 106–120.

Umar, I. and Mohammed, B.A. 2013. Effect of water hyacinth (*Eichhornia Crassipes* (Mart) Solms leaf extract on the juvenile mortality of *Meloidogyne Incognita*. J. Agric. Vet Sci. 4(2): 46–48.

Umar, I. and Mamman, A. 2014. Nematicidal potential of *Faidherbia albida* fruit against *Meloidogyne javanica* on cowpea. Pak. J. Nematol. 32(1): 77–83.

Umeh, A. and Ndana, R.W. 2010. Effectiveness of *Jatropha curcas* and *Jatropha gossypiifolia* plant extracts in the control of *Meloidogyne incognita* on okra. Int. J. Nematol. 20(2): 226–229.

Upgade, A. and Anusha, B. 2013. Characterization and medicinal importance of phytoconstiuents of Carica papaya from down south Indian region using gas chromatography and mass spectroscopy. Asian J. Pharm. Clinical Res. 6(4): 101–106.

Vetter, J. 2000. Plant cyanogenic glycosides, Toxi 38: 11–36.

Wang, L., Baldwin, E.A., Zhao, W., Plotto, A., Sun, X., Wang, Z., Bretch, J.K. and Bai, J. 2015. Suppression of volatile production in tomato fruit exposed to chilling temperature and alleviation of chilling injury by a pre-chilling heat treatment. Lebensm-Wiss Technol. 62: 115–121.

Wat, C.K., Prasad, S.K., Graham, E.A., Partington, S., Arnason, T., Towers, G.H.N. and Lam, J. 1981. Photosensitization of invertebrates by natural polyacetylenes. Biochem. Sys. Eco. 9: 59–62. doi: 10.1016/0305-1978(81)90060-0.

Widmer, T.L. and Abawi, G.S. 2002. Relationship between levels of cyanide in sudangrass hybrids incorporated into soil and suppression of *Meloidogyne hapla*. J. Nematol. 34: 16–22.

Wiratno, T.D., Van den Berg, H., Riksen, J.A.G., Rietjens, I.M.C.M., Djiwanti, S.R., Kammenga, J.E. and Murk, A.J. 2009. Nematicidal activity of plant extracts against root knot nematode *Meloidogyne incognita*. Open Nat. Products J. 2: 77–85.

CHAPTER 2

Green Technologies for Crop-Pest Control

Deepika Kumari,[1] Lucky Duhan,[1] Raman Manoharlal,[2]
G.V.S. Sai Prasad,[2] Marlia Mohd Hanafiah[3,4] and Ritu Pasrija[1,*]

1. Introduction

The Green revolution or the agricultural revolution started to safeguard food supply to the expanding world's population. It was achievable by allocating more land for farming and improving the yields from the area under farming. In order to achieve this target, many agronomical-practices (e.g., development of the high-yield varieties, setting up of seed-banks of resistant germplasm collection, upgraded irrigation, soil-management, crop-rotation, fertilizers, pesticides, and other advanced techniques) pioneered over the years. Despite these efforts, crop yield loss is still a big threat that causes huge economic losses across the world (Xing et al. 2014, Agarwal and Verma 2020, Oerke et al. 1994). In 2019, statistics of the Food and Agriculture Organization (FAO) of United Nations indicated spoilage swaying between 10–40% of the complete annual harvest, costing USD 220 billion at least. Moreover, multiple factors are responsible for loss in yields, categorised as either 'direct' or 'indirect'. Direct-losses, also known as primary-losses, are the pre-harvesting losses (e.g., diseases, shrivelled-grains and pest-infestation), whereas indirect-losses, also known as secondary-losses, are the losses that incur either during harvesting (e.g., broken grains, etc.) or post-harvesting losses (e.g., spillage, mould, or microbial-attack during storage, etc.). Among these, one of the prominent factors

[1] Department of Biochemistry, Maharshi Dayanand University, Rohtak, India.
[2] ITC Limited, ITC Life Science and Technology Centre (LSTC), Peenya Industrial Area, 1st Phase, Bengaluru - 560058, Karnataka, India.
[3] Center for Earth Sciences and Environment, Faculty of Science and Technology, Universiti Kebangsaan Malaysia, 43600 Bangi UKM, Selangor, Malaysia.
[4] Centre for Tropical Climate Change System, Institute of Climate Change, Universiti Kebangsaan Malaysia, 43600 Bangi UKM, Selangor, Malaysia.
* Corresponding author

affecting the crop-yield is pest-infestation. The term 'pests' is used for creatures that spoil our crops, forestry, and livestock and are largely referred to as phyto-pathogens, like bacteria, fungi, insects (viz. fruit-flies, termites, beetles, spiders, etc.), rodents, and weeds, which are accountable for spread of disease and decline in both crop yield and quality. The pests's mediated loss stands close to 1/3rd of total damage and economically translates to USD 70 billion. Among the different crops, the yield loss varies greatly and stand highest at 80% for cotton and sugar beet, 50% for barley, and variable for different crops like rice (30%), wheat (21%), maize (22%), soybean (21%), and potatoes (17%) (Savary et al. 2019). This yield loss due to pest-attack is an enormous burden on developing countries, like India, wherein majority of population depends on farming. Recently, an instance of locust's swarm attack in India made it to headlines in 2020. The earlier data on percentage yield loss by pests of some of the economical crop in India is shown in Table 1.

Thus, an efficient and effective protection mechanism of the agronomical-crops from pest strikes is a feasible option that can significantly escalate the quality and yields. For many decades, pest management has relied mainly on the use of chemicals such as pesticides, which are employed either as a single compound or a mixture of chemicals for killing, avoiding, repelling, or mitigating pests (Tiwari and Alam 2016). Indeed, the popularization of pesticides usage dates back to the early 1960s. Over the years, even after the employment of pesticides, the pre-harvest losses in different crops yields averages close to 30%, which otherwise could reach up to 50–80% in their absence (Oerke 2006). With an increasing land under agriculture and the use of pesticides, the production of major cereals (e.g., wheat, rice, and maize), has grown more than double. This accompanied unrestricted use of pesticides and chemical fertilizers for boosting agricultural productivity. Worldwide, pesticide

Table 1. Percentage harvest deficit because of pests infections in various Indian crops (Reddy and Zehr 2004).

Crop	Pests	Yield loss (%)
Rice (*Oryza sativa*)	Rice leaf roller (*Cnaphalocrocis medinalis*)	11–51
	Yellow stem borer (*Scirpophaga incertulas*)	11–47
	Asian rice gall midge (*Orseolia oryzae*)	9–49
	Rice whorl maggot (*Hydrellia philippina*)	21–31
Pigeon pea (*Cajanus cajan*)	Gram pod borer (*Helicoverpa armigera* Hubner)	13–102
	Legume pod borer (*Maruca testulalis*)	21–61
Cotton (*Gossypium barbadense*)	Asian spiny bollworm moth (*Earias vittella*)	31–41
	Corn earworm moth (*Helicoverpa armigera*)	21–81
Wheat (*Triticum aestivum*)	Oriental moth (*Mythimna separata*)	21–41
Cabbage (*Brassica oleracea var. capitata*)	Diamondback moth (*Plutella xylostella*)	21–51
Cauliflower (*Brassica oleracea var. botrytis*)	Diamondback moth (*Plutella xylostella*)	21–51
Brinjal (*Solanum melongena*)	Shoot borer moth (*Leucinodes orbonalis*)	26–93

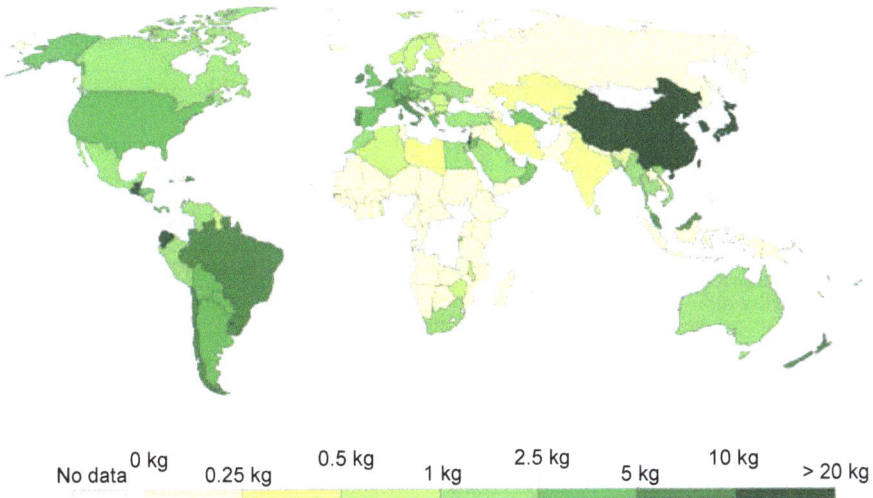

No data	0 kg	0.25 kg	0.5 kg	1 kg	2.5 kg	5 kg	10 kg	> 20 kg

Figure 1. Average pesticide application per unit of cropland, measured in kilograms per hectare (Source: https://ourworldindata.org/pesticides).

utilization has reached about 2 MT/year with an estimated market worth of USD 40 billion (Figure 1). A recent report published by MarketsandMarkets™ estimated the crop-protection chemicals market might reach USD 74.1 billion by 2025. In India, the average pesticide usage is 0.5 kg/ha, accounting for ~ 4% of global pesticide utilization (Agarwal et al. 2015). The high dependency on pesticides exists as in the case of a drop in production, food prices escalates drastically. In a rough estimate, every dollar spent in pesticide production increases overall agronomic production by 3–6 USD, and this gain is largely circulated among customers with a drop in the prices (Zilberman et al. 1991).

2. Impact and Consequences of Conventional Pesticides

Although the use of chemical pesticides did assist in pest management to a certain extent, its execution has been a double-edged sword (Isman 2019). It is not difficult to disregard the detrimental impressions on ecological and agrarian systems, biological diversity, civic well-being along with the spread of resilient strains (Sun et al. 2012). The aggressive usage of pest killers causes various ill bearings on our well-being as well as on the natural ecosystem (Ellgehausen et al. 1980). Pesticides enter into the human body either by straight contacts—such as consuming food (particularly fruits and vegetables) laden with chemicals—or through contaminated water and atmosphere (Mostafalou 2012). The misted fields when washed with rains result in the entry of these chemicals into soil and water-bodies. Both critical and prolonged illnesses can occur due to pesticide contact or consumption, including skin rashes, headache, impaired vision, body aches, nausea, dizziness, cardiovascular diseases, cramps, and in severe cases coma, cancer, and ultimately death (Mostafalou 2012, Nicolopoulou-Stamati et al. 2016). The data from a published study of 141 countries estimates annual 385 million cases of unintentional acute pesticide poisoning, out

Table 2. Clinical implications in persistently exposed person to pesticides.

Diseases	References
Birth defects	Mesnage et al. 2009, Winchester et al. 2009, Rappazzo et al. 2016
Hormonal imbalances	Bretveld et al. 2006, Leemans et al. 2019
Neurodegenerative diseases	Baldi et al. 2003, Baltazar et al. 2014
Respiratory diseases	Salameh et al. 2006, Mamane et al. 2015
Cardio-vascular disease	Sekhotha et al. 2016, Berg Zara et al. 2019
Diabetes	Swaminathan 2013, Juntarawijit and Juntarawijit 2018
Reproductive disorders	Bretveld et al. 2006, Frazier 2007
Cancer	Dich et al. 1997, Alavanja et al. 2013

of which 11,000 end up in fatalities (Boedeker et al. 2020). In addition, pesticides exposure could also result in many other long-term harmful effects (Table 2).

The pesticides along with their direct adverse reactions affect human well-being and also cause unfavourable effects on the ecosystem and diversity (Bürger et al. 2008, Mariyono 2008). Insecticides pollute soil, fields, and water bodies, which can be toxic to aquatic animals, birds, and can even instigate severe modifications in soil and crop-friendly earthworms, predators, and pollinators, thereby resulting in disturbances in their natural selection and evolution (Nicolopoulou-Stamati et al. 2016, Hassaan and El Nemr 2020). For instance, the pesticides (chlorfluazuron), sprayed fields grazed by cattle's lead to unsafe adulterated meat and rejection of exports. The birds feeding on dead animals, including cattle get pesticide exposure. However, the most notorious instance of prolonged chronic exposure is definitely through insecticide dichloro-diphenyl-trichloroethane (DDT, an organochlorine insecticide), which results in bird mortality and can reduce reproduction due to eggshell thinning (described as 'Biomagnification') that lead to its narrowed use from the 1970s onwards. This presses the utmost need to switch towards improved alternatives for pest management.

3. Green Pest Management (GPM)

To circumvent the aforementioned irreparable damages caused by pesticides and to meet the ever-increasing demand for organic and healthy food (free from the use of chemical pesticides) that are environmentally healthy and safe (e.g., soil fertility, Carbon foot-printing, fossil-fuel conservation, landscape and bio-diversity preservation, etc.), newer methods of pest-management can be adopted (El-Shafie 2019). These include the use of various 'Green Pest Management' (GPM), also called 'Integrated Pest Management' procedures. GPM implies controlling the pest attack/damage by embracing eco-friendly techniques with a positive impact on agro-ecosystem and sustainable agriculture (Kabir and Rainis 2015). These methods are moderately economical, highly target-specific, and prevent secondary pest out-breaks with negligible residual effect than the conventional pesticides, which are intrinsically less noxious and compatible when more than one method is applied

(Popp et al. 2013). Owing to these aforesaid properties, GPM in recent times has gained popularity and has attained its position as an efficient and effective method, particularly in controlling the pests that are otherwise challenging to restrain with chemical pesticides (Popp et al. 2013). GPM methods have also opened the avenues for use of natural/green pesticides over conventional ones, which still dominate our current agriculture practices. Broadly, the effective GPM strategies rely on the integration of various biological controls, use of resistant crop varieties, habitat manipulation, modification of cultural practices, mechanical, and microbial controls for achieving the goal of safe pest-management practices (Dara 2019). Describing in a more systematic way, GPM methods include (1) biological-control techniques, (2) allelopathic control (viz. natural phytotoxins, essential oils, sex pheromones, etc.), (3) microbial-control approaches, (4) genetic-manipulations such as production of genetically modified (GM) crops, (5) cultural-control (viz. crop rotation, inter-cropping, sowing, harvesting date manipulations, etc.), and last but not least (6) non-toxic physical- and chemical-approaches (viz. handpicking, mulching, heat treatment, etc.). Table 3 below is a compilation of various GPM methods and their underlying principle/mode of action.

With this background information, herein, we are trying to recapitulate some of these modern GPM technologies with the objective of discussing better, safer, and more economical crop yields for future generations.

Table 3. Various green strategies and their mode of action in the management of pest.

Type of Green Methods	Mode of Actions	References
Biological control	Minimize the pest population by the action of natural control agents (predators, parasitoidal insects, parasitic nematodes, etc.), allelopathy (pheromones, essential oil, etc.)	(Sanda and Sunusi 2016, Agarwal and Verma 2020)
Allelopathic control	Naturally occurring allelochemicals, phytotoxins, or bio-communicators, essential oils, sex pheromones, etc.	(Rice 1995)
Microbial pesticides	Naturally occurring or genetically modified organisms that produce species-specific toxins, resulting in the death of the host.	(Usta 2013)
Genetic manipulation	Methods reduce the fitness of the pest population by manipulating its genetic component.	(Alphey et al. 2010)
Cultural control	Methods reduce the pest population by adopting good agronomic practices.	(Smit and Matengo 1995)
Physical and mechanical treatments	Chemical-free method of pest control through the targeted application of handpicking, heat or solarization, etc.	(Hansen et al. 2011)

4. Green-Methods for Pest Control

GPM is an ecology-centred method, which is based on the principle of continuing crops shielding from pests and their harmful effects by engaging different techniques. A flow-chart explaining the different type of modern strategies employed in GPM are depicted in Figure 2.

The next section deals with the elaboration of these methods for better understanding:

4.1 Biological Control

The biological enemies can cause a significant reduction in pest density, which includes plant pathogens, invertebrates, weeds, etc. The practise of using natural rivals against targeted pests and reducing their population to a sizeable extent, thereby abolishing or curbing financial losses is counted under 'biological control' (Kaur and Garg 2014). Conventionally, the term 'biological control' refers to predators, parasitoids, nematodes, pathogens, etc., consuming other animals for sustenance. All of them exploit the range of varied trapping approaches plus manners in seizing or consuming their prey. Although an older practice, biological control has revamped attention because of inconveniencies faced with the use of chemical pesticides (Hajek and Eilenberg 2018). Currently, the three classes of usual-rivals of pests are discussed below.

Predators:

Distinct predators consume specific bugs. Among them, arthropods are commonly treated as biological control since they consume many insects as prey (Hajek and Eilenberg 2018, Nazir et al. 2019). They also have a short life-cycle, and thus their population fluctuates with variations in the crowdedness of their target. Some notable predators include dragon flies, mantids, lady beetles, scorpion flies, ground beetles, flower bugs, rove beetles, lacewings, true bugs, alder flies, damsel flies,

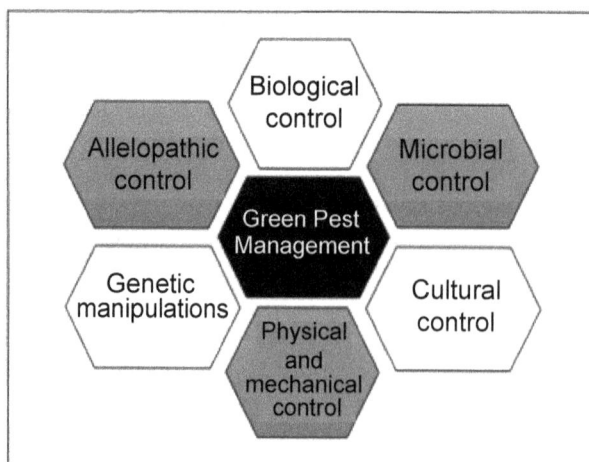

Figure 2. Flow chart explaining the different types of modern strategies employed in GPM.

feather-legged assassin bugs, hover flies, etc. (Croft 1990). In addition, spiders, nematodes, predatory mites, centipedes, slugs, predatory snails, planarian worms, and some mites sp. are also considered as predators of insects (Schöller and Flinn 2012). Here, the prey's death happens straight away after the capture.

Parasitoids:

Parasites are frequently tinier than their host and parasitoids are organisms in which larvae mature or else it grows inside the host, ultimately exterminating it (Colmenarez et al. 2018). These are narrow in their host range but are highly specialized. The grownups, however, are conventional non-parasitic and may be predators, which could survive on additional supplies too, including plant-nectar, honey-dew, pollen, etc. (Wang et al. 2019). Therefore, the accurate detection of the host and parasitoid is critical for use in biological control. Parasitoids comprise classes of wasps, beetles, worms (e.g., Gordian worms), and flies (e.g., tachinid flies) (Feener Jr and Brown 1997). Some such examples are rove beetle sp. which infects snake eggs, spiders, and wasps (Blouin-Demers and Weatherhead 2000, Korenko and Pekár 2011, Polidori et al. 2020). In these cases, the larva emerges and develops by nursing on the biological-fluids or the interior tissues of the host, ultimately released out of the host's body.

Pathogens:

Insects, like other creatures, are infested by disease-causing bacteria, fungi, protozoans, and viruses, which may shrink their proportion of eating, slower development, and even thwart their propagation or eradicate them. The insects might be stricken with nematodes sp. and their bacterial symbionts, which results in such outcomes. In suitable ecological conditions (like high insect density), the execution of a suitable biological control method could serve the purpose. One such example is *Entomophaga maimaiga*, an exclusive fungus pathogenic to the gypsy moth, *Lymantria dispar* (Hajek 1999). The fungal latent spores grow only when on this moth larvae. The moth maggots get disseminated by wind and while crawling on trees, in suitable rainfall, the fungus spores in their bodies grow, multiply, and spread to other worms. The infested young ones expire in sizeable figures and latent spores survive until the next cycle. The success of this method is largely dependent on rain at appropriate times (Tobin and Hajek 2012).

Looking at the choices of biological methods and their selection for effective application (host-specificity) in various geographical locations, monitoring, and partnerships, the International Code of Best Practices was formulated by the U.S. Department of Agriculture (USDA) to provide help in the appropriate selection of measures that mitigate threats and boost the constructive effect of 'Bio-control'. Apart from the previously mentioned biological method for pest control, the bio-herbicide based weed management system based on allelopathic approaches has been also gaining popularity in recent times (described in a following separate section).

4.2 Allelopathic Control

The existing inclination to discover a resolution to diminish the dangerous effects of conventional pesticides in agriculture has led to the inception of a concept termed 'allelopathy'. Allelopathy is a multi-dimensional phenomenon, wherein a diverse array of natural compounds (from either root, shoot, leave, or flower of a plant) are discharged into the environment, which in turn influences the agro-ecosystem dynamics in a decisive manner (Rice 1995). The compounds responsible for allelopathy are termed 'allelochemicals', 'phytotoxins', or 'bio-communicators'. These allelopathic molecules are created by a 'donor' and diffused to a 'receiver' is either 'wounded' or 'stirred', with later being a rare phenomenon. These allelochemicals are classified either as non-nutritional primary-metabolites (e.g., fatty-acids, non-protein amino acids, etc.) or secondary metabolites (e.g., benzoquinones, flavonoids, terpenoids, triketones, coumarins, strigolactones, phenolic acids, tannins, lignin, etc.) that are formed as side-shoots in the chief biochemical plants' pathways (He et al. 2019). According to their different structures and properties, allelochemicals can be broadly classified into 14 categories (Rice 1974):

1. Water-soluble organic acids, straight-chain alcohols, aliphatic aldehydes, and ketones

2. Long-chain fatty acids and polyacetylenes

3. Simple unsaturated lactones

4. Quinines (quinone and quinines)

5. Simple phenolics (benzoic acid and its derivatives)

6. Cinnamic acid and its derivatives

7. Flavonoids

8. Coumarins

9. Tannins

10. Steroids and terpenoids (sesquiterpene lactones, diterpenes, and triterpenoids)

11. Alkaloids and cyanohydrins

12. Amino acids and peptides

13. Purines and nucleosides

14. Sulphide and glucosinolates

Notably, in recent times, plant growth controllers, including gibberellic acid, salicylic acid, and ethylene are also counted as allelochemicals. Various such regular molecules have the likelihood to stimulate a varied array of biological outcomes and can deliver impressive gains in cultivation and weed controlling (Duke and Lydon 1987, Macías et al. 2006). Generally, allelochemicals penetrate the ground as existent plant-active molecules, e.g., momilactones, cyanamide, phenolic acids, heliannuols, etc. On the other hand, certain allelochemicals need to be altered into effective forms by microbes or by definite ecological conditions (temperature, pH, light, oxygen, moisture, etc.). For example, benzoxazolin-2-one, juglone, and 2-amino-3-H-phenoxazin-3-one. In one such instance, the plants of *Brassica* genus, in particular, are known to suppress the weeds by releasing the glucosinolates in soil, which upon conversion to isothiocyanate acts as a toxic metabolite to other plants (Petersen et al. 2001). Likewise, leguminosae plants such as pea (*Pisum sativum*), winter vetch

(*Vicia villosa*), velvet bean (*Mucuna pruriens*), alfalfa (*Medicago sativa*), soybean (*Glycine max*), chickpea (*Cicer arietinum*), red clover (*Trifolium pretense*), and cowpea (*Vigna unguiculata*) display allelopathic properties (Huber and Abney 1986, Qasem 1998, Yasmin et al. 1999, Akemo et al. 2000, Fujii 2001, Dang Xuan et al. 2003, Kato-Noguchi 2003, Kamo et al. 2006, Xiao et al. 2006, Yan and Yang 2008, Narwal and Haouala 2013).

Allelopathic compounds affect the growth and development of their receiver plants by interfering with various physiological processes, including respiration, photosynthesis, hormonal, and water balance. The primary reason for their activity is predominantly restraining enzymes activity (Nimbal et al. 1996, Dayan et al. 2009). This can also be exploited to eliminate weeds that remain unaffected by commercial weed killers with a similar mechanism. These features make them worth considering for probable use as weed killers or bio-herbicides. Table 4 is a compilation of different allelochemicals obtained from plants, which display inhibitory ability on seed sprouting and weed growth.

Overall, allelochemicals appear highly attractive and a new class of herbicides due to their variety of advantages. Being eco-friendly, safe, natural, relatively unstable with low shelf life coupled with structural and functional diversity, allelochemicals offer an effective and specific weed-control strategy. On the contrary, a number of limitations are also reported in their possible use. For example, high dose-dependent effect on acceptor plants, low recovery, rapid degradation, limited efficacy, and specificity of many allelochemicals are major concerns. It is worth mentioning that for any allelochemical to become a future herbicide, some crucial requirements need to be met. These include phytotoxic action on the scale of 10^{-5}–10^{-7} M, acknowledged chemical structure and mechanism of activity, and period of stay in the land likely to have an effect on microbial ecology and unintended plants, probable toxicity to our well-being and profitability during production in marketable dimension (Bhowmik and Inderjit 2003). Considering all these achievable obligations, it seems that allelochemicals could serve as an alternative in weed management strategy.

Besides, the strategy to change the behaviour of the pest can also be utilized to biologically control the pest through traps, baits, and mating interfering techniques with the help of sex pheromones (Morrison et al. 2016, Heinz et al. 1992, Foster and Harris 1997, Shorey and Gerber 1996). Baits enclosing venomous molecules fascinate and destroy pests after dispersal in the field or position in tricks. Pests are mesmerised to specific colours, light, odours of attractant, or pheromone. The strategies that use these chemicals attract them to get trapped or kill pests. Pheromones allure and puzzle mature insects, finally disturbing their coupling aptitude and reducing their progenies (Barclay and Judd 1995). Some of the successful combinations of pest and mating disrupting sex pheromones are listed in Table 5.

The Essential oils (EO) are volatile secondary metabolites, which are present in restricted families like Rutaceae, Myrtaceae, Apiaceae, Lauraceae, Lamiaceae, Asteraceae, Poaceae, Cupressaceae, Zingiberaceae and Piperaceae and also show allelopathic behaviour. EO and their derivative compounds are employed for restraining harmful insects because of their characteristic odour (Tripathi et al. 2009). Almost 300 different oils are used for the industrious purpose, including

Table 4. Plant sources of various allelopathic compounds and their inhibitory activity against different weeds, adapted from Soltys et al. (2013).

Allelopathic Compound	Plant Source	Weeds
Sorgoleone	Sorghum	Little seed canary grass (*Phalaris minor* Retz.), Swine wartcress (*Coronopus didymus* L.), Coco-grass (*Cyperus rotundus* L.), Blackberry nightshade (*Solanum nigrum* L.), Common tumbleweed (*Amaranthus retroflexus* L.), Common ragweed (*Ambrosia atrtemisiflora* L.), and Chinese senna (*Senna obtusifolia* L.).
Glucosinolates, Isothiocyanates	Mustard and Radish	Prickly sow-thistle (*Sonchus asper* L.), Wild chamomile (*Tripleurospermum inodorum*), Slim amaranth (*Amaranthus hybridus*), Cockspur (*Echinochloa cruss-galli* L. Beauv.), Black twitch (*Alopecurus myosuroides* Huds.), Morning glory (*Convolvulus arvensis*), Dodders (*Cuscuta* spp.), Wild carrot (*Daucus carota*), and Shortfruit hedgemustard (*Sisymbrium polyceratium*).
Artemisinin	Sweet wormwood	Redroot pigweed (*Amaranthus retroflexus*), White morning-glory (*Ipomoea lacunose*), and Common duckweed (*Lemna minor*).
Momilactone	Rice and Hypnum moss	Barnyard grass, (*Echinochloa colonum*), Livid amaranth (*Amaranthus lividus* L.), Purple crabgrass (*Digitaria sanguinalis* L.), and Poa (*Poa annua* L.).
Leptospermone	Bottle brush and Manuka	Barnyard millet (*Echinochloa cruss-galli* L. Beauv.), Hairy crabgrass (*Digitaria sanguinalis* L.), Yellow foxtail (*Setaria glauca* L.), and Yellow dock (*Rumex crispus* L.).
Sarmentine	Pepper	Barnyard millet (*Echinochloa cruss-galli* L. Beauv.), Redroot pigweed (*Amaranthus retroflexus* L.), Poa (*Poa annua* L.) and Yellow dock (*Rumex crispus* L.).

Table 5. Some cases of pest mating interference by sex pheromones in crops.

Crop	Pest	Sex pheromone	Reference
Grape	Grapevine Moth (*Lobesia botrana*)	(7E,9Z)-7,9-dodecadienyl acetate	(Schmitz et al. 1997)
		(E, Z)-7,9,-dodecadienol (Z)-9-dodecanyl acetate	(Torres-Vila et al. 2002)
Cotton	Pink Bollworm (*Pectinophora gossypiella*)	(Z, Z)/(Z,E)-7,11-hexadecadien-1-yl acetate	(Flint and Merkle 1984, Carde et al. 1998)
Pome fruit Stone fruit	Oriental Fruit Moth (*Grapholita molesta*)	(Z)-8-dodecenyl acetate and (E)-8-dodecenyl acetate, and (Z)-8-dodecen-1-ol	(Charlton and Carde 1981, Il'Ichev et al. 2006, Stelinski et al. 2007a)
Pome fruit	Codling Moth (*Cydia pomonella*)	(E,E)-8, 10-dodecadien-1-ol	(Pfeiffer et al. 1993)

Table 6. Essential oils of some plants having biological control activity.

Plant	Insect	References
Citrus limonum	*Tenebrio molitor*	(Wang et al. 2015)
Carum carvi	*Meligethes aeneus*	(Pavela 2011)
Artemisia scoparia	*Callosobruchus maculates*	(Negahban et al. 2006)
Ocimumbasilicum	*Aphis craccivora*	(Sammour et al. 2011)
Curcuma longa	*Agrotisipsilon*	(Abdelaziz et al. 2014)
Origanum vulgare	*Nezaraviridula*	(Gonzalez et al. 2011)
Artemisia judaica	*Ephestiakuehniella*	(Ayvaz et al. 2010)
Zataria multiflora	*Brevicorynebrassicae*	(Motazedian et al. 2014)
Anethum graveolens	*Callosobruchus chinensis*	(Chaubey 2008)
Mentha microphylla	*Sitophilus oryzae*	(Mohamed and Abdelgaleil 2008)
Laurelia sempervirens	*Triboliumcastaneum*	(Zapata and Smagghe 2010)
Cymbopogon nardus	*Euprosternaelaeasa*	(Hernandez-Lambrano et al. 2014)
Saturejakhuzistanica	*Leptinotarsa decemlineata*	(Saroukolai et al. 2014)
Citrus paradise	*Rhyzoperthadominica*	(Abbas et al. 2012)
Eucalyptus globules	*Agrotisipsilon*	(Jeyasankar 2012)
Ageratum conyzoides	*Callosobrochus maculates*	(Aboua et al. 2010)
Trachyspermumammi	*Triboliumcastaneum*	(Chaubey 2007)
Salvia bracteata	*Callosobruchus maculates*	(Shakarami et al. 2005)
Thymus kotschyanus	*Callosobruchus maculates*	(Akrami et al. 2011)

pesticides. Being lipophilic in structure, they interfere with various biochemical and physiological reactions in insects. They are exploited for their specificity and environmental neutral behaviour. Different oils have different mechanisms, some are larvicidal, delay development and adult emergence, induce antifeedant behaviour, or alter fertility. However, despite these advantageous properties, their volatile nature and poor solubility are the limitations in tapping their full potential. Some examples include linalool, menthol, geraniol, verbenol, citronellol, Ocimene, etc. The insect targeted with EO's includes orders of Lepidoptera, Coleoptera, Diptera, Isoptera, Hemiptera, etc. Cinnamon oil with cinnamaldehyde is used in *Cinnamite*, a commercial aphidicide/miticide/fungicide for glasshouse; *Valero* is a miticide/ fungicide in fields of grapes, berry crops, citrus, nuts, etc. EO has a herbal origin and thus are not always subject to strict testing procedures but may not always be safe for use. Biological control activities of some of the essential oils of plants are summarized in Table 6.

4.3 *Microbial-Control*

The process of controlling the pest population with the help of either natural or genetically altered microbes or their secretions comes under 'microbial-control'. It is included as a green method, as it is not only ecologically effective against pests, but also do not leave any traceable quantity of toxic residues in the food and fields

(Usta 2013). This is the reason that microbial pesticides are harmless for application on crops, even just before harvesting time. It involves the use of entomopathogenic bacteria (including the families *Xenorhabdus* and *Photorhabdus*), fungi (including *Beauveria bassiana*, *Metarhizium anisopliae*, and *Entomophagama imaiga*), and nematodes (including the families *Heterorhabditidae* and *Steinernematidae*) microsporidia or viruses. Even the fermentation by-products of some of the microbes against plant-parasitic nematodes, arthropod pests and plant-pathogens are also useful in pest-management (Paulitz and Belanger 2001, Hajek and Delalibera 2010, Dong and Zhang 2006). The mechanism of action of microbial entomopathogens involves invasion in the gut of pest followed by its multiplication there, which ultimately leads to the death of the pest (Kalha et al. 2014). Most of the secretions of microbes are peptide in nature but vary greatly in their structure, toxicity, and specificity (Burges 1981). Some of the microbes, their targets and mechanism of action are listed in Table 7. The advantage of using microscopic organisms for control lies in the fact that microbes offer target-specificity towards pests and do not harm non-targeted organisms, including humans. Besides that, the chances of emergence of pest resistance are almost negligible than conventional products (Qadri et al. 2020). It is generally employable in the form of a spray and effective against Colorado beetle (or elm leaf beetle) larvae and fungi gnat larvae. Among the various microbes used for pest control, the most well studied must be the *Bacillus* bacterium and discharged *Bt,* which is a protein/toxin (Cry protein) after being ingested by pests, gets proteolytically cleaved in the intestine, oligomerized and is inserted in the membrane to create pores. The purified Cry protein is crystallized to understand its structure and complete mechanism of action (Sanchis and Bourguet 2008).

In addition, microbial control also augments plant and root growth by enriching and stimulating the helpful soil microbes and thus, in addition to pest control, it also increases the crop yields (Usta 2013). A characteristic example involving the use of microbial-method is the development of a talc-based formulation of fluorescent *Pseudomonads,* which were first developed for the management of potato seed tubers and their development (Kloepper and Schroth 1981). Although the drawbacks of the method are irrefutable, like being species-specific, only a controlled quantity of pests can be restrained by the use of one type of microbe. Besides, they have a brief shelf life and sensitivity to surroundings (low moisture levels, temperature, UV exposure, etc.) is also a major concern (Lacey and Georgis 2012). Thereby, one cannot rely entirely on microbial control but have to employ other green methods in combination as well.

4.4 *Genetic Manipulations*

Internal host plant resistance toward pests, established through genetic-manipulations (GM) or biotechnological approaches, also comes under GPM technologies. GM approaches for harvest enhancement include initiating variations, choosing beneficial modifications, and studying the chosen descents and crossbreeds. With the establishment of GM procedures, it becomes feasible to introduce unusual genes of interest into the plant genetic material that bestows endurance against pests. Various genetic factors from microbes, like trypsin inhibitor, ribosome-inactivating proteins,

Table 7. The mechanism of action of some common pathogens against their targeted pest.

Pathogen	Target pest	Mechanism	Reference
Bacteria			
Bacillus thuringiensis (Bacillaceae)	Lepidoptera, Diptera, Coleoptera, etc.	Invasion of the midgut epithelium	(Sanchis 2010, Eski et al. 2017)
B. popilliae	*Popillia japonica* (Japanese beetle)	Penetration into the haemocoel leading to death	(Rippere et al. 1998, Stahly et al. 2006)
Serratia entomophila (Enterobacteriaceae)	*Costelytragiveni* (Grass grub)	Invasion of foregut and cessation of feeding	(Hurst et al. 2000, Wright et al. 2017)
Xenorhabdus nematophila	Lepidoptera, Coleoptera, Diptera	Symbiont in the entomopathogenic nematode	(Ji et al. 2004, Stilwell et al. 2018)
Fungi			
Verticillim lecanii	Greenhouse whitefly, aphids	Fungal toxicity and colonization of vital tissues	(Sinha et al. 2016)
Beauveria bassiana (Cordycipitaceae)	Greenfly, fungus gnats,	Fungal toxicity and colonization of vital tissues	(Lacey et al. 2015)
Metarhiziumanisopliae	Termites, mosquitoes, and cattle ticks; Various ticks and beetles; root weevils, flies, gnats, thrips.	Conidia land on the host and germinate into the cuticle	(Lacey et al. 2015, Aw and Hue 2017)
Lagenidiumgiganteum	Mosquito (larvae)	Biflagellate zoospores attach to the larval cuticle, proliferates in the host.	(Suh and Axtell 1999)
Viruses			
Nuclear polyhedrosis viruses (NPV)	Lepidoptera	Capsid dissolves in the alkaline midgut to release the virus particle causing cell lysis	(Chiu et al. 2012a)
Ns NPV (*Neodiprion sertifer*)	Pine sawfly	Capsid dissolves in the alkaline midgut to release the virus particle causing cell lysis	(Podgwaite et al. 1984, Chiu et al. 2012b)
Ag NPV (*Anticarsia gemmatalis*)	Velvet bean caterpillar	Capsid dissolves in the alkaline midgut to release the virus particle causing cell lysis.	(Castro et al. 1997, Chiu et al. 2012c)
Protozoa			
Nosema pyrausta	European corn borer (*Ostrinia nubilalis*)	Disrupts larval development, pupation, adult longevity, oviposition and fecundity.	(Gassmann and Clifton 2017)

Table 7 contd. ...

...Table 7 contd.

Pathogen	Target pest	Mechanism	Reference
Nosema locustae	European cornworm	Infects fat body tissues; disrupting host metabolism and energy storage	(Solter et al. 2012a)
Vairimorphanecatrix	Armyworm (Noctuidae)	Spore consumed by the larva, penetrating a midgut cell, and inoculating it with sporoplasm	(Solter et al. 2012b)

lectins, vegetative insecticidal proteins, secondary plant metabolites, and small RNAs are used for tolerance buildup in various plants (Sharma and Ortiz 2000). Most GM crops offer tolerance to lepidopteran and coleopteran pests effectively, and a large area around the world is covered by these GM crops (Figure 3). The most popular example of pest protection by GM method is imparted by insect killer toxin derived from *Bacillus thuringiensis* (*Bt*) gene that has been introduced into cotton, brinjal, soybean, and maize (Anderson et al. 2019). Transgenic *Bt* crops express insecticidal toxins (mainly Cry proteins) derived from *Bacillus thuringiensis* that is highly effective in dissolving the midgut of bug, thereby causing the death of pest, is discussed already (Tabashnik et al. 2013). Studies reveal that these crystalline toxins remain biological inactive until they are solubilised by the gut proteases (Gill et al. 1992). The earliest *Bt* crop cultivars were *Bt* toxin Cry1Ab producing corn and *Bt* toxin Cry1Ac generating cotton (Tabashnik et al. 2009). *Bt* corn can control pests, such as corn earworm, south-western corn borer, and European corn borer, whereas *Bt* cotton effectively controls the cotton bollworm, tobacco budworm, and pink bollworm (Mendelsohn et al. 2003).

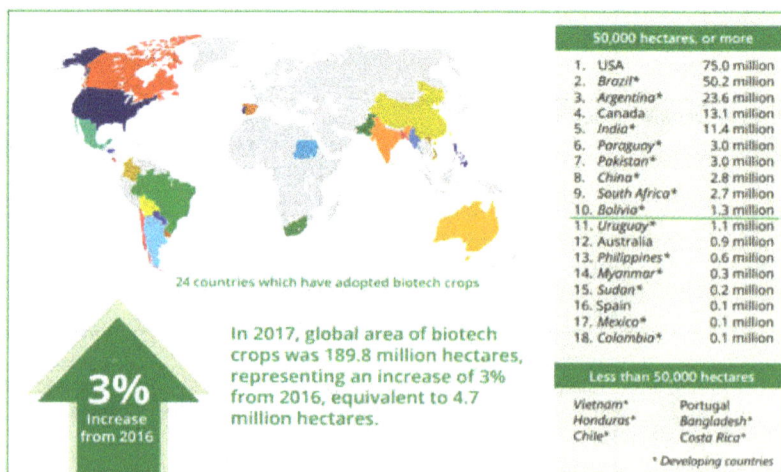

50,000 hectares, or more

1.	USA	75.0 million
2.	Brazil*	50.2 million
3.	Argentina*	23.6 million
4.	Canada	13.1 million
5.	India*	11.4 million
6.	Paraguay*	3.0 million
7.	Pakistan*	3.0 million
8.	China*	2.8 million
9.	South Africa*	2.7 million
10.	Bolivia*	1.3 million
11.	Uruguay*	1.1 million
12.	Australia	0.9 million
13.	Philippines*	0.6 million
14.	Myanmar*	0.3 million
15.	Sudan*	0.2 million
16.	Spain	0.1 million
17.	Mexico*	0.1 million
18.	Colombia*	0.1 million

24 countries which have adopted biotech crops

In 2017, global area of biotech crops was 189.8 million hectares, representing an increase of 3% from 2016, equivalent to 4.7 million hectares.

3% Increase from 2016

Less than 50,000 hectares

Vietnam*	Portugal
Honduras*	Bangladesh*
Chile*	Costa Rica*

* Developing countries

Figure 3. Global area (million hectares) of biotech crops, 1996 to 2017, by country, mega countries, and for the top ten countries (Source: ISAAA 2017).

In recent times, an innovative ribonucleic acid interference (RNAi) procedure—wherein double-stranded RNA (dsRNA) are used to suppress the expression of particular target genetic factors—can also be applied to provide protection from pest incidences (Gordon and Waterhouse 2007). The length, stability, and high levels of expression of dsRNA in the gut of the target pests determine the application of plant-mediated RNAi practices (Zhang et al. 2017). Lately, the advancement of SmartStax PRO maize by Bayer got approval for MON87411 in Canada and the USA to regulate *Diabrotica virgifera* is a landmark in the application of this expertise (Head et al. 2017).

Another GM based control method is the sterile insect technique (SIT), which effectively brings down the pest populace, by circulating the sterile insects (prepared by chromosomal disruption through genetic manipulation as well as through irradiation), which then contest for partners with the wild population. When the circulated sterile insects copulate with the wild types, the resultant eggs do not produce young ones due to the impairment in the genome in the parent's germ lines. Thus, it is clear that SIT is a mating based (biology) method and not chemistry for tackling pest populations. The effectiveness of this method lies in the release of an adequate number of sterile insects over a longer period. On the positive side, being species-specific, it does not cause any direct effect on non-targeted organisms in the environment (Alphey and Bonsall 2018). The *Cochliomyia hominivorax* (Coquerel) or New World screwworm was exterminated from the U.S.A., Costa Rica, Panama, Mexico, Belize, Guatemala, El Salvador, Honduras, Nicaragua, and some Caribbean islands by engaging the matching technique (Dyck et al. 2005).

Besides reduction in dependence on the use of pesticides, all these aforementioned methods are compatible with other green pest management tactics (Mohankumar and Ramasubramanian 2014). However, being target-specific, the practical implication of GM based pest-control method against the broad spectrum of pests is a key disadvantage. Also, the risk of horizontal gene transfer from GM organisms is a major concern (Keese 2008). Moreover, the development and use of transgenic plants and seeds are a costly, time-consuming, and legitimate affair that relies entirely on the necessary approval from the concerned regulatory and government bodies. The general public acceptability is another major issue that resists the switching to global use of GM crops.

4.5 Cultural Control

Cultural control is one of the most practical and convenient methods that rely on the adoption of good agronomical practices, including manipulation of the pest micro-environment in such a way that makes it less promising for pests to survive, although supporting the circumstances favourable for high crop productivity (All 2005). It is employed while commencing with sowing by choosing the clean and healthier seeds/plant material, thereby avoiding the possibility of introducing pests in the planted area (Schellhorn et al. 2000). Even alteration of sowing dates also helps in evading the pest infections or avoiding them during the most vulnerable stages of initial development. For instance, in Texas, USA sorghum is sown during early May month, instead of June, as it is premature and avoid the pest break due to panicle hindering pests (Archer et al. 1990). Similarly, shifting of harvesting date has been also found to impart promising results, as perceived with the early harvesting of cotton crop in Northern India, which lessens the pink bollworm mediated damage (Sundaramurthy and Gahukar 1998). The plant density or row-spacing is another factor that can be tinkered for achieving the desired results. For example, the high-density planting of canola has been demonstrated to reduce the root maggot (*Delia* sp.) attack in Canada (Dosdall et al. 2011). Likewise, modifying the irrigation drills, compost schedule, and other agronomical exercises have been found to create a suitable and conducive environment for reduced pest infestation. Hodson and Lampinen reported the reduction in the intensity of arthropods by increasing leaves' nitrogenous matter (Hodson and Lampinen 2019). Destroying crop residue is another good green cultural practice, which eliminates the existing reproducing spots and soil-dwelling phase of the pest. Leach et al. reported a significant reduction of spotted-wing drosophila (*Drosophila suzukii*) insects by the application of sanitation methods (Leach et al. 2018).

Along with that, crop-rotation with non-host tolerant/resistant host varieties also negatively influences the pest's life-cycle, thus bringing the much-needed pest control has been already practised in many parts of the world (Liebman and Dyck 1993, Curl 1963, Wright 1984). Inter-cropping with non-host plants traps, catch crops, or the ones that are unfavorable for pest growth commendably shields the main crop by diverting the pests away (Pretty and Bharucha 2015). The increase in numerous-cropping systems establishes the superiority and number of new natural enemy complexes (Landis et al. 2000). Like the tomato crops in India, a tall variety of African marigold substantially suppress the damage by the pest *Helicoverpa armigera* (Srinivasan et al. 1994). Thus, the cultural practices for pest control could be very beneficial as well economical in the long run.

4.6 Physical and Mechanical Methods

Various physical or mechanical strategies are also employed for pest elimination (Webb and Linda 1992). These include pest prohibition with the use of mesh or row-covers, hand-picking or vacuuming, or modifying the environmental conditions, such as heat, humidity levels in the green-houses, heat sterilization, or solarization which are some common methods employed (Gogo et al. 2014, Porto et al. 2017). One of the easiest methods of control is manual handpicking or hand pulling of different weeds and pests which works best when pests are clearly visible (Sorensen et al. 2016). Among the various other methods, heat treatment (HT) or thermal pest control is one of the most studied and diversified post-harvest methods for pest control. Porto et al. reported that HT (an air temperature > 45°C for 9 hours) to control pests in flour mills is possible by thermal analyzes and temperature trend models. However, extra caution should be given to the thermal bridges in this thermal envelope, so that the surface temperatures remain deadly to insects (Porto et al. 2017). The thermal sprays used for HT do not penetrate the hard outer shell, thus the success of this method relies on heat carried through the Entotherm system, which can effectively kill all the stages of insects, i.e., egg, larva, pupa, and adult through dehydration wherein temperature should not exceed beyond 56–60°C (Agarwal and Verma 2020).

Although it cannot be ruled out that the HT is a complex treatment, directed by the use of sophisticated computers and thermal sensors, which adds to the cost and is a deterrent factor in its execution (Hansen et al. 2011).

Intermittently, covering the soil with plastic sheets to raise the temperature (solarization), thereby killing various pests is another viable option. This is done at the pre-plant stage and captures solar radiations for 4–6 weeks during summers (Gill et al. 2017). Mulching through vegetative matter or polythene-film (plastic) reduces the evaporation in soil and restores the humidity, which enhances the number of beneficial soil microbes and suppresses the weed development. Phyto-sanitation focuses on the elimination and demolition of the diseased plant, which is a probable home for the spread of bugs or disease. A few successfully contained soil-borne microbes include *Rhizoctonia* sp., *Fusarium* sp., *Clavibacter michiganensis*, *Sclerotinia* sp., *Phytophthora* sp., *Macrophomina* sp., *Verticillium* sp., *Agrobacterium tumefaciens*, *Pythium* sp., and *Streptomyces scabies* (Ahmad et al. 1996, McGovern and Mcsorley 1997, Chellemi and Mirusso 2006, Gelsomino et al. 2006). It is pertinent to mention here that the achievement rate is dependent on this single method alone, which is not agreeable in pest management. Thus, different permutations and combinations of various suitable alternative methods would prove helpful in achieving the goal of healthy agricultural practices. Indeed, this is the reason that people are shifting towards organic farming and demand for organic products are gaining popularity (described in the following section).

5. Organic Farming and The Rise of Green Pesticides

'Organic Farming' or 'Organic Agriculture' is an ecologically unified production/farming method, which centres around maintaining soil fertility and biological diversity, crucial in achieving the sustainable-agriculture, and good human health (Eyhorn et al. 2019). In simpler term, organic farming involves the use of traditional, innovative, and scientific methods, which relies on green methods like the application of bio-fertilizers, biological pest control, microbial pest control, crop rotation, etc., without depending on chemical/synthetic fertilizers, insecticides, pesticides, weedicides, and growth promoters (El-Shafie 2019) (Figure 4). However, it should be noted that organic farming does not support GM crops. It is seen that the crop yield is relatively low in organic farming than in conventional farming but is more friendly to the pollinators and environment and provides more nutritious foods with substantially lesser to almost negligible pesticide residues (Seufert et al. 2012, Ponisio et al. 2015, Kennedy et al. 2013, Seufert and Ramankutty 2017). Instead of using conventional fertilizer, this method advocates the use of natural manures, bone meal, and shell-based fertilizers. Green pesticides are a core component of organic farming, occupied ~ 2% of the worldwide crop protection and pesticide sales in 2010 with an observed increasing trend of 7–10% every year (McDougall 2010). To increase yield in organic farming, the seeds pre-soaked or 'primed' with overnight soaking dissolves germination inhibitors and enhances the germination rate significantly. Similarly, pelleting done to escalate the seed size by adding an oxygen-permeable external layer (like clay) increases/streamlines the roundness of especially small seeds, helps in uniform sowing, and upsurges the plant ability of

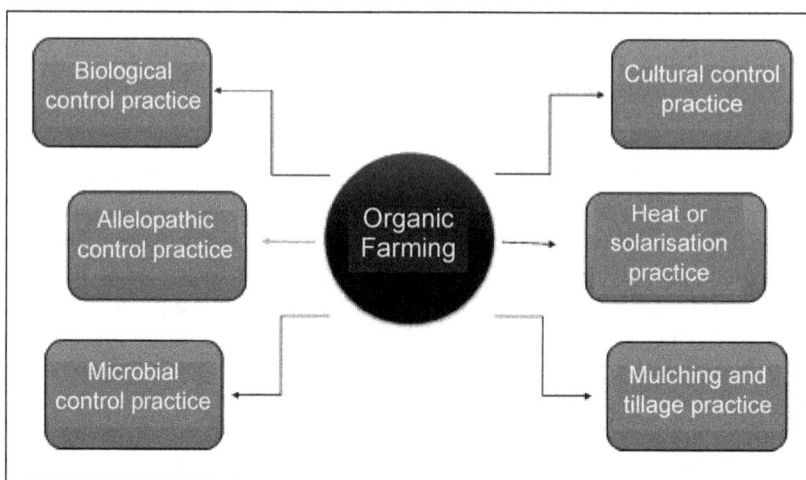

Figure 4. Various green practices involved in organic farming.

seeds. The controlled hot water treatment or diluted bleach (sodium hypochlorite) treatment is another option for disinfection and better results. Besides that, various commercial green pesticides, derived from EOs (as discussed in preceding-section) or microorganisms are permitted in organic farming.

Thus, it is understandable that the employment of these healthier green methods is the reason behind organic farming gaining immense popularity globally. Scientifically, organic farming enhances the biodiversity of the ecosystem and also harmoniously maintains ecological homeostasis (Gomiero et al. 2011). Among different countries, Australia has the largest farmland area under organic farming followed by Europe, China, and India. In India, particularly Sikkim has already achieved its ambition of converting to 100% organic farming back in 2016 itself. Whereas other states including Kerala, Goa, Mizoram, Rajasthan and Meghalaya have also announced their targets to shift to complete organic farming.

6. Conclusions, Challenges and Future Prospective

It is unanimously agreed that rampant use of chemical pesticides in pest management, although improved productivity significantly, has undesirable consequences on human health and the environment that often outweigh other benefits. These chemicals' entry into the food chains has reached the highest trophic levels along with the development of resistance in pests. The substantial decline in the density of beneficial organisms, such as predators, pollinators, earthworms, etc., lead to a disbalance among various microbes' populations in soil and water which is a serious concern.

In this scenario, the awareness and resulting enthusiasm for food produced through green methods and its economic viability at both farmer's and consumer's level make it conceivable to replace the crop produced with conventional practices. However, like two sides of the coin, limitations do exist in the implementation of

GPM strategies. For example, GPM cannot be employed in every situation but can be exercised conditionally, e.g., natural vegetal defiance is effective in some harvests where pest and infection impervious or tolerant varieties are available (Dara 2019). Additionally, green methods in the fields showed relatively slower impacts as compared to chemical pesticides. Moreover, it should be understood that although GPM-based food is generally considered safe and sustainable, its harvest is still not entirely free from harmful pesticides as certain green pesticides display negative impacts too (Dara 2019). A few green extracts, i.e., azadirachtin and pyrethrins, and microbe-derived toxic metabolites, i.e., avermectin and spinosad, are even though interpreted as biological (Lasota and Dybas 1991), their chemical structure is still similar to synthetic molecules and thus exhibit few safety threats (Larkin 2008). Furthermore, like conventional pesticides, pests can also acquire resistance to green pesticides (Dara 2019). In one such instance, arthropod resistance to abamectin and insects resistance to *Bacillus thuringiensis* (*Bt*) formulations or its *Cry* toxins have been already recorded in GM crops (Stumpf and Nauen 2002, Tabashnik et al. 2013). Likewise, insect resistance towards spinosad has been also testified (Scott 2008). An additional challenge with GM is possible in case of pest shift, i.e., increased population of a secondary pest that used to be collaterally reduced by chemical insecticides in traditional practice but not with GM crops farming. For example, widespread adoption of *Bt* cotton in China has led to lessened reliance on chemical insecticides, which had boosted the mirid bugs populace (Hemiptera: Miridae) in some fields (Lu et al. 2010). Mechanical pest control practices such as vacuuming or tilling utilize fossil fuels, which have a negative influence on the environment. All plant diseases and arthropod pests cannot be managed through green or non-chemical methods (Finckh et al. 2006). Manures used in organic farms can have a considerable carbon footprint and nitrate shortage in the root zone, and nitrate leaching are also frequently noticed in organic fields (Tal 2018). Tuomisto et al. reported both positive and negative impacts of organic farming and emphasize on a mix of both organic and chemical management for cutting the negative environmental impacts without affecting the yields (Tuomisto et al. 2012).

Food security for the growing world population is necessary for regulating the costs, thus new and better green methods in combating various pest diseases and producing healthier crops without compromising yields are need of the hour. The relative safety and better revenues of crops by green methods perfectly align with the popularity of organic farming. Firstly, their use can be done in a phase-wise manner, either one method at a time or separate ones in combination. To achieve this, scientists, extension workers, and policy planners need to work as a cohesive unit in consultation with farmers, who undoubtedly have a better understanding of their land and ways to improve the fertility of the soil. Furthermore, research and development are constantly needed along with publicising, spreading awareness, and giving guidance to the farmers to explain the short-term and long-term benefits of these options.

References

Abbas, S., Ahmad, F., Sagheer, M., Hasan, M., Yasir, M., Saeed, A. and Wali, M. 2012. Insecticidal and growth inhibition activities of *Citrus paradisi* and *Citrus reticulata* essential oils against lesser grain borer, *Rhyzopertha dominica* (F.) (Coleoptera: Bostrichidae). World J. Zool 7(4): 289–294.

Abdelaziz, N.F., Salem, H.A. and Sammour, E.A. 2014. Insecticidal effect of certain ecofriendly compounds on some scale insects and mealybugs and their side effects on antioxidant enzymes of mango nurslings. Arch. Phytopathol. Plant Prot. 47(1): 1–14.

Aboua, L.R.N., Seri-Kouassi, B.P. and Koua, H.K. 2010. Insecticidal activity of essential oils from three aromatic plants on *Callosobruchus maculatus* F. in Côte d'Ivoire. Eur. J. Sci. Res. 39: 243–250.

Agarwal, A., Prajapati, R. Singh, O.P., Raza, S.K. and Thakur, L.K. 2015. Pesticide residue in water—a challenging task in India. Environ. Monit. Assess. 187(2): 54.

Agarwal, M. and Verma, A. 2020. Modern technologies for pest control: a review. In Biotechnology in Mining and Metallurgical Industry. Intechopen Limited, London.

Ahmad, Y., Hameed, A. and Aslam, M. 1996. Effect of soil solarization on corn stalk rot. Plant Soil 179(1): 17–24.

Akemo, M.C., Regnier, E.E. and Bennett, M.A. 2000. Weed suppression in spring-sown rye (*Secale cereale*)–pea (*Pisum sativum*) cover crop mixes. Weed Technol. 14: 545–549.

Akrami, H., Moharramipour, S. and Imani, S. 2011. Comparative effect of *Thymus kotschyanus* and *Mentha longifera* essential oil on oviposition deterrence and repellency of *Callosobruchus maculates* F. I. J. Med. Aro. Plants. 27: 1–10.

Alavanja, M.C.R., Ross, M.K. and Bonner, M.R. 2013. Increased cancer burden among pesticide applicators and others due to pesticide exposure. CA. Cancer J. Clin. 63(2): 120–142.

All, J. 2005. Cultural control of insect pests. pp. 649–649. *In*: Encyclopedia of Entomology. Springer Netherlands, Dordrecht.

Alphey, N., Bonsall, M. and Alphey, L. 2010. Modeling resistance to genetic control of insects. J. Theor. Biol. 270(1): 42–55.

Alphey, N. and Bonsall, M.B. 2018. Genetics-based methods for agricultural insect pest management. Agric. For. Entomol. 20(2): 131–140.

Anderson, J.A., Ellsworth, P.C., Faria, J.C., Head, G.P., Owen, M.D.K., Pilcher, C.D., Shelton, A.M. and Meissle, M. 2019. Genetically engineered crops: importance of diversified integrated pest management for agricultural sustainability. Front. Bioeng. Biotechnol. 7: 24.

Archer, T., Losada, J. and Bynum, E. 1990. Influence of planting date on abundance of foliage-feeding insects and mites associated with sorghum. J. Agri. Entomol. 7(3): 221–231.

Aw, K.M.S. and Hue, S.M. 2017. Mode of infection of *Metarhizium* spp. fungus and their potential as biological control agents. J. Fungi Basel Switz. 3(2): 30.

Ayvaz, A., Sagdic, O., Karaborklu, S. and Ozturk, I. 2010. Insecticidal activity of the essential oils from different plants against three stored-product insects. J. Insect Sci. Online 10: 21.

Baldi, I., Lebailly, P., Mohammed-Brahim, B., Letenneur, L., Dartigues, J.-F. and Brochard, P. 2003. Neurodegenerative diseases and exposure to pesticides in the elderly. Am. J. Epidemiol. 157(5): 409–414.

Baltazar, M.T., Dinis-Oliveira, R.J., de Lourdes Bastos, M., Tsatsakis, A.M., Duarte, J.A. and Carvalho, F. 2014. Pesticides exposure as etiological factors of Parkinson's disease and other neurodegenerative diseases—a mechanistic approach. Toxicol. Lett. 230(2): 85–103.

Barclay, H. and Judd, G. 1995. Models for mating disruption by means of pheromone for insect pest control. Res. Popul. Ecol. 37: 239–247.

Berg Zara, K., Rodriguez Beatriz, Davis James, Katz Alan, R., Cooney Robert, V. and Masaki Kamal. 2019. Association between occupational exposure to pesticides and cardiovascular disease incidence: the Kuakini Honolulu heart program. J. Am. Heart Assoc. 8(19): e012569.

Bhowmik, P.C. and Inderjit. 2003. Challenges and opportunities in implementing allelopathy for natural weed management. Crop Prot. 22: 661–671.

Blouin-Demers, G. and Weatherhead, P.J. 2000. A novel association between a beetle and a snake: Parasitism of *Elaphe obsoleta* by *Nicrophorus pustulatus*. Écoscience 7(4): 395–397. Taylor & Francis.

Boedeker, W., Watts, M., Clausing, P. and Marquez, E. 2020. The global distribution of acute unintentional pesticide poisoning: estimations based on a systematic review. BMC Public Health 20: 1875.

Bretveld, R.W., Thomas, C.M., Scheepers, P.T., Zielhuis, G.A. and Roeleveld, N. 2006. Pesticide exposure: the hormonal function of the female reproductive system disrupted? Reprod. Biol. Endocrinol. 4: 30.

Bürger, J., de Mol, F. and Gerowitt, B. 2008. The "necessary extent" of pesticide use—Thoughts about a key term in German pesticide policy. Crop Prot. 27(3-5): 343–351.

Burges, H.D. 1981. Safety, safety testing and quality control of microbial pesticides. Microb. Control Pests Plant Dis. 1970–1980.

Cardé, R.T., Staten, R.T. and Mafra Neto, A. 1998. Behaviour of pink bollworm males near high-dose, point sources of pheromone in field wind tunnels: insights into mechanisms of mating disruption. Entomol. Exp. Appl. 89(1): 35–46.

Castro, M.E., Souza, M.L., Araujo, S. and Bilimoria, S.L. 1997. Replication of *Anticarsia gemmatalis* nuclear polyhedrosis virus in four lepidopteran cell lines. J. Invertebr. Pathol. 69(1): 40–45.

Charlton, R.E. and Cardé, R.T. 1981. Comparing the effectiveness of sexual communication disruption in the oriental fruit moth (*Grapholitha molesta*) using different combinations and dosages of its pheromone blend. J. Chem. Ecol. 7: 501–508.

Chaubey, M. 2007. Insecticidal activity of *Trachyspermum ammi* (Umbelliferae), *Anethum graveolens* (Umbelliferae) and *Nigella sativa* (Ranunculaceae) essential oils against stored-product beetle *Tribolium castaneum* Herbst (Coleoptera: Tenebrionidae). Afr. J. Agric Res. 2.

Chaubey, M.K. 2008. Fumigant toxicity of essential oils from some common spices against pulse beetle, *Callosobruchus chinensis* (Coleoptera: Bruchidae). J. Oleo Sci. 57(3): 171–179.

Chellemi, D.O. and Mirusso, J. 2006. Optimizing soil disinfestation procedures for fresh market tomato and pepper production. Plant Dis. 90(5): 668–674.

Chiu, E., Coulibaly, F. and Metcalf, P. 2012. Insect virus polyhedra, infectious protein crystals that contain virus particles. Curr. Opin. Struct. Biol. 22(2): 234–40.

Colmenarez, Y.C., Corniani, N., Jahnke, S.M., Sampaio, M.V. and Vásquez, C. 2018. Use of parasitoids as a biocontrol agent in the neotropical region: challenges and potential. In Horticulture, IntechOpen Limited.

Croft, B.A. 1990. Arthropod biological control agents and pesticides. Environ. Sci. Technol. USA 20(5): 1492.

Curl, E.A. 1963. Control of plant diseases by crop rotation. Bot. Rev. 29: 413–479.

Dang Xuan, T., Tsuzuki, E., Terao, H., Matsuo, M., Khanh, T., Murayama, S. and Hong, N. 2003. Alfalfa, rice by-products and their incorporation for weed control in rice. Weed Biol. Manag. 3: 137–144.

Dara, S.K. 2019. The new integrated pest management paradigm for the modern age. J. Integr. Pest Manag. 10(1): 12.

Dayan, F.E., Howell, J. and Weidenhamer, J.D. 2009. Dynamic root exudation of sorgoleone and its in planta mechanism of action. J. Exp. Bot. 60: 2107–2117.

Dich, J., Zahm, S.H., Hanberg, A. and Adami, H.O. 1997. Pesticides and cancer. Cancer Causes Cont. 8(3): 420–443.

Dong, L.Q. and Zhang, K.Q. 2006. Microbial control of plant-parasitic nematodes: a five-party interaction. Plant Soil 288: 31–45.

Dosdall, L.M., Herbut, M.J., Cowle, N.T. and Micklich, T.M. 2011. The effect of seeding date and plant density on infestations of root maggots, *Delia* spp. (Diptera: Anthomyiidae), in canola. Can. J. Plant Sci. 76(1): 169–177.

Duke, S.O. and Lydon, J. 1987. Herbicides from natural compounds. Weed Technol. 1: 122–128.

Durán-Lara, E.F., Valderrama, A. and Marican, A. 2020. Natural organic compounds for application in organic farming. Agriculture 10(2): 41.

Dyck, V., Hendrichs, J. and Robinson. 2005. Sterile insect technique principles and practice in area-wide integrated pest management.

Ellgehausen, H., Guth, J.A. and Esser, H.O. 1980. Factors determining the bioaccumulation potential of pesticides in the individual compartments of aquatic food chains. Ecotoxicol. Environ. Saf. 4(2): 134–157.

El-Shafie, H.A.F. 2019. Insect pest management in organic farming system. In Multifunctionality and Impacts of Organic Agriculture. IntechOpen, Limited.

Eski, A., Demir, İ., Sezen, K. and Demirbağ, Z. 2017. A new biopesticide from a local *Bacillus thuringiensis* var. *tenebrionis* (Xd3) against alder leaf beetle (Coleoptera: Chrysomelidae). World J. Microbiol. Biotechnol. 33(5): 95.

Eyhorn, F., Muller, A., Reganold, J.P., Frison, E., Herren, H.R., Luttikholt, L., Mueller, A., Sanders, J., Scialabba, N.E.-H., Seufert, V. and Smith, P. 2019. Sustainability in global agriculture driven by organic farming. Nat. Sustain. 2(April 2019): 253–255.

Feener Jr, D.H. and Brown, B.V. 1997. Diptera as Parasitoids. Annu. Rev. Entomol. 42: 73–97.

Finckh, M.R., Schulte-Geldermann, E. and Bruns, C. 2006. Challenges to organic potato farming: disease and nutrient management. Potato Res. 49(1): 27–42.

Flint, H.M. and Merkle, J.R. 1984. The pink bollworm (Lepidoptera: Gelechiidae): alteration of male response to *Gossyplure* by release of its component Z,Z- Isomer. J. Econ. Entomol. 77(5): 1099–1104.

Foster and, S.P. and Harris, M.O. 1997. Behavioral manipulation methods for insect pest-management. Annu. Rev. Entomol. 42(1): 123–146.

Frazier, L.M. 2007. Reproductive disorders associated with pesticide exposure. J. Agromedicine 12(1): 27–37.

Fujii, Y. 2001. Screening and future exploitation of allelopathic plants as alternative herbicides with special reference to hairy vetch. J. Crop Prod. 4: 257–275.

Gassmann, A.J. and Clifton, E. 2017. Current and potential applications of biopesticides to manage insect pests of maize. pp. 173–184. *In*: Microbial Control of Insect and Mite Pests: From Theory to Practice.

Gelsomino, A., Badalucco, L., Landi, L. and Cacco, G. 2006. Soil carbon, nitrogen and phosphorus dynamics as affected by solarization alone or combined with organic amendment. Plant Soil 279(1): 307–325.

Gill, H.K., Aujla, I.S., De Bellis, L. and Luvisi, A. 2017. The role of soil solarization in india: how an unnoticed practice could support pest control. Front. Plant Sci. 8.

Gill, S.S., Cowles, E.A. and Pietrantonio, P.V. 1992. The mode of action of *Bacillus thuringiensis* endotoxins. Annu. Rev. Entomol. 37(1): 615–634.

Gogo, E.O., Saidi, M., Ochieng, J.M., Martin, T., Baird, V. and Ngouajio, M. 2014. Microclimate modification and insect pest exclusion using agronet improve pod yield and quality of french bean. HortScience 49(10): 1298–1304.

Gomiero, T., Pimentel, D. and Paoletti, M.G. 2011. Environmental impact of different agricultural management practices: conventional vs. organic agriculture. Crit. Rev. Plant Sci. 30(1-2): 95–124.

Gordon, K.H.J. and Waterhouse, P.M. 2007. RNAi for insect-proof plants. Nat. Biotechnol. 25(11): 1231–1232.

Hajek, A.E. 1999. Pathology and epizootiology of *Entomophaga maimaiga* infections in forest Lepidoptera. Microbiol. Mol. Biol. Rev. 63(1): 814–835.

Hajek, A.E. and Delalibera, I. 2010. Fungal pathogens as classical biological control agents against arthropods. BioControl 55(1): 147–158.

Hajek, A.E. and Eilenberg, J. 2018. Natural Enemies: An Introduction to Biological Control. Cambridge University Press.

Hansen, J.D., Johnson, J. and Winter, D.A. 2011. History and use of heat in pest control: a review. Int. J. Pest Manag. 57: 267–289.

Hassaan, M.A. and El Nemr, A. 2020. Pesticides pollution: Classifications, human health impact, extraction and treatment techniques. Egypt. J. Aquat. Res. 46(3): 207–220.

He, S., Wang, S.Q., Wang, Q.Y., Zhang, C., Zhang, Y.M., Liu, T.Y., Yang, S.X., Kuang, Y., Zhang, Y.X., Han, J.Y. and Qin, J.C. 2019. Allelochemicals as growth regulators: a review. Allelopathy J. 48: 15–26.

Head, G.P., Carroll, M.W., Evans, S.P., Rule, D.M., Willse, A.R., Clark, T.L., Storer, N.P., Flannagan, R.D., Samuel, L.W. and Meinke, L.J. 2017. Evaluation of SmartStax and SmartStax PRO maize against western corn rootworm and northern corn rootworm: efficacy and resistance management. Pest Manag. Sci. 73(9): 1883–1899.

Heinz, K.M., Parrella, M.P. and Newman, J.P. 1992. Time-efficient use of yellow sticky traps in monitoring insect populations. J. Econ. Entomol. 85(6): 2263–2269.

Hernández-Lambraño, R., Caballero-Gallardo, K. and Olivero-Verbel, J. 2014. Toxicity and antifeedant activity of essential oils from three aromatic plants grown in Colombia against *Euprosterna elaeasa* and *Acharia fusca* (Lepidoptera: Limacodidae). Asian Pac. J. Trop. Biomed. 4: 695–700.

Hodson, A.K. and Lampinen, B.D. 2019. Effects of cultivar and leaf traits on the abundance of Pacific spider mites in almond orchards. Arthropod-Plant Interact. 13(3): 453–463.

Huber, D.M. and Abney, T.S. 1986. Soybean allelopathy and subsequent cropping. J. Agron. Crop Sci. 157: 73–78.

Hurst, M.R., Glare, T.R., Jackson, T.A. and Ronson, C.W. 2000. Plasmid-located pathogenicity determinants of *Serratia entomophila*, the causal agent of amber disease of grass grub, show similarity to the insecticidal toxins of *Photorhabdus luminescens*. J. Bacteriol. 182(18): 5127–5138.

Il'Ichev, A.L., Stelinski, L.L., Williams, D.G. and Gut, L.J. 2006. Sprayable microencapsulated sex pheromone formulation for mating disruption of oriental fruit moth (Lepidoptera: Tortricidae) in Australian peach and pear orchards. J. Econ. Entomol. 99(6): 2048–2054.

Isman, M.B. 2019. Challenges of pest management in the twenty first century: new tools and strategies to combat old and new foes alike. Front. Agron. 1.

Jeyasankar, A. 2012. Antifeedant, insecticidal and growth inhibitory activities of selected plant oils on black cutworm, *Agrotis ipsilon* (Hufnagel) (Lepidoptera: Noctuidae). Asian Pac. J. Trop. Dis. 2: S347–S351.

Ji, D., Yi, Y., Kang, G.-H., Choi, Y.-H., Kim, P., Baek, N.-I. and Kim, Y. 2004. Identification of an antibacterial compound, benzylideneacetone, from *Xenorhabdus nematophila* against major plant-pathogenic bacteria. FEMS Microbiol. Lett. 239(2): 241–248.

Juntarawijit, C. and Juntarawijit, Y. 2018. Association between diabetes and pesticides: a case-control study among Thai farmers. Environ. Health Prev. Med. 23.

Kabir, M. and Rainis, R. 2015. Do farmers not widely adopt environmentally friendly technologies? lesson from integrated pest management (IPM). Mod. Appl. Sci. 9(3): 208.

Kalha, C.S., Singh, P.P., Kang, S.S., Hunjan, M.S., Gupta, V. and Sharma, R. 2014. Entomopathogenic viruses and bacteria for insect-pest control. pp. 225–244. *In*: Abrol, D.P. (ed.). Integrated Pest Management. Academic Press, San Diego.

Kamo, T., Kato, K., Hiradate, S., Nakajima, E., Fujii, Y. and Hirota, M. 2006. Evidence of cyanamide production in hairy vetch *Vicia villosa*. Nat. Prod. Res. 20: 429–433.

Kato-Noguchi, H. 2003. Isolation and identification of an allelopathic substance in *Pisum sativum*. Phytochemistry 62: 1141–1144.

Kaur, H. and Garg, H. 2014. Pesticides: environmental impacts and management strategies.

Keese, P. 2008. Risks from GMOs due to horizontal gene transfer. Environ. Biosafety Res. 7: 123–149.

Kennedy, C.M., Lonsdorf, E., Neel, M.C., Williams, N.M., Ricketts, T.H., Winfree, R., Bommarco, R., Brittain, C., Burley, A.L., Cariveau, D., Carvalheiro, L.G., Chacoff, N.P., Cunningham, S.A., Danforth, B.N., Dudenhöffer, J.-H., Elle, E., Gaines, H.R., Garibaldi, L.A., Gratton, C., Holzschuh, A., Isaacs, R., Javorek, S.K., Jha, S., Klein, A.M., Krewenka, K., Mandelik, Y., Mayfield, M.M., Morandin, L., Neame, L.A., Otieno, M., Park, M., Potts, S.G., Rundlöf, M., Saez, A., Steffan-Dewenter, I., Taki, H., Viana, B.F., Westphal, C., Wilson, J.K., Greenleaf, S.S. and Kremen, C. 2013. A global quantitative synthesis of local and landscape effects on wild bee pollinators in agro ecosystems. Ecol. Lett. 16(5): 584–599.

Kloepper, J.W. and Schroth, M.N. 1981. Development of a Powder Formulation of [Plant Growth-Promoting] Rhizobacteria for Inoculation of Potato Seed Pieces. Phytopathol. USA.

Korenko, S. and Pekár, S. 2011. A parasitoid wasp induces overwintering behaviour in its spider host. PLOS ONE 6: e24628.

Lacey, L.A. and Georgis, R. 2012. Entomopathogenic nematodes for control of insect pests above and below ground with comments on commercial production. J. Nematol. 44(2): 218–225.

Lacey, L.A., Grzywacz, D., Shapiro-Ilan, D.I., Frutos, R., Brownbridge, M. and Goettel, M.S. 2015. Insect pathogens as biological control agents: Back to the future. J. Invertebr. Pathol. 132: 1–41.

Landis, D.A., Wratten, S.D. and Gurr, G.M. 2000. Habitat management to conserve natural enemies of arthropod pests in agriculture. Annu. Rev. Entomol. 45(1): 175–201.

Larkin, R.P. 2008. Relative effects of biological amendments and crop rotations on soil microbial communities and soilborne diseases of potato. Soil Biol. Biochem. 40(6): 1341–1351.

Lasota, J.A. and Dybas, R.A. 1991. Avermectins, A novel class of compounds: implications for use in arthropod pest control. Annu. Rev. Entomol. 36(1): 91–117.

Leach, H., Moses, J., Hanson, E., Fanning, P. and Isaacs, R. 2018. Rapid harvest schedules and fruit removal as non-chemical approaches for managing spotted wing Drosophila. J. Pest Sci. 91(11): 219–226.

Leemans, M., Couderq, S., Demeneix, B. and Fini, J.-B. 2019. Pesticides with potential thyroid hormone-disrupting effects: a review of recent data. Front. Endocrinol. 10.

Liebman, M. and Dyck, E. 1993. Crop rotation and intercropping strategies for weed management. Ecol. Appl. 3: 92.

Lu, Y., Wu, K., Jiang, Y., Xia, B., Li, P., Feng, H., Wyckhuys, K.A.G. and Guo, Y. 2010. Mirid bug outbreaks in multiple crops correlated with wide-scale adoption of Bt cotton in China. Science 328: 1151–1154.

Macías, F.A., Chinchilla, N., Varela, R.M. and Molinillo, J.M.G. 2006. Bioactive steroids from *Oryza sativa* L. Steroids 71: 603–608.

Mamane, A., Baldi, I., Tessier, J.-F., Raherison, C. and Bouvier, G. 2015. Occupational exposure to pesticides and respiratory health. Eur. Respir. Rev. 24: 306–319.

Mariyono, J. 2008. Direct and indirect impacts of integrated pest management on pesticide use: a case of rice agriculture in Java, Indonesia. Pest Manag. Sci. 64(10): 1069–1073.

McDougall. 2010. Phillips McDougall, AgriService, Industry Overview—2009 Market, Vineyard Business Centre Saughl and Pathhead Midlothian EH37 5XP Copyright 2010.

McGovern, R. and Mcsorley, R. 1997. Chapter 12 - Physical methods of soil sterilization for disease management including soil solarization. pp. 283–312. *In*: Rachcigl, N.A. and Rachcigl, J.E. (eds.). Environmentally Safe Approaches to Crop Disease Control. CRC Press.

Mendelsohn, M., Kough, J., Vaituzis, Z. and Matthews, K. 2003. Are Bt crops safe? Nat. Biotechnol. 21(9): 1003–1009.

Mesnage, R., Clair, E., Vendômois, J. and Seralini, G. 2009. Two cases of birth defects overlapping Stratton-Parker syndrome after multiple pesticide exposure. Occup. Environ. Med. 67: 359.

Mohamed, M.I.E. and Abdelgaleil, S.A.M. 2008. Chemical composition and insecticidal potential of essential oils from Egyptian plants against *Sitophilus oryzae* (L.) (Coleoptera: Curculionidae) and *Tribolium castaneum* (Herbst) (Coleoptera: Tenebrionidae). Appl. Entomol. Zool. 43: 599–607.

Mohankumar, S. and Ramasubramanian, T. 2014. Chapter 18 - Role of genetically modified insect-resistant crops in IPM: agricultural, ecological and evolutionary implications. pp. 371–399. *In*: Abrol, D.P. (ed.). Integrated Pest Management. Academic Press, San Diego.

Morrison, W.R., Lee, D.-H., Short, B.D., Khrimian, A. and Leskey, T.C. 2016. Establishing the behavioral basis for an attract-and-kill strategy to manage the invasive *Halyomorpha halys* in apple orchards. J. Pest Sci. 89(1): 81–96.

Mostafalou, S. 2012. Concerns of environmental persistence of pesticides and human chronic diseases. Clin. Exp. Pharmacol. 01.

Motazedian, N., Aleosfoor, M., Davoodi, A. and Bandani, A.R. 2014. Insecticidal activity of five medicinal plant essential oils against the cabbage aphid, *Brevicoryne brassicae*. J. Crop Prot. 3(2): 137–146.

Narwal, S. and Haouala, R. 2013. Role of allelopathy in weed management for sustainable agriculture. pp. 217–249. *In*: Allelopathy.

Nazir, T., Khan, S. and Qiu, D. 2019. Biological control of insect pest. pp. 1–14. *In*: Pests-Insects, Management, Control. IntechOpen Limited.

Negahban, M., Moharramipour, S. and Sefidkon, F. 2006. Insecticidal activity and chemical composition of *Artemisia sieben Besser* essential oil from Karaj, Iran. J. Asia-Pac. Entomol. 9(1): 61–66.

Nicolopoulou-Stamati, P., Maipas, S., Kotampasi, C., Stamatis, P. and Hens, L. 2016. Chemical pesticides and human health: the urgent need for a new concept in agriculture. Front. Public Health 4.

Nimbal, C.I., Yerkes, C.N., Weston, L.A. and Weller, S.C. 1996. Herbicidal activity and site of action of the natural product sorgoleone. Pestic. Biochem. Physiol. 54: 73–83.

Oerke, E.-C., Dehne, H.-W., Schoenbeck, F. and Weber, A. 1994. Chapter 3 - Crop production and crop protection: estimated losses in major food and cash crops. Amsterdam (Netherlands) Elsevier.

Oerke, E.-C. 2006. Crop losses to pests. J. Agric. Sci. 144(1): 31–43. Cambridge University Press.

Paulitz, T. and Belanger, R. 2001. Biological control in greenhouse systems. Annu. Rev. Phytopathol. 39: 103–33.

Pavela, R. 2011. Insecticidal and repellent activity of selected essential oils against of the pollen beetle, *Meligethes aeneus* (Fabricius) adults. Ind. Crops Prod. 34: 888–892.

Petersen, J., Belz, R., Walker, F. and Hurle, K. 2001. Weed suppression by release of isothiocyanates from turnip-rape mulch. Agron. J. 93: 37–43.

Pfeiffer, D.G., Kaakeh, W., Killian, J.C., Lachance, M.W. and Kirsch, P. 1993. Mating disruption to control damage by leafrollers in Virginia apple orchards. Entomol. Exp. Appl. 67(1): 47–56.

Podgwaite, J., Rush, P., Hall, D. and Walton, G. 1984. Efficacy of the *Neodiprion sertifer* (Hymenoptera: Diprionidae) nucleopolyhedrosis virus (*Baculovirus*) product, Neochek-S. J. Econ. Entomol. 77: 525–528.

Polidori, C., Ballesteros, Y., Wurdack, M., Asís, J.D., Tormos, J., Baños-Picón, L. and Schmitt, T. 2020. Low host specialization in the cuckoo wasp, *Parnopes grandior*, weakens chemical mimicry but does not lead to local adaption. Insects 11(2): 136.

Ponisio, L.C., M'Gonigle, L.K., Mace, K.C., Palomino, J., de Valpine, P. and Kremen, C. 2015. Diversification practices reduce organic to conventional yield gap. Proc. Biol. Sci. 282: 20141396.

Popp, J., Pető, K. and Nagy, J. 2013. Pesticide productivity and food security. A review. Agron. Sustain. Dev. 33(1): 243–255.

Porto, S.M.C., Valenti, F., Bella, S., Russo, A., Cascone, G. and Arcidiacono, C. 2017. Improving the effectiveness of heat treatment for insect pest control in flour mills by thermal simulations. Biosyst. Eng. 164: 189–199.

Pretty, J. and Bharucha, Z.P. 2015. Integrated pest management for sustainable intensification of agriculture in Asia and Africa. Insects 6(1): 152–182.

Qadri, M., Short, S., Gast, K., Hernandez, J. and Wong, A.C.-N. 2020. Microbiome innovation in agriculture: development of microbial based tools for insect pest management. Front. Sustain. Food Syst. 4: 547751.

Qasem, J.R. 1998. Chemical control of branched broomrape (*Orobanche ramosa*) in glasshouse grown tomato. Crop Prot. 17: 625–630.

Rappazzo, K.M., Warren, J.L., Meyer, R.E., Herring, A.H., Sanders, A.P., Brownstein, N.C. and Luben, T.J. 2016. Maternal residential exposure to agricultural pesticides and birth defects in a 2003–2005 North Carolina birth cohort. Birt. Defects Res. A. Clin. Mol. Teratol. 106(4): 240–249.

Reddy, K.V.S. and Zehr, U.B. 2004. Novel strategies for overcoming pests and diseases in India. New Dir. Diverse Planet 4th Int. Crop Sci. Congr. Conjunction 12th Aust. Agron. Conf. 5th Asian Crop Sci. Conf. 26 Sept.–1 Oct. 2004 Brisb. Aust. Australian Society of Agronomy Inc.

Rice, E.L. 1974. Chapter 1- Introduction in Allelopathy. Elsevier, Academic Press.

Rice, E.L. 1995. Biological Control of Weeds and Plant Diseases: Advances in Applied Allelopathy. University of Oklahoma Press.

Rippere, K.E., Tran, M.T., Yousten, A.A., Hilu, K.H. and Klein, M.G. 1998. *Bacillus popilliae* and *Bacillus lentimorbus*, bacteria causing milky disease in Japanese beetles and related scarab larvae. Int. J. Syst. Evol. Microbiol. 48(2): 395–402.

Salameh, P., Waked, M., Baldi, I., Brochard, P. and Saleh, B.A. 2006. Respiratory diseases and pesticide exposure: a case-control study in Lebanon. J. Epidemiol. Community Health 60(3): 256–261.

Sammour, E.A., El-Hawary, F.M.A. and Abdel-Aziz, N.F. 2011. Comparative study on the efficacy of neemix and basil oil formulations on the cowpea aphid *Aphis craccivora* Koch. Arch. Phytopathol. Plant Prot. 44(7): 655–670.

Sanchis, V. and Bourguet, D. 2008. *Bacillus thuringiensis*: applications in agriculture and insect resistance management. A review. Agron. Sustain. Dev. 28: 11–20.

Sanchis, V. 2010. From microbial sprays to insect-resistant transgenic plants: history of the biospesticide *Bacillus thuringiensis*. A review. Agron. Sustain. Dev. 31: 217–231.

Sanda, N. and Sunusi, M. 2016. Fundamentals of biological control of pests. Int. J. Chem Biol. Sci. 1(6).

Saroukolai, A.T., Nouri-Ganbalani, G., Hadian, J. and Rafiee-Dastjerdi, H. 2014. Antifeedant activity and toxicity of some plant essential oils to Colorado potato beetle, *Leptinotarsa decemlineata* Say (Coleoptera: Chrysomelidae). Plant Prot. Sci. 50(2014): 207–216.

Savary, S., Willocquet, L., Pethybridge, S.J., Esker, P.D., McRoberts, N. and Nelson, A.D. 2019. The global burden of pathogens and pests on major food crops. Nat. Ecol. Evol. 3: 430–439.

Schellhorn, N., Harmon, J. and Andow, D. 2000. Using cultural practices to enhance insect pest control by natural enemies. pp. 147–170. *In*: Insect Pest Management.

Schmitz, V., Renou, M., Roehrich, R., Stockel, J. and Lecharpentier, P. 1997. Disruption mechanisms of pheromone communication in the European grape moth *Lobesia botrana* Den & Schiff. III. Sensory adaptation and habituation. J. Chem. Ecol. 23(1): 83–95.

Schöller, M. and Flinn, P. 2012. Biological control: Insect pathogens, parasitoids, and predators. pp. 203–212. *In*: Stored Product Protection.

Scott, J.G. 2008. Unraveling the mystery of spinosad resistance in insects. J. Pestic. Sci. 33(3): 221–227.

Sekhotha, M.M., Monyeki, K.D. and Sibuyi, M.E. 2016. Exposure to agrochemicals and cardiovascular disease: a review. Int. J. Environ. Res. Public. Health 13(2).

Seufert, V., Ramankutty, N. and Foley, J. 2012. Comparing the yields of organic and conventional agriculture. Nature 485: 229–32.

Seufert, V. and Ramankutty, N. 2017. Many shades of gray- the context-dependent performance of organic agriculture. Sci. Adv. 3: e1602638.

Shakarami, J., Kamali, K. and Moharamipour, S. 2005. Fumigant toxicity and repellency effect on essential oil of *Salvia bracteata* on four species of warehouse pests. J. Entomol. Soc. Iran. 24: 35–50.

Sharma, H.C. and Ortiz, R. 2000. Transgenics, pest management, and the environment. Curr. Sci. 79(4): 421–437.

Shorey, H.H. and Gerber, R.G. 1996. Use of puffers for disruption of sex pheromone connnunication among navel orangeworm moths (Lepidoptera: Pyralidae) in almonds, pistachios, and walnuts. Environ. Entomol. 25(5): 1154–1157.

Sinha, K.K., Choudhary, A.Kr. and Kumari, P. 2016. Chapter 15 - Entomopathogenic fungi. pp. 475–505. *In*: Omkar (ed.). Ecofriendly Pest Management for Food Security. Academic Press, San Diego.

Smit, N.E.J.M. and Matengo, L.O. 1995. Farmers' cultural practices and their effects on pest control in sweetpotato in South Nyanza, Kenya. Int. J. Pest Manag. 41(1): 2–7.

Solter, L., Becnel, J. and Oi, D. 2012. Microsporidian entomopathogens. pp. 221–263. *In*: Insect Pathology.

Soltys, D., Krasuska, U., Bogatek, R. and Gniazdowska, A. 2013. Chapter 20 - Allelochemicals as bioherbicides—present and perspectives. pp. 517–542. *In*: Herbicides—Current Research and Case Studies in Use. IntechOpen Limited.

Sorensen, K., Subbarayalu, M. and Thangara, S.R. 2016. Physical, mechanical and cultural control of vegetable insects. pp. 131–148. *In*: Muniappan, R. and Heinrichs, E. (eds.). Integrated Pest Management of Tropical Vegetable Crops.

Srinivasan, K., Moorthy, P.N.K. and Raviprasad, T.N. 1994. African marigold as a trap crop for the management of the fruit borer *Helicoverpa armigera* on tomato. Int. J. Pest Manag. 40(1): 56–63.

Stahly, D.P., Andrews, R.E. and Yousten, A.A. 2006. The genus *Bacillus* - Insect pathogens. pp. 563–608. *In*: Dworkin, M., Falkow, S., Rosenberg, E., Schleifer, K.-H. and Stackebrandt, E. (eds.). The Prokaryotes: Volume 4: Bacteria: Firmicutes, Cyanobacteria. Springer US, New York, NY.

Stelinski, L.L., Gut, L.J., Haas, M., McGhee, P. and Epstein, D. 2007. Evaluation of aerosol devices for simultaneous disruption of sex pheromone communication in *Cydia pomonella* and *Grapholita molesta* (Lepidoptera: Tortricidae). J. Pest Sci. 80(4): 225–233.

Stilwell, M.D., Cao, M., Goodrich-Blair, H. and Weibel, D.B. 2018. Studying the symbiotic bacterium *Xenorhabdus nematophila* in individual, living *Steinernema carpocapsae* nematodes using microfluidic systems. mSphere 3(1).

Stumpf, N. and Nauen, R. 2002. Biochemical markers linked to abamectin resistance in *Tetranychus urticae* (Acari: Tetranychidae). Pestic. Biochem. Physiol. 72(2): 111–121.

Suh, C.P. and Axtell, R.C. 1999. *Lagenidium giganteum* zoospores: effects of concentration, movement, light, and temperature on infection of mosquito larvae. Biol. Control 15(1): 33–38.

Sun, B., Zhang, L., Yang, L., Zhang, F., Norse, D. and Zhu, Z. 2012. Agricultural non-point source pollution in china: causes and mitigation measures. AMBIO 41(4): 370–379.

Sundaramurthy, V.T. and Gahukar, R.T. 1998. Integrated management of cotton insect pests in India. Outlook Agric. 27(4): 261–269.

Swaminathan, K. 2013. Pesticides and human diabetes: a link worth exploring? Diabet. Med. J. Br. Diabet. Assoc. 30(11): 1268–1271.

Tabashnik, B.E., Van Rensburg, J.B.J. and Carrière, Y. 2009. Field-evolved insect resistance to Bt crops: definition, theory, and data. J. Econ. Entomol. 102(6): 2011–2025.

Tabashnik, B.E., Brévault, T. and Carrière, Y. 2013. Insect resistance to Bt crops: lessons from the first billion acres. Nat. Biotechnol. 31(6): 510–521.

Tal, A. 2018. Making conventional agriculture environmentally friendly: moving beyond the glorification of organic agriculture and the demonization of conventional agriculture. Sustainability 10(4): 1078.

Tiwari, D. and Alam, M. 2016. Electronic pest repellent: a review. pp. 435–439. *In*: International Conference on Innovations in Information Embedded and Communication Systems.

Tobin, P.C. and Hajek, A.E. 2012. Release, establishment, and initial spread of the fungal pathogen *Entomophaga maimaiga* in island populations of *Lymantria dispar*. Biol. Control 63(2012): 31–39.

Torres-Vila, L.M., Rodríguez-Molina, M.C. and Stockel, J. 2002. Delayed mating reduces reproductive output of female European grapevine moth, *Lobesia botrana* (Lepidoptera: Tortricidae). Bull. Entomol. Res. 92(2): 241–249.

Tripathi, A., Upadhyay, S., Bhuyan, M. and Bhattacharya, P. 2009. A review of essential oils as biopesticide in insect-pest management. J. Pharmacogn. Phytother. 1(5): 000–000.

Trumble, J.T. and Alvarado-Rodriguez, B. 1993. Development and economic evaluation of an IPM program for fresh market tomato production in Mexico. Agric. Ecosyst. Environ. 43(3): 267–284.

Tuomisto, H.L., Hodge, I.D., Riordan, P. and Macdonald, D.W. 2012. Does organic farming reduce environmental impacts?—A meta-analysis of European research. J. Environ. Manage. 112: 309–320.

Usta, C. 2013. Microorganisms in biological pest control—a review (Bacterial toxin application and effect of environmental factors). Curr. Prog. Biol. Res.

Wang, X., Hao, Q., Chen, Y., Jiang, S., Yang, Q. and Li, Q. 2015. The effect of chemical composition and bioactivity of several essential oils on *Tenebrio molitor* (Coleoptera: Tenebrionidae). J. Insect Sci. 15(1): 116.

Wang, Z., Liu, Y., Shi, M., Huang, J. and Chen, X. 2019. Parasitoid wasps as effective biological control agents. J. Integr. Agric. 18(4): 705–715.

Webb, S.E. and Linda, S.B. 1992. Evaluation of spunbonded polyethylene row covers as a method of excluding insects and viruses affecting fall-grown squash in Florida. J. Econ. Entomol. 85(6): 2344–2352.

Werdin González, J.O., Gutiérrez, M.M., Murray, A.P. and Ferrero, A.A. 2011. Composition and biological activity of essential oils from Labiatae against *Nezara viridula* (Hemiptera: Pentatomidae) soybean pest. Pest Manag. Sci. 67(8): 948–955.

Winchester, M.D., Huskins, P.J. and Ying, J. 2009. Agrichemicals in surface water and birth defects in the United States. Acta Paediatr. Oslo Nor. 98: 664–9.

Wright, D.A., Zydenbos, S.M., Wessman, P., O'Callaghan, M., Townsend, R.J., Jackson, T.A., van Koten, C. and Mansfield, S. 2017. Surface coating aids survival of *Serratia entomophila* (Enterobacteriaceae) in granules for surface application. Biocontrol Sci. Technol. 27(12): 1383–1399.

Xiao, C.-L., Zheng, J.-H., Zou, L.-Y., Sun, Y., Zhou, Y.-H. and Yu, J.Q. 2006. Autotoxic effects of root exudates of soybean. Allelopathy J. 18: 121–127.

Xing, K., Zhu, X., Peng, X. and Qin, S. 2014. Chitosan antimicrobial and eliciting properties for pest control in agriculture: A review. Agron. Sustain. Dev. 35: 569–588.

Yan, F. and Yang, Z. 2008. Allelochemicals in pre-cowing soils of continuous soybean cropping and their autointoxication. pp. 271–281. *In*: Zeng, R.S., Mallik, A.U. and Luo, S.M. (eds.). Allelopathy in Sustainable Agriculture and Forestry. Springer, New York, NY.

Yasmin, S., Saleem, B. and Irshad, A. 1999. Allelopathic effects of aqueous extract of chickpea (*Cicer arietinum*) and wheat (*Triticum aestivum* L.) on each other's growth and quality. Int. J. Agric. Biol. 1: 110–111.

Zapata, N. and Smagghe, G. 2010. Repellency and toxicity of essential oils from the leaves and bark of *Laurelia sempervirens* and *Drimys winteri* against *Tribolium castaneum*. Ind. Crops Prod. 32(3): 405–410.

Zhang, J., Khan, S.A., Heckel, D.G. and Bock, R. 2017. Next-generation insect-resistant plants: RNAi-mediated crop protection. Trends Biotechnol. 35(9): 871–882.

Zilberman, D., Schmitz, A., Casterline, G., Lichtenberg, E. and Siebert, J.B. 1991. The economics of pesticide use and regulation. Science 253(5019): 518–522.

CHAPTER 3

Green Fertilizer Technologies in Agriculture

Mohamed Abdelsattar,[1,]* *Vinay Kumar*[2] and
Mohamed A. Abdelwahed[1]

1. Introduction

Since the steam engine invention kicked off the industrial revolution, our world has undergone rapid climate change that involves increasingly severe droughts, depletion of freshwater supplies, seawater acidification, increasing seawater levels, the rapid spread of diseases and parasites, and species extinction. Also, the world's natural resources are limited, some of which have already been destroyed or damaged. Green technology provides us with the best hope to mitigate climate change and pollution impacts and overcome the shortage of resources.

Green technology, also known as sustainable technology, describes the use of technology and science to create environmentally friendly products to protect the environment and, in some cases, to restore past damage.

There are three principles to define sustainability in any resources, which is proposed by Herman Daly (Daly 2007):

i) The exploitation of resources should not be, at any rate, higher than their generation;

ii) The waste generation should not be, at any rate, higher than the ability of an ecosystem to process it;

iii) The depletion of nonrenewable resources should not be, at any rate, higher than the development of renewable substitutes.

[1] Agricultural Genetic Engineering Cotton Research Institute (AGERI), Agricultural Research Center (ARC), 12619 Giza, Egypt.
[2] Department of Community Medicine, Saveetha Medical College, Saveetha Institute of Medical and Technical Sciences, Chennai, India.
* Corresponding author: mteima81@gmail.com

2. Green Fertilizer Technology

Green Fertilizer Technology (GFT) is one of the important technologies in the agricultural sector, which is leading to improved fertilizer applications and sources. The yield of this technology helps in improving the environmental structure and increasing agricultural and food productivity, which participates in the sustainable development of the agriculture sector (Adnan et al. 2019).

Generally, the fertilizers industry can be classified into two main groups called controlled-release fertilizers (CRFs) and slow-release fertilizers (SRFs). Synthesis of CRF and SRF are including some matrix and chemical methods, such as encapsulating the fertilizer within a larger compound. CRF and SRF methods are slowing down the nutrient releases either by dissolution or diffusion. SRF is involved in the formation of organic and inorganic encapsulation.

Both CRFs and SRFs gradually supply nutrients to the plants. However, they vary in many ways, including the usage of technology, the mechanism of release, the stability, the factors controlling the release and much more.

Only soil temperature affects the nutrients release of controlled release fertilizers, while multiple factors (soil moisture, temperate, pH, etc.) affect the nutrients release of SRFs. Therefore, it is much more reliable to use CRFs to release nutrients to the plants.

2.1 Controlled-Release Fertilizers (CRFs)

CRFs are granulated fertilizers that when intercalated within carrier molecules can release the nutrients gradually into the soil. CRFs combined with state-of-art technology can achieve superior results in plant growth. Multi-coated CRFs are made of high-quality granular nutrients which are encapsulated in a multi-layer polymeric coating. Multi-coated granules absorb soil moisture that dissolves inside nutrients. CRF application provides continuous nutrients throughout the plant growth cycle. To consume fertilizer materials with maximum efficiency, nutrients are released at a rate that matches plant requirements and thus fertilizer losses are avoided and pollution of groundwater is prevented. Stable nutrient availability without hazardous excesses or shortages contributes to healthy growth.

2.1.1 Classification of Controlled-Release Fertilizers (CRFs)

Based on Shaviv's grouping, CRFs may be classified as shown in Figure 1.

2.1.2 Mechanism of Nutrients Release From CRFs

The main method in delaying the release of plant nutrients from fertilizers is covering a soluble fertilizer with a protective coating (encapsulation) of a substrate that is water-insoluble, semipermeable, or impermeable-with-pores. This regulates water penetration (hence, the dissolution rate) and ideally synchronizes the release of nutrients with the plants' needs. Most mechanisms show that nutrients released from CRFs are regulated primarily by the diffusion mechanism concerning temperature, coating material thickness, nutrient form, and the presence of the related soil microorganisms (Sempeho et al. 2014).

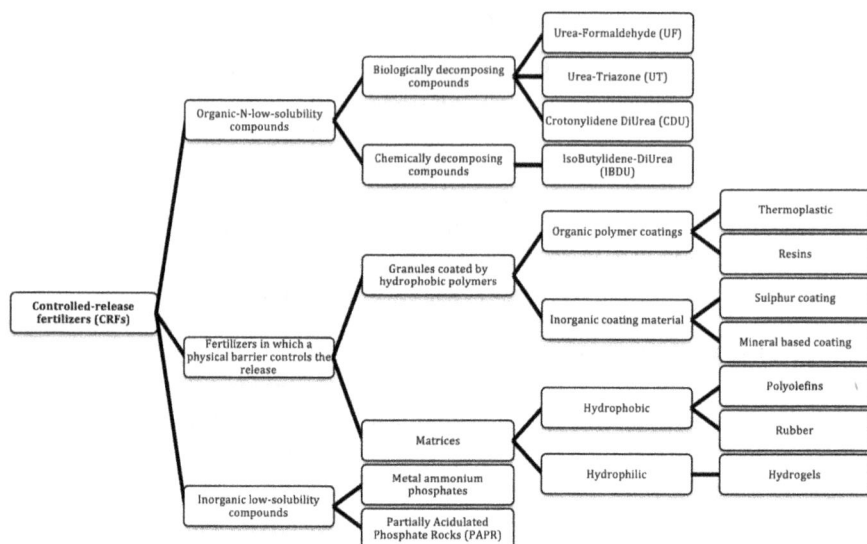

Figure 1. Classification of CRFs.

One of the newest technologies currently available, i.e., nano-technology, to be described later in this chapter, can be used to prepare nano-carriers to control the release of compounds or nutrients (Guo et al. 2018).

2.2 Slow-Release Fertilizers (SRFs)

SRFs are natural fertilizers that attenuate the release of nutrients over a long period and are uncoated like CRFs. The biologically and microbially decomposed nitrogen products, such as Urea-Formaldehyde are commonly referred to in the trade as SRFs and coated or encapsulated products as CRFs. SRFs slowly release the nutrients, but it is due to the chemical hydrolysis and/or microbial activity. The best example for the organic SRFs is compost, e.g., nitrogen products that are decomposed by microbe activity which are pointed to SRFs. SRFs can be organic or inorganic, and their availability is longer than quick-release fertilizers (QRFs), such as potassium chloride, ammonium phosphate, and ammonium nitrate (Liu et al. 2107). Natural SRFs include plant manures, such as green manure, seaweeds, all animal manures (cow, birds, and horse) and vermicompost (Mahmud et al. 2018). Due to seaweeds and vermicomposting technologies have become popular in recent years, so we will highlight their utility as SRFs in the agricultural sector.

2.2.1 The Composting Process Technology

The composting process depends on the activity of micro-organisms, such as bacteria and other organisms like insects and worms. However, this process needs some critical conditions that comprise temperature, moisture, air, acidity, and a type of organic material. The composting cascading process comprises three stages; heating phase or fermentation, cooling phase, and maturation phase.

i) **Heating Phase:**

To achieve this phase, the organic material is broken down by making a heap that helps to accelerate the fermentation process quickly by increasing the temperature from 60°C to 70°C. Also, the existing micro-organisms, oxygen, and water will accelerate the process.

ii) **Cooling Phase:**

In this stage, the temperature generated by the fermentation process is gradually slowed down from 50°C to 30°C. This cooling process helps to emerge new types of microorganisms to convert the organic materials into humus.

iii) **Maturation Phase:**

In this phase, the temperature of fermentation drops to soil temperature from 15°C to 25°C. On the other hand, a large quantity of soil fauna is active, such as earthworms, that are playing a good role in the vermicomposting process. Also, termites do the same role. Finally, the decomposition process faces a slow rate and is ready to be used as brown and/or black organic material as SRF.

2.2.2 Vermicompost as Organic Slow-Release Fertilizer (SRF)

Vermicompost is black colored organic compost product that is produced by utilizing earthworms to convert or turn the green-organic waste into high-quality compost (Ramnarain et al. 2019). Earthworms are classified into epigeic, endogeic, and anecic species based on ecological and trophic functions (Brown 1995, Bhatnagar and Palta 1996). Vermicomposting is a non-thermophilic and bio-oxidative process that earthworms associate with microbes to achieve the process (Pathma and Sakthivel 2012). Vermicomposting helps to convert organic wastes, such as domestic refuse, animal manure, and green agro-wastes (Gajalakshmi and Abassi 2004). The vermicompost is divided into peat-like material with a high-value structure, good aeration, microbial activity, water holding capacity, buffering capacity, and good aeration and thus can be used as a bio-fertilizer to enhance and increase plant growth and its physiological parameters. Also, it can efficiently improve soil fertility which results in a longer decomposing time compared to chemical fertilizers (Edwards et al. 2011, Mahmud et al. 2018). Vermicompost is a good source for some important nutrients for plants, such as N, P, K, Fe, Zn, Cu, Mg, Ca, and Zn when compared with soil (Table 1) (Ramnarain et al. 2019) that in turn increases the efficiency of exchangeable calcium, phosphates, and nitrates for plants (Jindo et al. 2016). Interestingly, vermicompost is a SRF as the mineralization for the materials can occur gradually by microbial decomposition or chemical hydrolysis (Morgan et al. 2009). Many studies investigated the positive effect of vermicompost and manure compost on a wide range of crop species, including legumes, cereals, and vegetables (Subler et al. 1998, Atiyeh et al. 2000). Vermicompost can work directly as enhancers for plant growth-regulating hormones and enzymes; on the other hand, it can indirectly control plant pathogens, pests, and nematodes (Pathma and Sakthivel 2012). To discover the bacterial community interactions in vermicompost, scientists have recently developed metagenomics tools that help to understand the hidden communications between the bacterial communities. These can help us to generate a

Table 1. Chemical characterization of vermicompost vs. soil.

Chemical Contents	Vermicompost	Soil
Mg	4.58 gKg^{-1}	0.9 cmolc/dm^3
P	2.4 gKg^{-1}	6.0 mg/dm^3
N%	1.3	0.06
C%	17.0	0.7
K	19.0 gKg^{-1}	48.0 mg/dm^3
pH	6.3	5.4

large database derived from various environments, especially in soil; vermicompost and plants interactions that reflect on the gene expression changes in host plant to improve plant behavior against abiotic and biotic stresses (Sobti et al. 2012).

2.2.3 Seaweeds as Organic SRF

Seaweeds are among the most important marine sources that are used on a wide commercial scale as fertilizer, pharmaceutical products, and food (Anandhan et al. 2011). Seaweeds are macroscopic marine algae and are present in the marine seabed in relatively shallow coastal waters (Ganapathy and Sivakumar 2013). Recently, seaweed extracts are among the latest natural organic fertilizers that contain a rich and highly effective nutritious source where it promotes plant growth rapidly and increases the yield of crops. Also, it helps the plants to resist some environmental changes (Ganapathy and Sivakumar 2013). As mentioned above, seaweed fertilizers contain a large number of highly active component fertilizers and their organic matter can be slow-release and improve the soil micro-community. There are different forms from seaweed extract, such as Liquid Fertilizers (LF), Seaweed Liquid Fertilizers (SLF), and powder form have been used as a biofertilizer (Patel et al. 2017) and can be used as SRFs (Ogawa and Fujita 2006). Seaweed compromises of macro-element and micro-element nutrients, amino acids, vitamins, and some regulatory and phytohormones, such as cytokinin, auxins, and abscisic acid (ABA) (Patel et al. 2017). Also, seaweed extracts comprise another range from osmoprotectants compounds, such as quaternary ammonium compounds, betaine, proline, sugar-alcohol, and mannitol (Calvo et al. 2007). All of these components are playing distinct roles to increase crop yield and enhancing the plant to tolerate environmental stresses (Patel et al. 2017).

There are many important economic crops and vegetables that could be improved using alginates, such as rice, peanut, tomato, and pepper. Alginates, depolymerized at concentrations of 0.5^{-1} mg ml^{-1}, enhance the growth rate of rice and peanut plants when cultivated under a hydroponic system (Hien et al. 2000). In particular, it has been shown that alginates, depolymerized using γ-radiation, at concentrations of 0.5^{-1} mg ml^{-1} using foliar spraying, increase the dry matter and enhance the biochemical, physiological functions, the growth rate of rice and peanut plants cultivated hydroponically (Hien et al. 2000). After 15 days of alginate (100 ppm), the net of photosynthesis is strong and can increase the growth promotion in peanuts. In tomato, liquid seaweed extracts from *U. lactuca* and *P. gymnospora* have enhanced the best growth seedling. The extracts were applied at lower concentrations than more concentrated extracts (Hernández-Herrera et al. 2014). On the other hand, the evaluation of the seaweeds commercial products deriving from *Ascophyllum nodosum* (An) and *E. maxima* (Em) on plant-parasitic root-knot nematodes (*Meloidogyne chitwoodi* and *Meloidogyne hapla*) infection; the products were affected on hatching, host location, and nematode penetration by second-stage juveniles (J2). *M. chitwoodi* egg masses were reduced from 50 to 100% which significantly reduced the final percentage hatch (Bruno Massa Ngala 2016). In *Brassica rapa* L. under nutrient stress conditions, the root application of *E. maxima* extracts improved the biomass production and nitrate content (NO3-N), antioxidant activity (Ascorbic acid), lipophilic (Trolox), total phenols (gallic acid), nutritional status (higher P and K and lower Na), photosynthetic rate, and chlorophyll content (Di Stasio et al. 2018).

In Glycine max, the seed germination, yield, pigment characteristics, and biochemical parameters were increased by *E. intestinalis* when treated as seaweed liquid fertilizer as follows: seed germination (100%), root (6 cm), shoot length (5 cm), carbohydrates (0.01 mg/g), protein (0.6 mg/g), chl a (0.44 mg/g), chl b (1.07 mg/g), carotenoids (4 mg/g), and phenol content (3 mg/g). Also, benzoic compounds could promote plant growth performance (Mathur et al. 2015). When *Vigna mungo* was treated with *U. reticulate* with 2% of SLF as a foliar system, the plants exhibited the maximum growth rate. Moreover, the number of epidermal and stomatal cells increased, and the size of stomata was higher on the abaxial side than adaxial (Ganapathy-Selvam and Sivakumar 2013).

So, according to everything mentioned above, we will provide a brief about three seaweed species that are used as bio-fertilizer.

Ecklonia maxma

E. maxima or sea bamboo is a Kelp species, which grows only in the clean and cold waters off the Atlantic coast of Southern Africa. *E. maxima* is compromising many important compounds that are helping plants to grow and carry over the environmental stresses, but the most important compound discovered in *E. maxima* is Alginic acid. Alginic acid, algin, or alginate is a hydrophilic polysaccharide compound and distributed in the cell walls of brown algae (Phaeophyceae) (Vera et al. 2011). Alginates are made up of three types of block polymers, namely poly (D-glucuronic acid) blocks (PG), poly (D-mannuronic acid) blocks (PM), alternating blocks of D-glucuronic, and D-mannuronic residues (GM) (Vera et al. 2011). Oligosaccharides are derived from depolymerization of alginates by acid hydrolysis, enzymatic degradation of sodium alginate, or degradation radiation techniques (Woods and Pikaev 1994). When the alginate is used as a stimulator in medicinal plants, such as *Mentha arvensis*, the growth performance, physiological activities, yields, quality, and essential oil content (Naeem et al. 2012). Also, *Catharanthus roseus* is a source of indole alkaloids that are used in cancer chemotherapy, the yield of vinblastine, vincristine, and vindoline contents, which increased when treated under foliar application system with Gamma irradiated sodium alginate (ISA) (Naeem et al. 2015). So, the quality and yield production of several medicinal and aromatic plants can be improved using alginate derivatives.

Interestingly, alginates can improve the structure and condition of soil due to their work as natural chelation. Alginate binds to metal ions that form complex polymers. These polymers can absorb moisture and increase soil aeration and water-holding capacity. Also, alginate improves the environmental conductivity between the plant and rhizosphere growth activity (Spinelli et al. 2010, Battacharyya et al. 2015, Illera-Vives et al. 2015).

Ascophyllum nodosum

A. nodosum, knotted kelp, egg wrack or knotted wrack is a brown alga, which grows on the coast of the Northern Atlantic Ocean, Northwestern coast, Northeastern of North America, and East Greenland. *A. nodosum* is a good source of nutrients and minerals, such as calcium, iron, potassium, lipids, and Ash. Also, *A. nodosum* is a rich source of some important phenolic compounds, such as fucoidans (11%), laminarin (4.5%), mannitol (7.5%), and alginic acid (28%) (Yuan and Macquarrie 2015, Moreira et al. 2016).

Enteromorpha intestinalis

E. intestinalis, sea lettuce, grass kelp, or gutweed, is a green alga in the family Ulvaceae. *E. intestinalis* may be considered to be a good source for the nutrient content (Metin and Baygar 2018).

3. Nanotechnology in Green Technology

Nanotechnology is one of the most important modern science, and its applications are burgeoning day by day. Nanotechnology has a prospect to innovate the feed and agricultural and food sectors (Chunga et al. 2017). Nanotechnology includes nanoparticles, nanotubes, and nanowires (Ramkumar et al. 2016). Metal nanoparticles (NPs) have many applications in different fields, such as environmental monitoring, pharmacy, medicine, and agriculture (Rai et al. 2008, Kumar and Yadav 2009). In the last years, many marine bio-resources have been used in the field of nanotechnology (Vijayan et al. 2014, Chanthini et al. 2015). The NPs can be synthesized by chemical, physical, and biological methods. Synthesis of NPs can be achieved by biological agents, such as microorganisms, plants, and marine organisms (Ramkumar et al. 2016). Interestingly, Seaweeds can be used as NPs-machine that develop NPs with high yield and low cost and are more stable and the rate of synthesis is faster (Ramkumar et al. 2016). AgNPs that extracted from *Codium capitatum, Enteromorpha flexuosa, Caulerpa scalpelliformis* and *U. lactuca* were exhibited promising agricultural, biomedical, and antimicrobial and mosquitocidal activity applications (Ponnuchamy and Jacob 2016).

4. Interplay Between Green Fertilizers and Gene Expression Pathways in Plants

The root application of *A. nodosum* on Arabidopsis can enhance the gene expression of endogenous phytohormone pathways. The endogenous phytohormones in the extract can induce ABA biosynthesis genes, such as 9-cis-epoxycarotenoid dioxygenase 3

(NCED3) and xanthoxin dehydrogenase (ABA2) that play as distinct markers for regulating ABA biosynthesis.

Following these ABA signaling cascades, the transcript abundance of ABA-responsive gene RD29a can increase upon *A. nodosum* extract.

Cytokinenes (CKs) are important plant growth regulators that control the emerge of axillary buds, photomorphogenic development, growth rate and development of shoot apical meristem, and plant cell division (Stirk et al. 2003, Werner and Schmulling 2009).

Exogenous treatment of CK can increase seed set, leaf size, and delay of senescence. The *A. nodosum* contains CK, therefore the exogenous treatment application of seaweed extract can exhibit the same effect of CK. Ck signaling can induce Arabidopsis Histidine kinase 2 (AHK2) which mediates phosphorylate Arabidopsis histidine phosphotransfer proteins (AHPs) that are phosphorylated response regulators (ARRs) (Wally et al. 2013).

Seaweed extracts contain high levels of auxins and auxins-like compounds (Kumari et al. 2011, Yokoya et al. 2010). Auxins play important roles as coordinators of many growth and behavioural processes and plant body development during its life cycles. Auxins play a clear role to promote the elongation of primary roots and initiation and development of lateral roots (Aloni et al. 2006, Nibau et al. 2008, Osmont et al. 2007). Signaling cascades of auxin are very complicated with five biosynthetic pathways (Zhao et al. 2010) and cross-talking with many phytohormones. The most important transcription factor that controls cellular expansion in the Auxin pathway is auxin response factors (ARF) that mediate Pin-formed 1 (PIN1) proteins and activate auxin-responsive genes. Seaweeds containing Salicylic acid (SA), jasmonic acid (JA), and ethylene (ET) play an important role in mediating the gene activation of pathogenesis-related genes (PRs) and systemic acquired resistance (SAR). When the infected tomato and sweet pepper, which were infected with *Xanthomonas campestris*, was treated with *Ascophyllum nodosum* extract, it resulted in a significant reduction in plant disease levels and an increase in crop yield. PR1a. PINII, and ETR-1 as defense pathway markers of salicylic acid, jasmonic acid, and ethylene (ET) presented high expression levels. Also, hormonal biosynthesis genes, such as IPT, IAA and Ga2Ox, were shown high transcript levels (Ali et al. 2019).

In the last 20 years, there have been small regulatory transcripts that scientists discovered their roles to control the regulatory plant and animal pathways. These regulatory are called small RNA (miRNA). The miRNAs are short non-coding RNAs and can be regulated at transcriptional and post-transcriptional cleavage. Interestingly, miRNAs are involved in the regulation of many biological processes, such as development, intra- and extracellular stress response, and immune-defense strategies (Baldrich and Segundo 2016, Calil and Fontes 2017). When the salt-stressed Arabidopsis plants were treated by *Ascophyllum nodosum* extract, the expression levels of miRNAs families that integrated with salt stress were changed in response to salt stress (Shukla et al. 2018). The extract has improved the ability of plants for phosphorus (P) uptake under salinity conditions.

5. Conclusion

The use of science and technology to create environmental-friendly products is needed to protect our environment. These products should be sustainable to be regenerated and processed by the ecosystem at high rates. Green fertilizers will assist the planet in reducing toxicity and increasing plant production. CRFs and SRFs are the main groups of fertilizers used in industry to gradually supply nutrients to the plants. The vermicompost and seaweed extracts are the most used green fertilizers which help plants to grow healthily.

References

Adnan, N., Nordin, S.M. and Rasli, A.M. 2019. A possible resolution of Malaysian sunset industry by green fertilizer technology: factors affecting the adoption among paddy farmers. Environmental Science and Pollution Research 26: 27198–27224.

Ali, O., Ramsubhag, A. and Jayaraman, J. 2019. Biostimulatory activities of *Ascophyllum nodosum* extract in tomato and sweet pepper crops in a tropical environment. PLoS One 14(5): e0216710.

Aloni, R., Aloni, E., Langhans, M. and Ullrich, C.I. 2006. Role of cytokinin and auxin in shaping root architecture: Regulating vascular differentiation, lateral root initiation, root apical dominance and root gravitropism. Ann. Bot. (Lond.) 97: 883–893.

Anandhan, S. and Sorna, k.H. 2011. Biorestraining potentials of marine macroalgae collected from Rameshwaram. Tamil Nadu. J. Res. in Boil. 5: 385–392.

Atiyeh, R., Subler, S., Edwards, C., Bachman, G., Metzger, J. and Shuster, W. 2000. Effects of vermicomposts and composts on plant growth in horticultural container media and soil. Pedobiologia 44: 579–590.

Badenes, M.L., Martí, F.I., Ríos, G. and Rubio-Cabetas, M.J. 2016. Application of genomic technologies to the breeding of trees. Front. Genet. 7: 198.

Battacharyya, D., Babgohari, M.Z., Rathor, P. and Prithiviraj, B. 2015. Seaweed extracts as biostimulants in horticulture. Sci. Hortic. 196: 39–48.

Bhatnagar, R.K. and Palta, R.K. 1996. Earthworm-Vermiculture and Vermicomposting. Kalyani Publishers, New Delhi.

Brown, G.G. 1995. How do earthworms affect microfloral and faunal community diversity? Plant Soil 170: 209–231.

Calil, P.I. and Fontes, B.P.E. 2017. Plant immunity against viruses: antiviral immune receptors in focus. Ann. Bot. 119(5): 711–723.

Calvo, P., Nelson, L. and Kloepper, J.W. 2014. Agricultural uses of plant biostimulants. Plant Soil 383: 3–41.

Chanthini, AB., Balasubramani, G., Ramkumar, R., Sowmiya, R., Balakumaran, M.D., Kalaichelvan, P.T. and Perumal, P. 2015. Structural characterization, antioxidant and *in vitro* cytotoxic properties of seagrass, Cymodocea serrulata (R. Br.) Asch. & Magnus mediated silver nanoparticles. Journal of Photochemistry and Photobiology B: Biology 153: 145–152.

Chunga, I.M., Rajakumara, G., Gomathib, T., Parka, S.K., Kima, S.H. and Thiruvengadama, M. 2017. Nanotechnology for human food: Advances and perspective. Frontiers in Life Science 10(1): 63–72.

Daly, H. 2007. Ecological Economics and Sustainable Development Selected Essays of Herman Daly; Edward Elgar Publishing: Cheltenham, UK.

Deng, Y., Yao, J., Wang, X., Guo, H. and Duan, D. 2012. transcriptome sequencing and comparative analysis of *Saccharina japonica* (Laminariales, Phaeophyceae) under blue light induction. PLoS ONE 7(6): e39704.

Di Stasio, E., Rouphael, Y., Colla, G., Raimondi, G., Giordano, M., Pannico, A., El-Nakhel, C. and De Pascale, S. 2018. The influence of Ecklonia maxima seaweed extract on growth, photosynthetic activity and mineral composition of *Brassica rapa* L. subsp. sylvestris under nutrient stress conditions. Eur. J. Hortic. Sci. 82(6): 286–293.

Edwards, C.A., Subler, S. and Arancon, N. 2011. Quality criteria for vermicomposts. pp 287–301. *In*: Edwards, C.A., Arancon, N.Q. and Sherman, R.L. (eds.). Vermiculture Technology: Earthworms, Organic Waste and Environmental Management. CRC Press, Boca Raton.

Gajalakshmi, S. and Abassi, S.A. 2004. Earthworms and vermicomposting. Int. J. Biotechnol. 3: 486–494.

Ganapathy, S.G. and Sivakumar, K. 2013. Effect of foliar spray form seaweed liquid fertilizer of *Ulva reticulata* (Forsk.) on *Vigna mungo* L. and their elemental composition using SEM-energy dispersive spectroscopic analysis. A. Pac. J Rep. 2(2): 119–125.

Garcia-Jimenez, P., Llorens, C., Roig, F.J. and Robaina, R.R. 2018. Analysis of the transcriptome of the red seaweed *Grateloupia imbricata* with emphasis on reproductive potential. Mar. Drugs. 16(12): 490.

Guo, H., White, C.J., Wang, Z. and Xing, B. 2018. Nano-enabled fertilizers to control the release and use efficiency of nutrients. Current Opinion in Environmental Science & Health 6: 77–83.

Hashim, M.M.A., Yusop, M.K., Othman, R. and Wahid, S.A. 2017. Field evaluation of newly-developed controlled release fertilizer on rice production and nitrogen uptake. Sains Malays 46(6): 925–932.

Hernández-Herrera, R.M., Santacruz-Ruvalcaba, F., Ruiz-López, M.A., Norrie, J. and Hernández-Carmona, G. 2013. Effect of liquid seaweed extracts on growth of tomato seedlings (*Solanum lycopersicum* L.). J. Appl. Phycol. 26.

Hien, N.Q., Nagasawa, N., Tham, L.X., Yoshii, F., Dang, H.V., Mitomo, H., Makuuchi, K. and Kume, T. 2002. Growth promotion of plants with depolymerised alginates by irradiation. Radiat. Phys. Chem. 59: 97–101.

Illera-Vives, M., Labandeira, S.S., Brito, L.M., López-Fabal, A. and López-Mosquera, M.E. 2015. Evaluation of compost from seaweed and fish waste as a fertilizer for horticultural use. Sci. Hortic. 186: 101–107.

Jindo, K., Chocano, C., De Aguilar, J.M., Gonzalez, D., Hernandez, T. and Garcia, C. 2016. Impact of compost application during 5 years on crop production, soil microbial activity, carbon fraction, and humification process. Commun. Soil Sci. Plant Anal. 47: 1907–1919.

Jonas Collén, Betina Porcel, Wilfrid Carré, Steven G. Ball, Cristian Chaparro, Thierry Tonon, Tristan Barbeyron, Gurvan Michel, Benjamin Noel, Klaus Valentin, Marek Elias, François Artiguenave, Alok Arun, Jean-Marc Aury, José F. Barbosa-Neto, John H. Bothwell, François-Yves Bouget, Loraine Brillet, Francisco Cabello-Hurtado, Salvador Capella-Gutiérrez, Bénédicte Charrier, Lionel Cladière, Mark Cock, J., Susana M. Coelho, Christophe Colleoni, Mirjam Czjzek, Corinne Da Silva, Ludovic Delage, France Denoeud, Philippe Deschamps, Simon M. Dittami, Toni Gabaldón, Claire M.M. Gachon, Agnès Groisillier, Cécile Hervé, Kamel Jabbari, Michael Katinka, Bernard Kloareg, Nathalie Kowalczyk, Karine Labadie, Catherine Leblanc, Pascal J. Lopez, Deirdre H. McLachlan, Laurence Meslet-Cladiere, Ahmed Moustafa, Zofia Nehr, Pi Nyvall Collén, Olivier Panaud, Frédéric Partensky, Julie Poulain, Stefan A. Rensing, Sylvie Rousvoal, Gaelle Samson, Aikaterini Symeonidi, Jean Weissenbach, Antonios Zambounis, Patrick Wincker and Catherine Boyen. 2013. Chondrus crispus shed light on evolution of the Archaeplastida. PNAS 11: 5247–5252.

Kalaichelvan and Perumal, P. 2015. Structural characterization, antioxidant and *in vitro* cytotoxic properties of seagrass, Cymodocea serrulata (R.Br.) Asch. & Magnus mediated silver nanoparticles, J. Photoch Photobio B. 153: 145–152.

Kumari, R., Kaur, I. and Bhatnagar, A.K. 2011. Effect of aqueous extract of *Sargassum johnstonii* Setchell & Gardner on growth, yield and quality of *Lycopersicon esculentum* Mill. J. Appl. Phycol. 23(3): 623–633.

Kumar, V. and Yadav, S.K. 2009. Plant-mediated synthesis of silver and gold nanoparticles and their applications. J. Chem. Technol. Biotechnol. 84: 151–157.

Lee, J.M. Lee, Yang, E.C., Graf, L., Yang, J.H., Qiu, H., Udi Zelzion, Cheong Xin Chan, Timothy G. Stephens, Andreas P.M. Weber, Ga Hun Boo, Sung Min Boo, Kyeong Mi Kim, Younhee Shin, Myunghee Jung, Seung Jae Lee, Hyung-Soon Yim, Jung-Hyun Lee, Debashish Bhattacharya and Hwan Su Yoon. 2018. Analysis of the draft genome of the red seaweed Gracilariopsis chorda provides insights into genome size evolution in rhodophyta. Mol. Biol. Evol. 35(8): 1869–1886.

Liu, G., Zotarelli, L., Li, Y., Dinkins, D., Wang, Q. and Hampton, M.O. 2017. Controlled-Release and Slow-Release Fertilizers as Nutrient Management Tools. http://edis.ifas.ufl.edu.

Liu, T., Wang, X., Wang, G., Jia, s., Liu, G., Shan, G., Chi, S., Zhang, J., Yu, Y., Xue, T. and Yu, J. 2019. Evolution of complex Thallus alga: Genome sequencing of *Saccharina japonica*. Front. Genet. 10: 378.

Mahmud, M., Abdullah, R. and Yaacob, J.S. 2018. Effect of vermicompost amendment on nutritional status of sandy loam soil, growth performance, and yield of pineapple (*Ananas comosus* var. MD2) under field conditions. Agronomy 8: 183; doi:10.3390/agronomy8090183.

Manea, A.I. and Abbas, K.A.U. 2018. Influence of seaweed extract, organic and inorganic fertilizer on growth and yield broccoli. International Journal of Vegetable Science 24(6): 550–556.

Mathur, C., Rai, S., Sase, N., Krish, S. and Jayasri, M.A. 2015. Enteromorpha intestinalis derived seaweed liquid fertilizers as prospective biostimulant for *Glycine max*. Braz. Arch. Biol. Technol. 58: 813–820.

Metin, C. and Baygar, M. 2018. Determination of nutritional composition of Enteromorpha intestinalis and investigation of its usage as food. Ege Journal of Fisheries and Aquatic Sciences 35(1): 7–14.

Moreira, R., Chenlo, F., Sineiro, J., Arufe, S. and Sexto, S. 2017. Water sorption isotherms and air drying kinetics of fucus vesiculosus brown seaweed. Journal of Food Processing and Preservation 41: 1–10.

Morgan, K.T., Cushman, K.E. and Sato, S. 2009. Release mechanisms for slow- and controlled-release fertilizers and strategies for their use in vegetable production. Horttechnology 19(1): 10–12.

Naeem, M., Idrees, M., Aftab, T., Khan, M.M.A. and Varshney, L. 2012. Irradiated sodium alginate improves plant growth, physiological activities and active constituents in *Mentha arvensis* l. Journal of Applied Pharmaceutical Science 2(5): 28–35.

Naeem, M., Idrees, M., Aftab, T., Alam, M.M., Khan, M.M.A. and Moinuddin. 2015. Radiation processed carrageenan improves plant growth, physiological activities and alkaloids production in *Catharanthus roseus* L. Advances in Botany.

Ngala, B.M., Valdes, Y., dos Santos, G., Perry, R.N. and Wesemael, W.M.L. 2016. Seaweed-based products from *Ecklonia maxima* and *Ascophyllum nodosum* as control agents for the root-knot nematodes Meloidogyne chitwoodi and Meloidogyne hapla on tomato plants. J. Appl. Phycol. 28(3): 2073–2082.

Nibau, C., Gibbs, D.J. and Coates, J.C. 2008. Branching out in new directions: The control of root architecture by lateral root formation. New Phytol. 179: 595–614.

Ogawa, H. and Fujita, M. 2006. The effect of fertilizer application on farming of the seaweed Undaria pinnatifida (Laminariales, Phaeophyta). Physiological Research. doi.org/10.1111/j.1440-1835.1997. tb00070.x.

Osmont, K.S., Sibout, R. and Hardtke, C.S. 2007. Hidden branches: Developments in root system architecture. Annu. Rev. Plant Biol. 58: 93–113.

Othman, R., Rasib, A.A.A., Ilias, M.A., Murthy, S., Ismail, N. and Hanafi, N.M. 2019. Transcriptome data of the carrageenophyte Eucheuma denticulatum. Data in Brief 24.

Patel, R.V., Pandya, K.Y., Jasrai, R.T. and Brahmbhatt, N. 2017. Review: scope of utilizing seaweed as a biofertilizer in agriculture. Int. J. Adv. Res. 5(7): 2046–2054.

Pathma, J. and Sakthivel, N. 2012. Microbial diversity of vermicompost bacteria that exhibit useful agricultural traits and waste management potential. SpringerPlus. 1(26): 1–19.

Ponnuchamy, K. and Jacob, J.A. 2016. Metal nanoparticles from marine seaweeds—a review. Nanotechnol. Rev. 5(6): 589–600.

Rai, M., Yadav, A. and Gade, A. 2008. Current trends in phytosynthesis of metal nanoparticles. Crit. Rev. Biotechnol. 28: 277–284.

Ramkumar, V.S., Prakash, S., Ramasubburayan, R., Pugazhendhi, A., Kumar, G., Kannapiran, E. and Rajendran, R.B. 2016. Seaweeds: A resource for marine bionanotechnology. Enzyme and Microbial Technology. doi.org/10.1016/j.enzmictec.2016.06.009.

Ramnarain, Y.I., Ansari, A.A. and Ori, L. 2019. Vermicomposting of diferent organic materials using the epigeic earthworm Eisenia foetida. International Journal of Recycling of Organic Waste in Agriculture 8: 23–36.

Rasyid, A. 2017. Evaluation of nutritional composition of the dried seaweed *Ulva lactuca* from pameungpeuk waters, Indonesia. Tropical Life Sciences Research 28(2): 119–125.

Ratana-arporn, P. and Chirapart, A. 2006. Nutritional evaluation of tropical green seaweeds *Caulerpa lentillifera* and *Ulva reticulata*. Kasetsart Journal (Natural Science) 40: 75–83.

Sempeho, S.I., Kim, H.T., Mubofu, E. and Hilonga, A. 2014. Meticulous overview on the controlled release fertilizers Hindawi publishing. Advances in Chemistry. doi.org:10.1155:2014:363071.

Sempeho, S.I., Kim, H.T., Mubofu, E. and Hilonga, A. 2014. Meticulous overview on the controlled release fertilizers. Advances in Chemistry 363071: 16.

Shanmuganathan, B. and Devi, K.P. 2016. Evaluation of the nutritional profile and antioxidant and anti-cholinesterase activities of *Padina gymnospora* (Phaeophyceae). Eur. J. Phycol. 1–9.

Shaviv, A. 2001. Advances in controlled-release fertilizers. Advances in Agronomy 71: 1–49.

Shukla, P.S., Borza, T., Critchley, A.T., Hiltz, D., Norrie, J. and Prithiviraj, B. 2018. *Ascophyllum nodosum* extract mitigates salinity stress in *Arabidopsis thaliana* by modulating the expression of miRNA involved in stress tolerance and nutrient acquisition. PLoS ONE 13(10): e0206221.

Sobti, N., Juneja, S.K., Yadav, P.K., Chibbar, R.N. and Behl, R.K. 2012. Vermicompost application for improving grain yield and quality in cereals harnessing metagenomics and induced gene expression changes. Annals of Biology 28(2): 73–77.

Spinelli, F., Fiori, G., Noferini, M., Sprocatti, M. and Costa, G. 2010. A novel type of seaweed extract as a natural alternative to the use of iron chelates in strawberry production. Sci. Hortic. 125: 263–269.

Stirk, W.A., Novaik, O., Strnad, M. and van Staden, J. 2003. Cytokinins in macroalgae. Plant Growth Regul. 41(1): 13–24.

Subler, S., Edwards, C. and Metzger, J. 1998. Comparing vermicomposts and composts. Biocycle. 39: 63–66.

Trenkel, M.E. 2010. Slow- and Controlled-Release and Stabilized Fertilizers: An Option for Enhancing Nutrient Use Efficiency in Agriculture, International Fertilizer Industry Association (IFA).

Vera, J., Castro, J., Gonzalez, A. and Moenne, A. 2011. Seaweed polysaccharides and derived oligosaccharides stimulate defense responses and protection against pathogens in plants. Mar Drugs 9: 2514–2525.

Vijayan, S.R., Santhiyagu, P., Singamuthu, M., Kumari Ahila, N., Jayaraman, R. and Ethiraj, K. 2014. Synthesis and characterization of silver and gold nanoparticles using aqueous extract of seaweed, Turbinaria conoides, and their antimicrofouling activity. The Sci. World J. e938272.

Wally, O.S.D., Critchley, A.T., Hiltz, D., Craigie, J.S., Han, X., Zaharia, L.I., Abrams, S.R. and Prithiviraj, B. 2013. Regulation of phytohormone biosynthesis and accumulation in arabidopsis following treatment with commercial extract from the marine macroalga *Ascophyllum nodosum*. J. Plant Growth Regul. 32: 324–339.

Werner, T. and Schmülling, T. 2009. Cytokinin action in plant development. Curr. Opin. Plant Biol. 12: 527–538.

Woods, R.J. and Pikaev, A.K. 1993. Applied Radiation Chemistry: Radiation Processing. R. Changotra et al. Journal of Cleaner Production 242.

Yokoya, N.S., Stirk, W.A., Van, S.J., Novak, O., Tureckova, V., Pencik, A. and Strnad, M. 2010. Endogenous cytokinins, auxins, and abscisic acid in red algae from Barazil. J. Phycol. 46(6): 1198–1205.

Yuan, Y. and Macquarrie, D. 2015. Microwave assisted extraction of sulfated polysaccharides (fucoidan) from *Ascophyllum nodosum* and its antioxidant activity. Carbohydrate Polymers 129(20): 101–107.

Zhang, M., Yang, Y.C., Song, F.P. and Shi, Y.X. 2005. Study and industrialized development of coated controlled release fertilizers. Journal of Chemical Fertilizer Industry 32: 7–12.

Zhao, Q., Wang, H., Yin, Y., Xu, Y., Chena, F. and Dixon, R.A. 2010. Syringyl lignin biosynthesis is directly regulated by a secondary cell wall master Switch. Proc. Natl. Acad. Sci. 107(32): 14496–14501.

CHAPTER 4

Green Chemistry in Organic Farming

Goutam B. Hosamani,[1,*] *S.S. Chandrashekhar,*[1] *Pooja C.A.*[1] and
Sivarama Krishna Lakkaboyana[2]

1. Introduction

Green chemistry is a requisite to generate greener inputs for agricultural production. Green chemistry alternatives are sprightly ways to sustainably produce agricultural goods without continued dependence on toxic pesticides and chemicals of concern. Sustainable development has caught the imagination and action all over the world. According to FAO, sustainable agricultural development is the management and conservation of the natural resource base and the orientation of technological and institutional change in such a manner as to ensure the attainment and continued satisfaction of human needs for present and future generations. As per International resource data from the Research Institute of Organic Agriculture (FiBL) and the International Federation of Organic Agriculture Movements (IFOAM) Statistics 2020, India stands at 9th position in terms of certified agricultural land with 1.94 million ha (2018–19). Sikkim is one of the North-Eastern states of India that became the first fully organic state of India.

Organic farming is one of the methods for achieving the goal of sustainable agriculture. Many organic farming techniques, such as intercropping, mulching, and crop-livestock integration, are not new to various agriculture systems, including conventional agriculture in countries like India. In terms of environmental protection, organic farming plays a defensive role. Organic farming is thought to be less detrimental to the environment because it does not allow the use of synthetic

[1] College of Agriculture Science, University of Agricultural Science, Dharwad – 580005, Karnataka – India.

[2] Department of Chemistry, Vel Tech Rangarajan Dr. Sagunthala R&D Institute of Science and Technology, Avadi, Chennai 600062, India.

* Corresponding author: goutam17hosamani@gmail.com

pesticides, the majority of which have the ability to damage water, soil, and local terrestrial and aquatic wildlife (Suryatapa et al. 2020). Furthermore, due to crop rotation methods, organic farms are better at maintaining biodiversity than traditional farms. When opposed to conventionally farmed soil, organic farming increases physico-biological properties, such as more organic matter, biomass, higher enzyme, greater soil stability, improved water percolation, holding capacities, less water, and wind erosion.

ZBNF (Zero Budget Natural Farming) is a form of farming that removes external inputs and uses local resources to rejuvenate soils and restore ecosystem health through complex and multi-layered cropping systems. It uses 10% less water and 10% less energy than chemical and organic farming. Cow dung microorganisms decompose the dried biomass on the soil and turn it into plant-ready nutrients.

Organic farming is also in its infancy in India. According to the Union Ministry of Agriculture and Farmers Welfare, approximately 2.78 million hectares of farmland were under organic cultivation in March 2020. This is 2% of the country's overall net sown area of 140.1 million ha.

2. Scope of Organic Farming

It consists of a set of practices evolved around several principles based on the widely varying character of resources. The totality of each ecosystem and mechanism of delicate balance available between components of each ecosystem would ultimately decide which principle would weigh more or less and ultimately which practice should be predominant in organic farming. The criteria of assessment of the scope of organic farming vary according to the soil, climate, cropping system, vegetation irrigation, allied activities (animal husbandry, poultry, fisheries, etc.) as well as harmony achieved between different components of each eco-system. Justifying examples are cited here:

1. Temperate climate does not favour faster decomposition of organic residues, as compared to tropical climate due to cold temperatures prevailing in a temperate climate for most of the year.

2. The scope for preparing farmyard manure by a farmer with large livestock is certainly brighter than that of a farmer maintaining a few cattle.

3. An irrigating farmer can easily adopt green manuring practice or the practice of returning the crop residues for the enrichment of the soil, as the availability of moisture can facilitate the decomposition. But a dry land farmer is not able to do the same so easily.

4. A farm nearer urban agglomerations has the advantage of obtaining decomposed compost or poultry manure at cheaper rates compared to interior rural areas.

The scope of adopting organic farming exists in every farming situation to a varying degree. It can be adopted for every crop and in every situation. Each farming situation offers a unique scope for adopting some principles of organic farming.

3. Principles of Organic Farming

At a theoretical level, organic farming aims at achieving the regeneration and continuance of natural processes of plant growth in a given ecosystem by making the ecosystem a self-supporting and self-sustainable system. Organic farming is indeed not restricted to the single practice of supplying organic manures or green manuring alone or even controlling pests by one of the non-chemical methods alone. It is an integrated evolution of all possible organic sources and practices synthesized for a particular ecosystem with the sole aim of strengthening crop husbandry.

The Basic Principles of Organic Farming Practices are as Follows:

1. A crop should be able to grow and yield successfully using the nutrients supplied by a given soil under ideal conditions provided by increased microbial activity for the buildup of fertility.

2. Natural enemies, predators, bio-control agents, the use of natural products, or bio-extracts should be mainly used to control crop diseases and pests.

3. To supply the plant nutrients, all available organic sources in nature could be used, whether digested, semi-digested, or undigested. This could lead to the introduction of manure composting methods outside of the agricultural sector or even *in situ* decomposition in the field.

4. The natural resistance of some crops to pests and diseases should be conveniently exploited for the benefit of crop production. No practice, which would suppress the natural resistance, is encouraged in organic farming.

IFOAM Listed the Principles as Follows:

- The Principle of health
- The Principle of Ecology
- The Principle of Fairness
- The Principle of Care

These four principles are the roots of the development of organic agriculture:

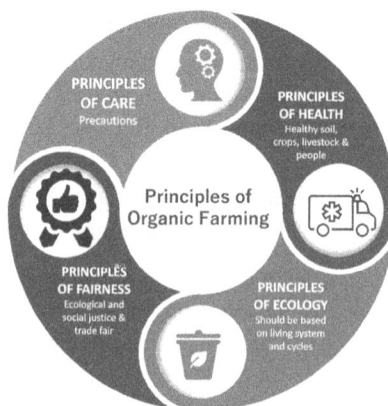

Figure 1. Principles of organic farming (adapted from IFOAM 1998).

4. Characteristics of Organic Farming

The most important characteristics of organic farming are as follows:

1. Optimal yet long-term use of local capital resources.
2. Maintaining the soil-water-nutrient-humus continuum's essential biological functions.
3. Maintaining a diversity of plant and animal species as a basis for ecological balance and economic stability.
4. Creating an appealing overall landscape that provides residents with a sense of satisfaction.
5. To reduce the risk, increasing crop and animal diversity through polyculture, agroforestry systems, integrated crop, livestock systems, and other methods.

Biocontrol agents are used mainly for pest and weed control with crop rotation, organic manuring, natural predators, and minimal chemical inputs that have been thrown in for good measure.

5. Advantages of Organic Farming

1. Organic manures provide ideal soil conditions for high yields and high-quality crops.
2. They have all of the nutrients that the plants need.
3. They help plants grow faster and work better physiologically.
4. They enhance the physical properties of the soil, such as granulation and tilth, allowing for better aeration, root penetration, and water holding capability.
5. They encourage beneficial chemical reactions by enhancing soil chemical properties, such as source supply and retention of soil nutrients.
6. They reduce the need for inputs that must be purchased.
7. The majority of organic manures are wastes or byproducts that, when accumulated, can cause contamination. Pollution is reduced when they are used for organic farming.
8. It helps to avoid chain reactions in the environment from chemical sprays and dust.
9. It helps to prevent environmental degradation and can be used to regenerate degraded areas.
10. Organic fertilizers are considered complete plant food. Organic matter restores the pH of the soil which may become acid due to the continuous application of chemical fertilizers.

6. Limitations of Organic Farming

1. Uniform practices of organic farming cannot be developed for all situations and all crops. Organic materials used in organic farming also differ vastly in various places. Packages of organic farming practice involving different components, which will have to be developed separately for each ecosystem.

2. Dramatic and large benefits in terms of yields and returns cannot be expected in a short time from organic farming practices. The increase in the yields and returns, commensurate with the resource structure of each ecosystem, can be gradually increased, stabilized, and sustained over a longer period of three to eight years.

3. Many organic manures are bulky with low concentrations of nutrients. Their carriage to distant places may pose problems of increased cost and labour. This will be a major limitation, where *in situ* decomposition of bio-mass due to moisture constraints are not possible and pre-digested manure has to be inevitably used.

4. Many more efforts are needed to identify natural enemies, and predators of some pests. Wherever natural enemies and predators of pests are absent in an ecosystem, the pests have to be managed by cultural and breeding methods, which may involve more time and labor.

Most of these limitations can pose threat to the adoption of organic farming partially or fully. In addition, the practice of organic farming is prioritized more than an option. It is just not possible to desist from practicing organic farming due to limitations.

6.1 Major Problems in Marketing Indian Organic Products: (Manchala et al. 2017)

- Price expectations are too high in comparison to quality.
- Low consistency of quality.
- Slow shipment and restrictions for importing Indian organic products.
- Time-consuming and complicated paperwork while dealing with export authorities.
- The poor customer service from the Indian traders after sales is the major problem in export marketing.
- Lack of proper marketing network for implementation.
- Less effort to develop domestic markets.
- An increase in biological activity makes lower depth nutrients availability possible.
- Increases water holding capacity of the soil.
- Improves texture and structure of the soil.

6.2 Market for Organically Grown Food

Concerns about high levels of saturated fats, sugarcane, salt, additives, and pesticide residues in foods as well as the risks of additives and pesticide residues have fueled the demand for healthy foods, especially organic foods. Furthermore, the environmental harm caused by modern agricultural techniques, especially agrochemicals, is becoming more widely recognized. Although the above trends have led to the development of the organic food industry, it is worth noting that there have been no significant organic food marketing initiatives. However, the media has been supportive of organic farming, which has partially compensated for the lack of product marketing by commercial advertisement outlets. Marketing concepts may be dominant in this context, but they cannot fully dominate it. As a result, marketing is a crucial component of organic farming growth.

7. Organic Farming and Sustainable Agriculture

Sustainable agriculture is described as food production that does not deplete Earth's resources or pollute the atmosphere. Sustainable agriculture is characterized as crop and animal production management practises that ensure long-term ecological sustainability without depleting natural resources or jeopardising human health. Sustainable agricultural management practices include maintenance of soil organic matter (e.g., conservation tillage and residue management), selection of crops adapted to local climate regimes, enhancement of agro-biodiversity (e.g., intercropping and agroforestry), prevention of soil erosion (e.g., windbreaks and terracing) and strengthening biogeochemical cycles (e.g., efficient crop rotation and adoption of proper irrigation and drainage techniques) and protection of environmental health (e.g., organic farming, integrated pest management and minimization of the use of synthetic fertilizers and biocides).

Environmental health, agricultural profitability, and social and economic justice are also central to the idea of sustainable agriculture. The philosophy of sustainability is based on the premise that we must address existing needs without jeopardising future generations' capacity to meet their own (Sarda et al. 2020). The following are the fundamentals of organic farming for the environment's long-term sustainability:

1. Improvement and maintenance of the natural landscape and agro-ecosystem.
2. Avoidance of overexploitation and pollution of natural resources.
3. Minimization of the consumption of non-renewable energy resources.
4. Exploitation synergies that exist in a natural ecosystem.
5. Maintenance and improve soil health by stimulating activity or soil organic manures and avoiding harming them with pesticides.
6. Optimum economic returns with a safe, secure and healthy working environment.
7. Acknowledgement of the virtues of indigenous know-how and traditional farming system.

8. Requirements of Organic Farming

In an organic farming system, certain minimum requirements are to be met so that farm is certified as organic (Seilan 2020).

8.1 Conversion

Conversion occurs as a farmer transitions from a traditional to an organic farming method. The conversion cycle is the interval between the start of organic management and certification. Measures should be taken to preserve bio-diversity, such as swamps, gross fields, and woodland, during the transfer process.

8.2 Mixed Farming

In addition to organic farming, animal husbandry, poultry, and fisheries should be practised. It is not permitted to change cultivation.

8.3 Cropping Pattern

If annual crops are cultivated, crop rotation should be practised. When growing seasonal crops, intercropping should be achieved. Green manure and feed crops should both be used in crop rotation.

8.4 Planting

Cultivated species and hybrids should be suited to the soil and climatic environments as well as insect and disease tolerant. Seeds and planting materials should be obtained from a supplier that is sustainable. Chemically untreated seeds or planting materials should be used once if chemically treated seeds or planting materials are not available. It is not permissible to use genetically modified seeds or planting products, such as tissue culture, pollen culture, or transgenic plants.

8.5 Manuring

Green manure crops, leguminous crops, and other crops should be raised to preserve or improve soil fertility. Bean residues should be recycled into the soil as far as possible after harvest. As manures, biodegradable products of microbial, plant, or animal origin must be used (e.g., vermicompost, farmyard manure, sheep penning, etc.). It is not allowed to use biological or chemical fertilizers. Mineral-based materials, such as rock phosphate, gypsum, lime, and other similar materials may also be used in small amounts where a solution is needed.

  In the organic fields, the following items may be used as a source of nutrients or for soil conditioning:

A. Farmyard manure, slurry, green manures, crop residues, straw, and other mulches from own farm.
B. Sawdust and wood-shaving from untreated wood.
C. Calcium chloride, limestone, gypsum and chalk, Magnesium rock, and Sodium chloride.

D. Bacterial preparations (Bio-fertilisers), e.g., Azospirillum, Rhizobium.

E. Bio-dynamic preparations.

F. Plant preparation and extracts, e.g., neem cake.

G. Vermicompost.

8.6 Processing

Solar drying, freeze-drying, and hot air chambers are also allowed processing methods. The use of ionizing radiation on farm products is banned. During fermentation, no industrial chemicals or days need to be used.

8.7 Labeling

The logo should be straightforward and correct facts about the product's organic status (i.e., organic or in-progress conversion). Different colored stickers for organic and conversion-in-progress goods should be distinguishable. The mark must contain information, such as the product's name, amount, producer's name and address, inspection agent name, certification, lot number, and so on. In the event of pollution, the lot number may be used to track down the plant, especially the field number where it was grown. The seed, region, field number, harvest date, and production year should all be included in the lot number.

8.8 Packaging

Recycling and recycled products, such as clean jute containers, should be used for packaging. Biodegradable fabrics must be used as well. Unnecessary packaging should be stopped at all costs. Except when labeled, organic and non-organic materials must not be processed or shipped together.

8.9 Documentation

Land charts, field history sheets, operation registers, input/output records, harvest records, stock, sales records, pest control records, movement records, machine cleaning, and marking records are all documents/records that must be preserved.

8.10 Certification Process

Organic farms must be certified to assure customers that the food is fully organic. The certification department inspects that the basic standards for organic cultivation are fulfilled and issues a certificate. Communication is made with the certifying entity by the manufacturer. Standards, fees, application, testing, evaluation, and appeal procedures are also covered by the licencing body. The producer then submits an application that includes field history, a shape chart, and a record-keeping method, among other things. The Inspector submits an inspection report to the agency with his decision, and the agency either approves or denies the credential. Since the certificate is only valid for the current year's crop, annual certification is required.

9. Impacts of Organic Farming

Agriculture has created environmental pressures in many parts of the world, including soil erosion, water use, and greenhouse gas emissions. The following are some of the basic effects of agriculture on the global climate (Ortiz and Hue 2007).

1. During the past 40 years, almost one-third of the world's cropland has been abandoned because of erosion and degradation.

2. Agriculture accounts for 80% of deforestation and 40% of the world's population live in regions where water resources are over-drafted and stressed and where users compete for water.

3. Methane (CH_4) and nitrous oxide (N_2O) emissions from agriculture in the EU amounted to 383 mt of carbon dioxide (CO_2) equivalent in the year 2000, which corresponds to approximately 10% of the total EU greenhouse gas emissions.

Agriculture's increased environmental burden is unlikely to abate anytime quickly, as the human population continues to outstrip global food production, and diets begin to move toward animal products.

Transitioning to organic farming could be a feasible option for lowering energy consumption and greenhouse gas emissions. Synthetic pesticides and fertilisers use a lot of resources, and switching to organic production, which is less dependent on these inputs, will help to mitigate these effects. According to FAO (1998), organic farming would have long-lasting, mostly beneficial effects on such important areas as:

9.1 Long-Term Productivity of the Land

Protecting soils and improving their fertility will ensure future generations' productive potential. Farmers often cite declining soil quality as one of the key factors for moving to organic farming. As a result, it is reasonable to believe that farmers who followed organic management methods find ways to increase the quality of their soil under the current management scheme or at the very least halted the degradation. The success of this mission hinges on the security of land tenure. Farmers have no reason to invest in a method that could only offer them profits in the future rather than immediate benefits if protection is not assured.

9.2 Food Security and Stability

Organic agriculture also grows a variety of crops and keeps a variety of livestock. Since different crops respond differently to climatic and edaphic changes or have different growth cycles, diversification reduces the likelihood of production variation (both in the time of the year and in the length of the growth period). Consumer appetite for organic food and premium rates open up new export markets for developing-world producers, allowing them to become more self-sufficient. Organic farming can help with local food security in a variety of ways. Organic farmers do not have high start-up costs, so they borrow less capital.

Synthetic inputs are not used because they are unaffordable to an increased number of resource-poor farmers due to reduced subsidies and the need for foreign currency. For resource-poor, small-scale farmers, organic soil enhancement might be the only economically viable alternative.

9.3 Environmental Impact

Synthetic fertilisers are not used by organic producers. Most certification programmes also limit the use of mineral fertilisers, which can only be used to replace organic matter generated on the farm to a certain degree that is possible. Where non-renewable resources are needed, there are environmental benefits. Leaching and fossil oil consumption are often limited. Instead, farmers improve soil productivity by using crop residues (e.g., corn stover, rice straw), legumes and green manures, and other natural fertilisers (although the type of manure and how it is handled have a major impact on N quality, the improper use can cause leaching problems) (e.g., rock phosphate, seaweed, guano, wood ash). Dairy processing systems are the main source of CH_4 and N_2O emissions in the agriculture sector, and therefore can have a significant capacity for greenhouse gas mitigation. There are some drawbacks to not using synthetic chemicals. If thermal and mechanical weeding or extensive soil tillage are used, energy needs can increase. Many resource-poor farmers lack access to livestock manure, which is an essential component of fertility. Immature composts, which can include bacteria and other toxins, are often used. Finally, certain areas in tropical countries may need synthetic inputs due to low soil fertility. Soil erosion, compaction, salinization, and deterioration are also combated by organic agriculture's soil conservation strategies (e.g., terracing in the tropical tropics, cover crops, etc.), which involve crop rotations and organic materials that enhance soil fertility and structure (including beneficial microbial influence and soil particle aggregation).

Using trees and shrubs in the agricultural system conserves soil and water while also protecting against adverse weather conditions, like winds, droughts, and floods. Organic farming techniques also aim to mitigate water waste and save water on the field. Although the benefits (both real and perceived) of organic farming and organic food are many, the potential negative effects should also be noted including the risk of contamination for human consumption. For example, nitrate leaching may contaminate groundwater used for drinking or organic livestock might be contaminated with disease-causing microorganisms from manure and animal parasites.

9.4 Social Impact

The social impact of organic farming is considerable, as mentioned in the IFOAM's principal aims. The main benefit according to some organic farmers in developing countries (e.g., China and India) is that they now have better standards of living. Good product prices and low unemployment dropped rural emigration and reduced health risks from chemicals.

10. Crop Protection

Rajasekhara (2003) observed a lower pest population of *Spodoptera litura* (1.18 and 1.09 larvae/m. row) in organically manured groundnut crop with neem cake (770 kg ha⁻¹) and vermicompost (3.75 t ha⁻¹), respectively.

Giraddi (2007) documented a higher population of *Chrysoperla carnea* and *Menochilus sexmaculatus* in organic, nourished crop plots of chilli, receiving organic manures and cakes in comparison to conventional agricultural practices.

In the groundnut ecosystem, organic treatment recorded less number of pests (aphids, leafhoppers, thrips, and defoliators) and a higher number of natural enemy (coccinellids and syrphids) populations compared to integrated and inorganic treatments. Similarly, a higher number of mycosed larvae due to *Metarhizium rileyi* on *Spodoptera litura* was noticed in the organic field (Kavitha 2009).

For control of Armyworm, Aphids, Cotton Semilooper, Green Leafhopper, Mites, Powdery Mildew, Pulse Beetle and Rice Weevil:

1 kg Turmeric + 4 liters of cows urine with 20 liters of water

For control of American ball worms, Aphids, Pulse Beetle Whitefly, etc.:

- 2 kg ginger paste and 30 12/13/06 Spray the filtrate in half an acre.
- Ginger, Garlic, and chilly extract
- Make 500 gm garlic paste in 100 mi kerosene + 100 gm chilly paste in 50 ml water + 100 gm ginger paste
- Add all the paste into 30 liter of water along with an emulsifier.
- Spray in the field over half an acre.

For control of aphids and beetles:

- Custard apple seed powder is an insecticide and antifeedant. It is a contact poison to flies, aphids, and several beetles.
- Take 500 gm in 2 liters of water and boil till 500 ml solution remains. Mix it with 15 liters of water and spray over the crop.
- A 2 kg custard apple leaves fresh juice in 500 ml water + 500 gm of chilly water extract + 1 kg Neem seed extract in 2 liters of water.
- Dilute it with 60 liters.

Neem Kernel Aqueous Extract:

The simple method of Neem Kernel Aqueous Extract preparation consist of the following steps:

- Take dried neem seed. Decorticate (removal of seed coat) with the help of mortar and pestle or any mechanical decorticator. Clean the neem kernel and seed coat mixture by winnowing the seed coat.
- Weigh 1 kg of clean neem kernel and make powder of grain size like fine tea powder. It should be pounded in such a way that no oil comes out. Soak in about

10 liters of clean water. Add 10 ml of neutral pH adjuvant (mixture of emulsifier, spreader, etc.) and stir the mixture. Finely ground soapnut powder is known to make a good natural emulsifier. Keep the mixture overnight and filter it on the next day with a clean muslin cloth. Add fresh water to the residue and repeat the extraction 2–3 times. Use spent residue as manure.

Cow Urine:

Cow urine alone is a good liquid fertilizer and can be used directly for spraying the crop. Dilute 1 liter of cow urine with 100 liter of water and use it as a foliar spray. For one acre of crop, 200 liter of such dilute suspension will be sufficient. This can be used in any crop in all seasons.

Vermiwash as Growth Promoter:

Vermiwash alone or mixed with cow urine can also be an excellent growth promoter. Dilute one liter of vermiwash or 0.5 liters of vermiwash + 0.5 lit of cow urine in 20 liters of water and use as a foliar spray. Three to four applications are needed for excellent results.

11. Conclusion

Organic farming is focused on a mutually beneficial relationship between soil, minerals, water, plants, microflora, insects, livestock, and people. It cultivates sustainable ecosystems while simultaneously managing food production and environmental security. Organic management enlists the help of local people and expertise to improve natural resource systems while being mindful of ecological carrying capacities. Ecosystem stability is improved, food security is enhanced, and additional incomes are created by reducing dependency on off-farm inputs and generating more sustainable nutrient and energy flows.

Going organic is a perfect way to eliminate pesticides and protect our health and the climate, but it comes with a lot of obstacles. People are not yet ready to embrace the use of organic food because of its high price. The other problem in organic farming is fulfilling the world's demand for food, as organic crop development is weak. Organic farming responds favorably to both sustainable agriculture and rural development objectives, thus assisting in the preservation of soil fertility, crop productivity, and farmer socio-economic conditions. Organic farming has been discovered to have beneficial effects on the climate and biodiversity. It helps farmers in terms of both economic and environmental benefits, allowing them to improve their living standards. Organic agriculture needs a robust national strategy to enforce. Considering India's current organic status, there is a lot of space for extending organic practises in order to achieve long-term agricultural production.

References

FAO, Food and Agriculture Organization. 1998. Evaluating the potential contribution of organic agriculture to sustainability goals. Environment and Natural Resources Service. http://www.fao.org/DOCREP/003/AC116E/AC116E00.

Giraddi, R.S. 2007. Effect of organic amendments on the activity of *Menochilus sexmaculatus* (Fabricius) and *Chrysoperla carnea* (Stephens) in chilli, (*Capsicum annum* L.). Pest Manage. Horti. Eco. 13(2): 38–43.

International Federation of Organic Agriculture Movements (IFOAM). (1998). The IFOAM basic standards for organic production and processing. General Assembly, Argentina, November, IFOAM, Germany. Organic Food Production Act of 1990 (U.S.C.) s. 2103.

Kavitha, A.S. 2009. Ecofriendly Practices Against Major Pests in Different Cropping Systems with Special Reference to Groundnut. M.Sc. (Agri.) Thesis, Univ. Agric. Sci., Dharwad, Karnataka (India).

Manchala, S., Chandramohan, R.G. and Sangwan, P.S. 2017. A review on organic farming—sustainable agriculture development. Int. J. Pure App. Biosci. 5(4): 1277–1282.

Ortiz, E.M. and Hue, N.V. 2007. Current developments in organic farming. Recent Res. Devel. Soilsci. 2(1): 29–62.

Rajasekhara, R.K. 2003. Influence of host plant nutrition on the incidence of *Spodoptera litura* (Fab.) and *Helicoverpa armigera* (Hub.) on groundnut. Indian J. Entomol. 65: 386–392.

Sarda, K.D., Wangkhem, T.C. and Priyanka, I. 2020. Organic farming for sustainable development of agriculture. AgriCos e-Newsletter 1(02): 48–51.

Seilan. 2011. Organic farming and sustainable agriculture. Environment and Rural Development 2(1): 415–427.

Suryatapa, D., Annalakshmi, C. and Tapan, K.P. 2020. Organic farming in India: a vision towards a healthy nation. Food Quality and Safety 4(1): 69–76.

CHAPTER 5

Ecosafe Farming with Microbes

Ali Samy Abdelaal,[1,*] *Abeer Mohamed Mosalam,*[2]
Sameh H. Youseif[3,4] and *Amal Samy Abdelaal*[5]

1. Introduction

Farmers have used high inputs of mineral fertilizers and chemical pesticides for decades to produce higher yields and protect crops from diseases and pests. The rising costs of chemical inputs along with raising awareness of health and environmental issues have led to intense attention to substitute hazardous agrochemicals (mineral fertilizers and pesticides) with environment-friendly microbes, which can reduce costs, improve the nutrition of crops and livestock, and protect them from environmental stresses as well as human health protection (Yang et al. 2009). Microbes or microorganisms are mostly microscopic small organisms that are placed in different groups such as bacteria, fungi, microalgae, protozoa, and viruses. These organisms live in water, soil, food, animal intestines, and other different habitats. Microbes are found in the soil to improve agricultural productivity and spreading them on crops have become a successful strategy for providing nutrients to plants and protecting them against pathogens. Moreover, microbes are found in the digestive system of livestock to improve digestion and immune response (Uyeno et al. 2015).

Over the last years, the low-cost DNA sequencing and other technologies have helped the scientific researchers to specifically identify and study every microbe of the microbial community surrounding the plant and inside the animal digestive system to understand various microbes' behavior under the different environmental conditions in order to use and modify them to enhance their impact in farming.

[1] Department of Genetics, Faculty of Agriculture, Damietta University, Damietta, Egypt.
[2] Agronomy Department, Faculty of Agriculture, Damietta University, Damietta, Egypt.
[3] National Gene Bank and Genetic Resources, Agricultural Research Center, Giza, Egypt.
[4] Faculty of Biotechnology, MSA University, 6th of October City 12451, Egypt.
[5] Agronomy Department, Cotton Research Institute, Agricultural Research Center, Giza, Egypt.
* Corresponding author: asamy82@gmail.com

1.1 Soil Health and Fertility

Soil health, also known as soil quality, is biological, chemical and physical features to long term, and sustainable agricultural productivity with minimal environmental impact (Arias et al. 2005). For the importance of soil management to be sustainable for future generations, we should always remember that soil contains living organisms that perform functions required to produce food and fiber when the necessities of life like food, shelter, and water are provided. The soil is not an inert growing medium, rather it is a combination of billions of bacteria, fungi, and other microbes that form the foundation of an elegant symbiotic ecosystem.

There are several strategies of soil health regeneration, including soil construction for plant health and nutrition, bio-remediation, and the introduction of beneficial soil organisms. All these approaches complement each other and follow the same soil ecology principles.

Soil fertility is the capacity of soil to sustain plant growth and other beneficial processes that occur in soil. Soil fertility is the combined effect of three main interacting components, the chemical, physical, and biological properties of the soil. The soil fertility consists of soil organic matter (including microbial biomass), soil texture, soil structure, soil depth, the content of nutrients, adsorption capacity, soil reactions and the absence of toxic elements.

1.2 Rhizosphere

The term 'rhizosphere' was originated in 1904 to describe the region of soil close to plant roots associated with intense microbial activity (Hiltner 1904). The health and productivity of a plant are depended on the presence of microbes in the rhizosphere. Bacteria are closely associated with roots and even inside roots. Plants can determine the distribution of microbes through the food released. A specific range of microbes is recruited by each plant species (Berendsen et al. 2012a). Such microbes do not act independently but communicate with each other in a complex pattern of exchange of food and information. Plants direct microbes near their roots through the release of flavonoids and strigolactones to allow microbes to fix nitrogen or biocontrol of pathogens. Basically, the health of a plant is determined by the whole rhizomicrobiome (Garbeva et al. 2004, Berendsen et al. 2012b, Oldroyd 2013, Bonfante and Anca 2009, Somers et al. 2004, Garbaye 1994).

1.3 Microorganisms in Soil

Microorganisms in soils include bacteria, fungi, actinomycetes, and microalgae. Almost all of them can only be seen under a microscope similar in size to the clay particles. Microbial community structure data improve the prediction of soil function compared to normal soil tests alone.

Different environmental factors, such as aeration conditions, temperature, pH value, nutrients and energy substrates, influence microbial activity in the soil. Oxidation-reduction conditions have a major impact on microbes in the soil. In general, an oxidation state (Eh > 50 mV) environment encourages aerobic microorganisms, such as nitrifying bacteria and iron-oxidizing bacteria, while a reduction state (–Eh) encourages anaerobic microorganisms such as sulfate-reducing bacteria, iron-reducing bacteria as well as methane-producing bacteria. Facultative microorganisms can grow in both aerobic and anaerobic environments (Eh from 0 to 50 mV).

The pH value is one of the most important factors that influence the presence of microorganisms in the soil. The optimum pH values for microorganisms are between 7.4 and 8.6. However, some microorganisms can grow under extremely low or high pH.

Anaerobic functions produce organic acids which may shift the pH and generate bio-gas (e.g., methane, hydrogen, and carbon dioxide). Microbial activity is accompanied by the accumulation of microbial mass in soil and gas generation. About 99% of microbial cells exist within biofilms (Dalton and March 1998, Nikolaev and Plakunov 2007) and are found attached to a solid surface by biofilms with a complex layered structure at the nanoscale (Petrova and Sauer 2012). The nutrients available and the hydrodynamic regime control the ability of a microbial mass to be attached to a solid surface of particles. Nutrients have a positive effect on the volume of biofilm, while hydraulic conductivity negatively affects biofilms. Biofilms are protected against antibiotic killing (Stewart 2002, Costerton et al. 1995).

1.3.1 *Impact of Microbial Activity on Soil Properties*

Microbial activity is one of the most important factors which can significantly change soil properties. The activity of soil microbes can affect soil properties in two main factors: pH and soil structure.

a. **pH value:** Soil pH affects the availability of nutrients in the soil as well as the types and activity of the soil microbes. Most soil microbes and plants live at a neutral pH. Anaerobic microorganisms produce organic acid by anaerobic fermentation in a deep layer of soil. Furthermore, aerobic microorganisms also generate proton ions that alter the soil pH (Sylvia 2005). The low pH suppresses the availability of phosphorus in the soil and increases the availability of aluminum ions which might reduce crop yields. Lime may be used to control pH by increasing the availability of nutrients from the soil and also providing calcium and magnesium for soil microorganisms and plants.

b. **Soil Structure:** The arrangement of particles and associated pores in soils is important to provide flow pathways for water and nutrients. Microorganisms play an important role in the aggregation of soil particles to determine the soil structure, which can promote aggregation by extracellular polysaccharides, glomalin, and hyphae. Also, soil microbes can bind soil particles to contribute to the formation of soil structure. Moreover, the products of soil microorganisms are central factors for soil aggregation (Sylvia 2005). The activity of soil microbes influences the soil structure by means of altering the soil grain size, the permeability of the soil, gas generation, and soil strength and deformability (Dashko and Shidlovskaya 2016).

1.3.2 Factors Affecting Microbial Communities in Agriculture

Soil microbial community structure is largely driven by soil conditions. The vast array of microorganisms that exist around plants can be changed by growing different plants, availability of nutrients, and human activities.

a. **Crop Rotation:** Crop rotation is one of the oldest agricultural methods that utilize different types of crops in the same field in different periods. It helps in increasing biodiversity and soil nutrients by using crops with different nutrient demands. It is also beneficial for pest and pathogen control.

b. **Fertilization:** Crop fertilization is an important factor in the soil nutrient pools (Stevenson and Cole 1999). When high yield plants are grown, they consume a lot of nutrients from the soil. In order to obtain high yields, fertilizers are needed. The soil fertility and availability of nutrients influence soil microbial growth and activity (Broeckling et al. 2008, Wei et al. 2008).

c. **Tillage:** Tillage is a human activity that overturns the soil surface to provide a suitable environment for seed germination and root growth of plants and affect soil microbial communities (Sylvia 2005, Muruganandam et al. 2010). When tillage is minimized, lower mineralization and nitrification rates combined with increased nitrogen immobilization and a higher denitrification potential will cause a decrease in available nitrogen which affect soil microbes.

2. Farming with Microbes

Microbes used in farming are isolated from the soil. There are about 8,000,000 bacteria, 30,000 fungi, 800,000 actinomycetes, 16,000 algae, and 10,000 protozoa in one gram of soil (Waksman 1932). The average number of bacteria split between 4,000–5,000 species can be 10 billion per gram (Raynaud and Nunan 2014). Soil also includes invertebrates and nematodes. Protozoa, hyperparasitic fungi, and nematodes consume bacteria and fungi. Fungi and larger species, including soil-dwelling mites, consume nematodes. The effect is a nutrient stream and knowledge describing the network of soil nutrition. Soil organisms are active, make full use of the food supply available, rarely fall asleep, and are always hungry in human terms. Pathogens and nematodes can invade plant roots if plants are introduced into this area, but beneficial fungi such as mycorrhizae colonize plant roots, supplying nutrients, and pathogens biocontrol. Plants release 10–40% of their produced food as root exudates, containing carbohydrates and amino acids. In the rhizosphere area, bacteria grow near plant roots to feed. Many soil bacteria are beneficial to the crop, many of which are antagonistic to pathogens.

Continuous cultivation of soil has resulted in a gradual decline in soil health that can be preserved and sustained to a greater extent through the use of organic manures. In this respect, further progress was observed through the incorporation of microorganisms in organic manures to produce bioorganic fertilizers (Chakradhar and Jauhri 2004). Therefore, the use of microbial inoculants for sustainable agri-industrial applications has been subjected to several recent reviews to exploit rhizosphere conditions through revolutionary techniques for improved plant growth and health (Bloemberg and Lugtenberg 2001).

Farmers use naturally occurring microbes to develop biofertilizers and bio-pesticides to enhance plant growth and control diseases and pests.

2.1 Biofertilizers

In nature, several useful soil microbes can help plants to absorb nutrients and resist diseases and pests. Biofertilizer is a group of living fertilizers compounds of efficient microbial inoculants, such as bacterial, fungal, and algae origin, which can fix atmospheric nitrogen or solubilize phosphorus and decompose organic material or oxidize sulfur in the soil. Hellriegel and Wilfarth demonstrated that fixation of atmospheric nitrogen takes place in legumes nearly 100 years ago (Burris 1977). In 1888, Beijerinck isolated Rhizobium from root nodules of legumes and at that time many other scientists have defined it (Beijerinck 1888). Biofertilizer is a low cost, effective, and sustainable plant nutrient source to replace chemical fertilizers. Gluconacetobacter diazotrophicus (Gd), a non-nodulating, non-rhizobial, and nitrogen-fixing bacterium, which is isolated from the intercellular juice of sugarcane inoculated under specific conditions to intracellularly colonize the roots and shoots of the cereals as well as other crops (Dent and Cocking 2017). To maximize their function in the soil, microorganisms that can be used as biofertilizers are applied to the rhizosphere region. Sustainable crop production is highly dependent on good soil health. The conservation of soil health requires an optimal mixture of organic and inorganic soil components. Repetitive use of chemical fertilizers destroys the biota of soil. Biofertilizers help plants use all of the nutrients available in the environment, thus allowing farmers to reduce the use of chemical fertilizers. This helps preserve the environment for a better life.

2.1.1 Classification of Biofertilizers

Several microorganisms are being exploited in the production of biofertilizers. They can be categorized in different ways according to their nature and function.

a. **Rhizobium:** Rhizobium species can fix molecular nitrogen from the atmosphere in association with plants. It varies from free-living condition to the bacteroid of nodules. It is the most efficient biofertilizer according to the amount of fixed nitrogen. Commercially, inoculation of legumes with Rhizobium spp. bacteria started in 1900 (Bashan 1998).

b. **Azotobacter:** Azotobacter species normally fix molecular nitrogen from the air without symbiotic relations with plants. However, some other Azotobacter species are associated with plants (Kass et al. 1971). In the presence of available nitrogen sources, nitrogen fixation is inhibited (Burgmann et al. 2003). Azotobacters are used as biofertilizers since 1902. Besides, its ability to fix atmospheric nitrogen, Azotobacters can exude plant hormones, solubilize phosphates and can counteract plant pathogens. Azotobacters have been found beneficial for all types of plants, such as cereals, oilseeds, vegetables, fruits, fiber crops, and trees (Das 2019).

c. **Azospirillum:** Azospirillum brasilense and Azospirillum lipoferum are primary inhabitants of soil, and the rhizosphere and intercellular are spaces of the root cortex of Gramineae plants. They develop a symbiotic relationship with Gramineae. In addition to nitrogen fixation, they help in growth-promoting substance production (IAA), disease resistance, and drought tolerance.

d. **Cyanobacteria:** Both free-living and symbiotic cyanobacteria (blue-green algae) can be used in rice cultivation as a source of nitrogen.

e. **Azolla:** A free-floating water fern that fixes nitrogen in association with Anabaena azollae, a blue-green algae. Azolla can be used as an alternate nitrogen source or as a supplement to commercial nitrogen fertilizers. Also, it can be used as a biofertilizer for wetland rice.

f. **Phosphate Solubilizing Microbes:** Phosphate plays an important role in crop maturity, quality, stress tolerance, and has a direct or indirect role in nitrogen fixation. About 80% of phosphate fertilizer applied to field crops is wasted. It binds to the soil by cations of aluminum or iron to form insoluble salts that cannot be used by plants (Richardson et al. 2009). Several soil bacteria and fungi, notably species of Bacillus, Streptomyces, Pseudomonas, Aspergillus, Penicillium, etc., secrete acids and other metabolites leading to lower pH and release of phosphate from organic sources, such as nucleic acids and inositol phosphates or inorganic complexes to be available for plants (Srivastav et al. 2004, Kkan et al. 2009).

g. **Arbuscular Mycorrhiza Fungi (AM Fungi):** The transfer of nutrients mainly phosphorus, nitrogen, sulfur, and micronutrients from the soil to the cells of the root cortex is mediated by intracellular obligate fungi that possess vesicles for storage of nutrients and arbuscules for pouring these nutrients into the root system. By far, the commonest genus appears to be Glomus, which has several species distributed in soil. The genera Glomus, Acaulospora, Gigaspora, Sclerocysts, and Endogone are examples of this kind of fungi, whereas Glomus is the most widespread genus found in soil.

h. **Silicate Solubilizing Bacteria (SSB):** Microbes can degrade silicates and aluminium silicates. Many organic acids are produced during the metabolism of microbes, and these have a dual role in weathering of silicate. These provide the medium with hydrogen ions to promote hydrolysis and organic acids, such as citric acid, Keto acids, oxalic acid, and hydroxy carbolic acids which form complexes with cations, promote retention of silicate in a dissolved state in the medium.

i. **Potash Mobilizing Bacteria:** The microbe 'Frateuria aurentia' is a species of Proteobacteria (Johansen et al. 2005) that can mobilize available Potash near to the roots of the plants. It works in all types of soil especially, low potassium content soil.

j. **Plant growth-promoting rhizobacteria (PGPR):** Group of bacteria that colonize roots or rhizosphere soil. The name 'plant growth-promoting rhizobacteria' has been given to these microbes because the yield of plants can be increased after their treatments with rhizobacteria (Burr 1978, Weller 1988, Weller 2007,

Kloepper 1989, Lugtenberg and Kamilova 2009). These bacteria promote growth through suppression of pathogen by producing antibiotics to kill plant pathogen, induce systemic acquired resistance (SAR), or release phytohormones which enhance the number of fine roots that increase the absorptive surface of plant roots to get more water and nutrients. Plants synthesize phytohormones, including Auxin, Gibberellins, Cytokinins, Ethylene, and Abscisic acid but under environmental stress conditions, plants may not be able to synthesize sufficient concentrations to sustain growth and development. Recently, it was shown that plant growth can be improved when specific microbial strains are used to treat plants due to the microbial production of phytohormones.

2.1.2 Types of Biofertilizers

Biofertilizers introduce macro-nutrients and micro-nutrients through the natural processes of atmospheric nitrogen fixation, phosphate solubilization, potassium mobilization, and plant growth stimulation through the production of growth-promoting substances. Their nature and function can be defined in different ways as summarized in Figure 1.

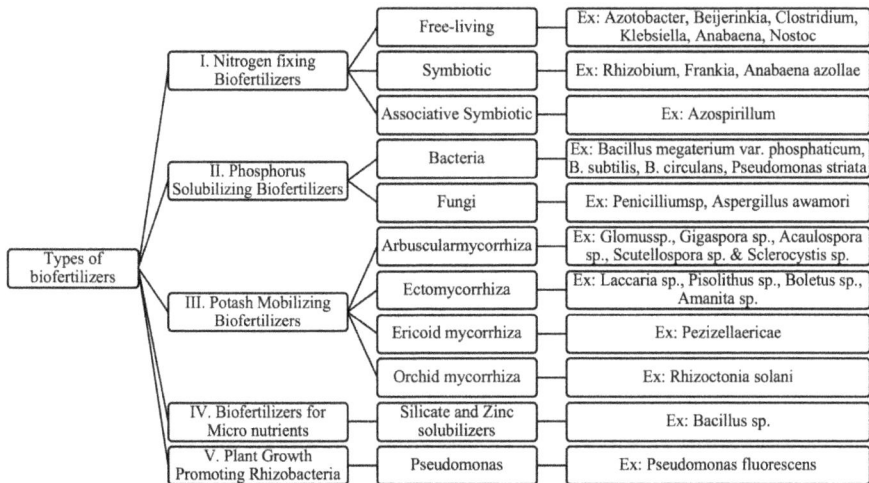

Figure 1. Types of biofertilizers. The types of biofertilizers include nitrogen fixing biofertilizers, phosphorus solubilizing biofertilizers, potash mobilizing biofertilizers, biofertilizers for micronutrients, and plant growth promoting rhizobacteria.

2.1.3 Methods of Application of Biofertilizers

Organic farmers have inoculated microbes with plants to increase the yields using different methods. Microbial inoculants treatment helps to enrich carbon starved soils, decrease mineral fertilizer and pesticide applications, and increase yields (O'Callaghan 2016).

a. **Seedling Root Dip:** This method is used before transplanting by dipping plant roots in a solution of microbes. For rice plants, recommended microbes are mixed in the water and the roots of seedlings are dipped for 8–10 hours and then transplanted.

b. **Soil Treatment:** The recommended biofertilizers are mixed with compost and kept overnight. This mixture is added to the soil near plant roots at the time of sowing or planting.

c. **Seed Treatment:** Seed treatments need less microbial material which reduces costs (Harman 2000). Seeds can be added to slurries of microbes in a carrier or a saline solution of inoculant and then dried in a process called 'biopriming'. Microbes can also be sprayed in a polymeric coating onto seeds. Biopriming is the best microbial survival process. Spore forming bacteria, such as Bacillus subtilis, survive better than non-spore forming bacteria, such as Pseudomonas spp. (Deaker et al. 2004, O'Callaghan 2016). Seeds can be treated before planting by the farmer or seed distributor as used with Trichoderma inoculants or microbes can be applied by the seed company with other additives (Harman 2000, O'Callaghan 2016).

2.1.4 Advantages of Using Biofertilizers

Many benefits associated with using biofertilizers in agriculture, which can be enumerated in these ten points:

1) They are both eco-friendly and cost-effective.
2) Their use contributes to soil enrichment and soil quality increases over time.
3) The findings seen over time are impressive, even though they do not produce immediate results.
4) These fertilizers capture and deliver atmospheric nitrogen directly to the crops.
5) They increase the phosphorous content of the soil by solubilizing and releasing unavailable phosphorous.
6) Biofertilizers improve root proliferation due to the release of growth-promoting hormones.
7) Microorganism converts complex nutrients into simple nutrients available for plants.
8) Biofertilizer includes microbes that promote the adequate supply of nutrients to the host plants and ensure their proper growth development and physiological regulation.
9) They help to increase the crop yield by 10–25%.
10) Biofertilizers can protect plants against soil born diseases.

2.1.5 Limitations in Biofertilizer Technology

While biofertilizer technology is a low cost and environmentally friendly technology, the application or implementation of the technology is limited by several constraints. The drawbacks could be as follows:

1) Technological limitations such as unavailability of carrier material with good quality and lack of qualified technically qualified staff.
2) Infrastructural limitations such as lack of essential facilities and equipment.
3) Financial limitations such as the non-availability of sufficient funds.
4) Environmental limitations, such as seasonal demand for biofertilizers, simultaneous cropping operations, and limited sowing/planting time in a specific location.
5) No visual difference in the crop growth immediately as that of inorganic fertilizers.
6) Marketing limitations, such as lack of the right inoculant at the right place at the right time and lack of retail outlets or the producers' distribution network.

2.2 Biocontrol of Plant Pathogens and Diseases

Microorganisms found in the soil are not all friendly. These microbes can cause disease or kill pathogens. Scientists used these disease-causing microbes to naturally control plant pathogens (Pal and McSpadden Gardener 2006). The use of beneficial microbes, such as specialized bacteria and fungi, to attack and control plant pathogens and the diseases they cause through direct competition with the target organism. This mechanism is called 'antibiosis' in which the biocontrol agent secretes an antibiotic or toxin that kills/harms the target organism or it can induce resistance of the infected plant by keeping it from being infected with other pathogens (Köhl et al. 2019). Biocontrol microbes were selected to be compatible with plants and were initially isolated from the soil where the plants grew.

Bacterial pathogens used for biocontrol of insects are in the genus Bacillus, which are spore-forming, rod-shaped bacteria. They commonly occur in soils, and most insecticide bacterial strains have been isolated from soil. Bacilli are well-studied bacteria that have the ability to produce spores when environmental conditions are unfavorable, such as high temperatures, extremely high or low pH, lack of nutrients or water, etc. (Piggot and Hilbert 2004). In order to be effective, bacterial insecticides have to be eaten; they are not poisons of contact. Insecticidal products from a single Bacillus species may be effective against only one species or an entire order of insects (Usta 2013).

Bacillus thuringiensis (Bt) is an aerobic, gram-positive, spore-forming soil bacterium capable to produce endogenous various kinds of crystals protein during its sporulation. *Bacillus thuringiensis* has been widely used in agriculture for the control of insect pests since it was discovered in 1901 as a microbial insecticide. Its main feature is the synthesis of a crystalline inclusion, containing Cry proteins, which have insecticidal properties (Schnepf et al. 1998). The inclusions of the crystal protein consist of toxins called δ-dotoxins or insecticidal crystal proteins that are either one or more crystal (Cry) and cytolytic (Cyt) toxins. Some of these toxins are harmful to certain insects and harmless to beneficial insects and humans. Since their insecticidal effect has been discovered, it has been marketed worldwide as a biocontrol of many important plant pests, mainly caterpillars of the Lepidoptera. The industrial Bt products contain a mixture of dried spores and crystals of toxins. They are added to the leaves or other areas where the insect larvae feed. The Bt genes of toxin crystals have been genetically engineered into several crop plants. The usage, mode of action, and host range of these Bt toxins may differ among other insecticide species of Bacillus. In the current era of transgenic technology, Bt toxins assume significant importance in the production of insect-resistant crops, such as cotton, maize, rice, potato, etc.

Fungi can be used as biological insecticides to control several pests such as Termites, Grasshoppers, Whiteflies, Aphids, and many other insects. The spores of the fungus are sprayed on the affected crop as an insecticide. The fungus can infect insects that come in contact with it using its spores that germinate and begin to grow after attaching to the cuticle (skin) of the insect. The fungus kills the insect when penetrating its exoskeleton and grow rapidly inside the insect and drain the insect of its nutrients. The infected insect can also infect other insects that get in contact with it. The fungi Metarhizium anisopliae, Beauveria bassiana, and Verticillium lecanii are a few examples of insecticide fungi. Also, fungi are used to control the growth of nematodes. They can be used as a nematicide by infecting nematode's eggs, juveniles, and adult females. Nematicide fungi are found in many kinds of soils, which include *Paecilomyces lilacinus* and *Arthrobotrys* spp. Interestingly, fungi can be used as fungicide fungi that prevent the crops from fungal diseases, such as *Trichoderma viride*.

2.3 Bio-Herbicides

Weeds cause major problems for farmers. They compete with crops for water, nutrients, sunlight, and space; block irrigation and drainage systems; host insect and disease pests; and keep weed seeds into crop harvests. Bioherbicides are biological control agents that are useful for controlling weeds. They are microbial phytotoxins or phytopathogenic microorganisms useful for targeting weeds. Most commonly the organism used in biological weed control is a fungus; therefore, the term 'mycoherbicide' is often used (Pacanoski 2015). Bioherbicides are used to target weeds with no effect on crops. The microbes possess virulent genes that can attack the defense genes of the weeds and kill them. The use of weed control plant pathogens was first recorded in the early 1900s but after the Second World War, the idea of using bioherbicides to control weeds sparked widespread interest among weed scientists and plant pathologists. The first experiment was performed in Hawaii, which involved *Fusarium oxysporum* Schlecht to control prickly pear cactus (*Opuntia ficus-indica* (L.) Mill). In the 1950s, a team from Russia applied the spores of *Alternaria cuscutacidae* Rudakov to the parasitic weed dodder (*Cuscata* spp.). In 1963, another team from China used a different fungus (*Colletotrichum gloeosporioides* f. sp. *cuscutae*) against the same weed (*Cuscata* spp.). Since the early 1980s, the number of research articles on bioherbicide has enlarged excessively (Pacanoski 2015).

2.4 Enhancing Farm Animal Health and Production with Microbes

Farmers have many ways to improve livestock health and production. They have used antibiotics to improve the production of their chickens, pigs, and cows (McEwen and Fedorka-Cray 2002, Landers et al. 2012). Also, they have used different breeds of animals. Moreover, they have added enzymes to digest feed.

Microbes have the potential to improve animal health and production with the lowest price and without using antibiotics that contribute to antibiotic resistance, which have severe effects on human health.

The term 'probiotics' has been modified by the FAO/WHO to "Live microorganisms, which, when administered in adequate amounts, confer a health benefit on the host" (Fuller 1989).

Some lactic acid bacteria (LAB), including species belonging to the Lactobacillus, Bifidobacterium, and Enterococcus genera are considered beneficial to the host and have been used as probiotics. Probiotics are capable of enhancing intestinal health by stimulating the development of a beneficial microbiota, preventing enteric pathogens from colonizing the intestine, increasing digestive capacity, reducing pH, and improving mucosal immunity (Uyeno et al. 2015). Probiotics as live microorganisms can help livestock to improve digestion and enhance nutrient absorption to gain weight, also they outcompete undesirables microbes, improve immune response, and damp down the inflammation. Probiotics produce feasible enzymes, like amylases and proteases, to digest the carbohydrates and proteins in animal feed. Probiotics also increase the height of intestinal villi that increase the surface area for better nutrient absorption. Probiotics can also change the population of gut microbes by outcompeting pathogenic microbes or knocking them out with antimicrobial substances. The probiotics also adhere to the intestinal lining to prevent pathogens from sneaking into the bloodstream. They keep the immune system active against any infections (Lopez 2000, Chaucheyras-Durand and Durand 2010).

Ruminants depend mainly on microbes that live in the rumen to break down the fibrous plant material in their feed. Those microbes produce enzymes that break down cellulose and hemicellulose, producing energy from the feed for the ruminant (Woodman 2009). The bodies of the proliferating microbes are also used as a source of protein. They enable the ruminant to produce high-value milk from the low-value feed.

3. Plant Improvement for Biological Nitrogen Fixation

Molecular genetics modification of a host plant led to the improvement of nitrogen fixation in association with microbes which including:

a. **Host Transformation to Modify Host Range:** Transgenic Lotus plants transformed with the soybean lectin gene became susceptible to infection by Bradyrhizobium japonicum. Rhijn et al. 1998 induced the soybean lectin gene into Lotus corniculatus, and it was found that nodule like outgrowths developed on transgenic L. corniculatus plant roots in response to Bradyrhizobium japonicum (van Rhijn et al. 1998).

b. **Host Modification to Synthesize Opines:** Because Rhizobium strains vary in their ability to use opines, genetic engineering of legumes or other plants for opine synthesis may result in the enhancement growth of rhizosphere organisms with the ability to utilize this substrate. Savka and Farrand (1997) examined the influence of a novel substrate produced by a transgenic plant on root colonization by near-isogenic bacteria, differing in their ability to use the resource, and concluded that resources produced and exuded by a plant host can confer a selective advantage to microbes that use the substrate (Savka and Farrand 1997).

c. **Genetic Transformation of Plants for Enhancement Malate Dehydrogenase (MDH) Synthesis in Roots and Nodules:** Malate is the primary plant carbon source used by bacteroids and is also a factor in plant adaptation to phosphate and aluminum stress. Alfalfa transformed with a MDH gene having high efficiency in malate synthesis, exuded more organic material into the rhizosphere, and fixed more nitrogen than the wild type in initial studies (Tesfaye et al. 2001). Whether this also translates into enhanced phosphate uptake and aluminum balance remains to be determined. Tesfaye et al. 2001 produced transgenic plants using nodule-enhanced forms of malate dehydrogenase and reported an increase in malate dehydrogenase enzyme-specific activity in root tips and an increase in root concentration as well as in root exudation of citrate, oxalate, malate, succinate, and acetate compared with control alfalfa plants. The degree of aluminum tolerance by transformed plants in hydroponic solutions and in naturally acid soil corresponded with their patterns of organic acid exudation and supports the concept that enhancing organic acid synthesis in plants may be an effective strategy to cope with soil acidity and aluminum toxicity (Tesfaye et al. 2001).

4. Use of Microbial Genetics in Agriculture

4.1 Identification of Microorganisms

In the early 1980s, the first use of molecular biology tools to identify microorganisms with the occurrence of polymerase chain reaction (PCR) to amplify microbial DNA and identify organisms based on their rRNA. Berlanas et al. 2019 characterized the rhizosphere bacterial and fungal microbiota across different grapevine rootstock genotypes cultivated in the same soil using 16S rRNA gene and ITS high-throughput amplicon sequencing for bacteria and qPCR for fungal pathogens associated with black-foot disease, and they found that grapevine rootstock genotypes in the mature vineyard were associated with different rhizosphere microbiomes (Berlanas et al. 2019).

4.2 Microbial Genomics and Proteomics Studies

Different advanced technologies have been developed for high-throughput genomics and proteomics uses to study genomes and proteomes of microbes with high precision. Kwak et al. (2018) comparatively analyzed the rhizosphere metagenomes from resistant and susceptible tomatoes against soil-borne pathogens to identify and assemble the flavobacterial genome (Kwak et al. 2018). Hara et al. 2019 used 'omics' approaches to identify functional nitrogen-fixing bacteria associated with field-grown sorghum. Bacterial samples from roots were analyzed by metagenome and proteome and found that major functional nitrogen-fixing bacteria in sorghum roots are unique bradyrhizobia (Hara et al. 2019). Espenberg et al. (2018) sequenced the metagenomes of soil bacteria and archaea to investigate the community structure of soil bacteria and archaea and their ability to perform nitrogen transformation processes. They found significant dissimilarities in the structure of soil bacterial and archaeal communities (Espenberg et al. 2018). Wolinska et al. (2017) analyzed the composition of the bacterial communities using next-generation sequencing (NGS) to determine the diversity of the potential nitrogen-fixing (PNF) bacteria in the soil. They realized that soil agricultural management and soil formation processes are the most conducive factors for PNF bacteria (Wolinska et al. 2017). Nash et al. (2018) used metagenomic analysis to investigate the bacterial community structure across different fields. They identified symbiotic nitrogen fixers through the presence of root associated diazotrophs. They also identified fermentative and sulfur cycling bacteria, halophiles, and anaerobes (Nash et al. 2018).

4.3 Genetically Modified Microbes in Agriculture

Microorganisms used as biofertilizers, biopesticides, or probiotics have been genetically modified for further improvement.

4.3.1 Use of Genetically Engineered Microbes as Biofertilizers

Sashidhar and Podile (2009) enhanced mineral phosphate solubilization and growth-promoting activity of sorghum, when seeds were bacterized with the transgenic Azotobacter vinelandii using glucose dehydrogenase (gcd) gene from *Escherichia coli* (Sashidhar and Podile 2009). Barney et al. (2015) increased nitrogen release by engineered Azotobacter vinelandii with gene deletions of urease gene complex ureABC and ammonium transporter gene ΔamtB (Barney et al. 2015). Ambrosio et al. (2017) metabolically engineered Azotobacter vinelandii by replacing the native promoter of the glutamine synthase gene with trc inducible promoter to improve ammonium release (Ambrosio et al. 2017). Bageshwar et al. (2017) have deleted the negative regulatory gene nifL and constitutively expressed the positive regulatory gene nifA in Azotobacter chroococcum. The nitrogen fixation efficiency was improved when wheat seeds were inoculated with the engineered Azotobacter strain (Bageshwar et al. 2017).

Van Dommelen et al. (2009) improved wheat growth through nitrogen fixation using genetically modified Azospirillum brasilense strain with a point mutation in the ammonium binding site of glutamine synthetase (Van Dommelen et al. 2009).

Similar mutants of Azospirillum, Kosakonia, Pseudomonas, and Azotobacter (Zhang et al. 2012, Setten et al. 2013, Geddes et al. 2015, Ambrosio et al. 2017, Bageshwar et al. 2017) demonstrated capable of promoting plant growth.

4.3.2 Use of Genetically Engineered Microbes as Biopesticides

Bacillus thuringiensis strains were engineered via *in vitro* modification and recombination of different genes of Bt insecticidal crystal protein (ICP) with various insecticidal activities to develop products with broader insecticidal spectra (Klier et al. 1983, Crickmore et al. 1990, Yue et al. 2005a, Yue et al. 2005b) and for high insecticidal activity (Lu et al. 2000, Hu et al. 2009, Perez-Garcia et al. 2010).

4.4 Nontraditional Solutions for Agricultural Challenges

With recent advances in biotechnology, environmental biotechnologists now are able to use microbial communities for detoxifying wastewater or soil to use in agriculture and producing renewable energy from agricultural waste.

Bioremediation refers to the use of biological processes to detoxify contaminants from soil and water. Bacteria, fungi, or their enzymes can be used to eliminate or reduce hazardous wastes from the contaminated area. The use of recombinant bacteria and fungi to remove specific metals or hazards from contaminated water and soil is currently being investigated. Delgadilloab et al. (2015) expressed the copAB genes from a Cu-resistant *Pseudomonas fluorescens* strain in *E. medicae*. They found that the strain of *E. medicae* is able to alleviate the toxic effect of Cu in Medicago truncatula. Moreover, nodules elicited by this strain were able to accumulate Cu more than the wild type strain (Delgadillo et al. 2015). Glandorf et al. (2001) released two genetically modified *Pseudomonas putida* with the phz biosynthetic gene of *P. fluorescens* into the rhizospheres of wheat plants to produce the antifungal compound phenazine-1-carboxylic acid (PCA). They also found that the presence of any of these bacterial strains transiently changed the rhizosphere fungal microflora (Glandorf et al. 2001).

Lignocellulosic biomass, including rice straw, wheat straw, corn stover, and sugarcane bagasse, is the major agricultural waste, which is used for biofuel production. This lignocellulosic biomass contains cellulose and hemicellulose that can be hydrolyzed to monosaccharides (mainly glucose and xylose) and are converted into biofuels using microbes. Abdelaal et al. (2019) engineered butanol production genes in the genome of two different *E. coli* strains using CRISPR/Cas9 technology to produce bio-butanol from glucose and xylose (Abdelaal et al. 2019).

5. Conclusion

Farming is a complex network of plant-microbe or animal-microbe interactions. There is an increasing demand for eco-friendly techniques for farming that can provide the growing human population with sufficient amounts of nutrients by improving the quality and quantity of farm products.

The use of beneficial microbes in farming would be an important alternative to some traditional farming techniques. The main ecological strategy of integrated management practices, such as nutrient management, disease, and pesticide management is farming with microbes to reduce the use of chemicals in the field and increase productivity.

References

Abdelaal, A.S., Jawed, K. and Yazdani, S.S. 2019. CRISPR/Cas9-mediated engineering of *Escherichia coli* for n-butanol production from xylose in defined medium. J. Ind. Microbiol. Biotechnol. 46: 965–975.

Ambrosio, R., Ortiz-Marquez, J.C.F. and Curatti, L. 2017. Metabolic engineering of a diazotrophic bacterium improves ammonium release and biofertilization of plants and microalgae. Metab Eng. 40: 59–68.

Arias, M.E., Gonzalez-Perez, J.A., Gonzalez-Vila, F.J. and Ball, A.S. 2005. Soil health—a new challenge for microbiologists and chemists. Int. Microbiol. 8: 13–21.

Bageshwar, U.K., Srivastava, M., Pardha-Saradhi, P., Paul, S., Gothandapani, S., Jaat, R.S., Shankar, P., Yadav, R., Biswas, D.R., Kumar, P.A., Padaria, J.C., Mandal, P.K., Annapurna, K. and Das, H.K. 2017. An environmentally friendly engineered Azotobacter strain that replaces a substantial amount of urea fertilizer while sustaining the same wheat yield. Appl. Environ. Microbiol. 83.

Barney, B.M., Eberhart, L.J., Ohlert, J.M., Knutson, C.M. and Plunkett, M.H. 2015. Gene deletions resulting in increased nitrogen release by Azotobacter vinelandii: application of a novel nitrogen biosensor. Appl. Environ. Microbiol. 81: 4316–28.

Bashan, Y. 1998. Inoculants of plant growth-promoting bacteria for use in agriculture. Biotechnology Advances 16: 729–770.

Beijerinck, M.W. 1888. Die bacterien der Papilionaceenknollchen (The root nodule bacteria). Botanische Zeitung 46: 725–804.

Berendsen, R.L., Pieterse, C.M. and Bakker, P.A. 2012a. The rhizosphere microbiome and plant health. Trends Plant Sci. 17: 478–86.

Berendsen, R.L., Pieterse, C.M.J. and Bakker, P.A.H.M. 2012b. The rhizosphere microbiome and plant health. Trends in Plant Science 17: 478–486.

Berlanas, C., Berbegal, M., Elena, G., Laidani, M., Cibriain, J.F., Sagues, A. and Gramaje, D. 2019. The fungal and bacterial rhizosphere microbiome associated with grapevine rootstock genotypes in mature and young vineyards. Front Microbiol. 10: 1142.

Bloemberg, G.V. and Lugtenberg, B.J. 2001. Molecular basis of plant growth promotion and biocontrol by rhizobacteria. Curr. Opin. Plant Biol. 4: 343–50.

Bonfante, P. and Anca, I.A. 2009. Plants, mycorrhizal fungi, and bacteria: a network of interactions. Annual Review of Microbiology 63: 363–383.

Broeckling, C.D., Broz, A.K., Bergelson, J., Manter, D.K. and Vivanco, J.M. 2008. Root exudates regulate soil fungal community composition and diversity. Appl. Environ. Microbiol. 74: 738–44.

Burgmann, H., Widmer, F., Sigler, W.V. and Zeyer, J. 2003. mRNA extraction and reverse transcription-PCR protocol for detection of nifH gene expression by Azotobacter vinelandii in soil. Appl. Environ. Microbiol. 69: 1928–35.

Burr, T.J., Schroth, M.n. and Suslow, T. 1978. Increased potato yields by treatment of seedpieces with specific strains of *Pseudomonas fluorescens* and *P. putida*. Phytopathology 68: 1377–1383.

Burris, R.H. 1977. Overview of nitrogen fixation. Basic Life Sci. 9: 9–18.

Chakradhar, T. and Jauhri, K.S. 2004. Development and evaluation of bioorganic fertilizer. J. Indian Microbiology 44(4): 291–294.

Chaucheyras-Durand, F. and Durand, H. 2010. Probiotics in animal nutrition and health. Beneficial Microbes 1: 3–9.

Costerton, J.W., Lewandowski, Z., Caldwell, D.E., Korber, D.R. and Lappin-Scott, H.M. 1995. Microbial biofilms. Annu. Rev. Microbiol. 49: 711–45.

Crickmore, N., Nicholls, C., Earp, D.J., Hodgman, T.C. and Ellar, D.J. 1990. The construction of *Bacillus thuringiensis* strains expressing novel entomocidal delta-endotoxin combinations. Biochem. J. 270: 133–6.

Dalton, H.M. and March, P.E. 1998. Molecular genetics of bacterial attachment and biofouling. Curr. Opin. Biotechnol. 9: 252–5.

Das, H.K. 2019. Azotobacters as biofertilizer. Adv. Appl. Microbiol. 108: 1–43.

Dashko, R. and Shidlovskaya, A. 2016. Impact of microbial activity on soil properties. Canadian Geotechnical Journal 53: 1386–1397.

Deaker, R., Roughley, R.J. and Kennedy, I.R. 2004. Legume seed inoculation technology—a review. Soil Biology & Biochemistry 36: 1275–1288.

Delgadillo, J., Lafuente, A., Doukkali, B., Redondo-Gomez, S., Mateos-Naranjo, E., Caviedes, M.A., Pajuelo, E. and Rodriguez-Llorente, I.D. 2015. Improving legume nodulation and Cu rhizostabilization using a genetically modified rhizobia. Environ. Technol. 36: 1237–45.

Dent, D. and Cocking, E. 2017. Establishing symbiotic nitrogen fixation in cereals and other non-legume crops: The Greener Nitrogen Revolution. Agriculture & Food Security, 6.

Espenberg, M., Truu, M., Mander, Ü., Kasak, K., Nõlvak, H., Ligi, T., Oopkaup, K., Maddison, M. and Truu, J. 2018. Differences in microbial community structure and nitrogen cycling in natural and drained tropical peatland soils. Scientific Reports, 8.

Fuller, R. 1989. Probiotics in man and animals. J. Appl. Bacteriol. 66: 365–78.

Garbaye, J. 1994. Helper bacteria: a new dimension to mycorrhizal symbiosis. New Phytol. 128: 197–210.

Garbeva, P., Van Veen, J.A. and Van Elsas, J.D. 2004. Microbial diversity in soil: Selection of microbial populations by plant and soil type and implications for disease suppressiveness. Annual Review of Phytopathology 42: 243–270.

Geddes, B.A., Ryu, M.H., Mus, F., Garcia Costas, A., Peters, J.W., Voigt, C.A. and Poole, P. 2015. Use of plant colonizing bacteria as chassis for transfer of N(2)-fixation to cereals. Curr. Opin. Biotechnol. 32: 216–222.

Glandorf, D.C., Verheggen, P., Jansen, T., Jorritsma, J.W., Smit, E., Leeflang, P., Wernars, K., Thomashow, L.S., Laureijs, E., Thomas-Oates, J.E., Bakker, P.A. and Van Loon, L.C. 2001. Effect of genetically modified *Pseudomonas putida* WCS358r on the fungal rhizosphere microflora of field-grown wheat. Appl. Environ. Microbiol. 67: 3371–8.

Hara, S., Morikawa, T., Wasai, S., Kasahara, Y., Koshiba, T., Yamazaki, K., Fujiwara, T., Tokunaga, T. and Minamisawa, K. 2019. Identification of nitrogen-fixing bradyrhizobium associated with roots of field-grown sorghum by metagenome and proteome analyses. Frontiers in Microbiology, 10.

Harman, G.E. 2000. Myths and dogmas of biocontrol changes in perceptions derived from research on *Trichoderma harzinum* T-22. Plant Dis. 84: 377–393.

Hiltner, L. 1904. Uber neure Efarhungen und Probleme auf dem Gebiet der Bodenbakteriologie und unter besonderer Berucksichtigung der Grundingung and Brace. Arb. Dtsch. Landwirtsch. Gesellschaftswiss, 98.

Hu, S.B., Liu, P., Ding, X.Z., Yan, L., Sun, Y.J., Zhang, Y.M., Li, W.P. and Xia, L.Q. 2009. Efficient constitutive expression of chitinase in the mother cell of *Bacillus thuringiensis* and its potential to enhance the toxicity of Cry1Ac protoxin. Appl. Microbiol. Biotechnol. 82: 1157–67.

Johansen, J.E., Binnerup, S.J., Kroer, N. and Molbak, L. 2005. Luteibacter rhizovicinus gen. nov., sp. nov., a yellow-pigmented gammaproteobacterium isolated from the rhizosphere of barley (Hordeum vulgare L.). Int. J. Syst. Evol. Microbiol. 55: 2285–91.

Kass, D.L., Drosdoff, M. and Alexander, M. 1971. Nitrogen fixation by Azotobacter paspali in association with Bahiagrass (Paspalum notatum)1. Soil Science Society of America Journal, 35.

Kkan, A., Akhtar, M., Naqvi, S.M.S., Rasheed, M., Khan, A., Jilani, G., Akhtar, M., Muhammad, S., Naqvi, S. and Rasheed, M. 2009. Phosphorus solubilizing bacteria: occurrence, mechanisms and their role in crop production. J. Agric. Biol. Sci. 1.

Klier, A., Bourgouin, C. and Rapoport, G. 1983. Mating between *Bacillus subtilis* and *Bacillus thuringiensis* and transfer of cloned crystal genes. Molecular and General Genetics MGG 191: 257–262.

Kloepper, J.W., Lifshitz, R. and Zablotowicz, R.m. 1989. Free-living bacteria inocula for enhancing crop productivity. Trends in Biotechnology 7: 39–44.

Köhl, J., Kolnaar, R. and Ravensberg, W.J. 2019. Mode of action of microbial biological control agents against plant diseases: relevance beyond efficacy. Frontiers in Plant Science, 10.

Kwak, M.J., Kong, H.G., Choi, K., Kwon, S.K., Song, J.Y., Lee, J., Lee, P.A., Choi, S.Y., Seo, M., Lee, H.J., Jung, E.J., Park, H., Roy, N., Kim, H., Lee, M.M., Rubin, E.M., Lee, S.W. and Kim, J.F. 2018. Rhizosphere microbiome structure alters to enable wilt resistance in tomato. Nat. Biotechnol.

Landers, T.F., Cohen, B., Wittum, T.E. and Larson, E.L. 2012. A review of antibiotic use in food animals: perspective, policy, and potential. Public Health Reports 127: 4–22.

Lopez, J. 2000. Probiotics in animal nutrition. Asian-Australasian Journal of Animal Sciences 13: 12–26.

Lu, S.Q., Liu, Z.D. and Yu, Z.N. 2000. The characterization of *Bacillus thuringiensis* strain YBT833 and its transformants that containing different ICP genes. Acta Genet. Sin 27: 839–844.

Lugtenberg, B. and Kamilova, F. 2009. Plant-growth-promoting rhizobacteria. Annu. Rev. Microbiol. 63: 541–56.

Mcewen, S.A. and Fedorka-Cray, P.J. 2002. Antimicrobial use and resistance in animals. Clinical Infectious Diseases 34: S93–S106.

Muruganandam, S., Israel, D.W. and Robarge, W.P. 2010. Nitrogen transformations and microbial communities in soil aggregates from three tillage systems. Soil Science Society of America Journal 74: 120–129.

Nash, M.V., Anesio, A.M., Barker, G., Tranter, M., Varliero, G., Eloe-Fadrosh, E.A., Nielsen, T., Turpin-Jelfs, T., Benning, L.G. and Sanchez-Baracaldo, P. 2018. Metagenomic insights into diazotrophic communities across Arctic glacier forefields. Fems Microbiology Ecology, 94.

Nikolaev, Y.A. and Plakunov, V.K. 2007. Biofilm - "City of microbes" or an analogue of multicellular organisms? Microbiology 76: 125–138.

O'callaghan, M. 2016. Microbial inoculation of seed for improved crop performance: issues and opportunities. Applied Microbiology and Biotechnology 100: 5729–5746.

Oldroyd, G.E.D. 2013. Speak, friend, and enter: signalling systems that promote beneficial symbiotic associations in plants. Nature Reviews Microbiology 11: 252–263.

Pacanoski, Z. 2015. Bioherbicides. Herbicides, Physiology of Action, and Safety.

Pal, K.K. and Mcspadden Gardener, B. 2006. Biological control of plant pathogens. The Plant Health Instructor.

Perez-Garcia, G., Basurto-Rios, R. and Ibarra, J.E. 2010. Potential effect of a putative sigma(H)-driven promoter on the over expression of the Cry1Ac toxin of *Bacillus thuringiensis*. J. Invertebr. Pathol. 104: 140–6.

Petrova, O.E. and Sauer, K. 2012. Sticky situations: key components that control bacterial surface attachment. Journal of Bacteriology 194: 2413–2425.

Piggot, P.J. and Hilbert, D.W. 2004. Sporulation of *Bacillus subtilis*. Curr. Opin. Microbiol. 7: 579–86.

Raynaud, X. and Nunan, N. 2014. Spatial ecology of bacteria at the microscale in soil. Plos One, 9.

Richardson, A.E., Barea, J.M., Mcneill, A.M. and Prigent-Combaret, C. 2009. Acquisition of phosphorus and nitrogen in the rhizosphere and plant growth promotion by microorganisms. Plant and Soil 321: 305–339.

Sashidhar, B. and Podile, A.R. 2009. Transgenic expression of glucose dehydrogenase in Azotobacter vinelandii enhances mineral phosphate solubilization and growth of sorghum seedlings. Microb Biotechnol. 2: 521–9.

Savka, M.A. and Farrand, S.K. 1997. Modification of rhizobacterial populations by engineering bacterium utilization of a novel plant-produced resource. Nat. Biotechnol. 15: 363–8.

Schnepf, E., Crickmore, N., Van Rie, J., Lereclus, D., Baum, J., Feitelson, J., Zeigler, D.R. and Dean, D.H. 1998. *Bacillus thuringiensis* and its pesticidal crystal proteins. Microbiol. Mol. Biol. Rev. 62: 775–806.

Setten, L., Soto, G., Mozzicafreddo, M., Fox, A.R., Lisi, C., Cuccioloni, M., Angeletti, M., Pagano, E., Diaz-Paleo, A. and Ayub, N.D. 2013. Engineering *Pseudomonas protegens* Pf-5 for nitrogen fixation and its application to improve plant growth under nitrogen-deficient conditions. PLoS One 8: e63666.

Somers, E., Vanderleyden, J. and Srinivasan, M. 2004. Rhizosphere bacterial signalling: A love parade beneath our feet. Critical Reviews in Microbiology 30: 205–240.

Srivastav, S., Yadav, K.S. and Kundu, B.S. 2004. Prospects of using phosphate solubilizing Pseudomonas as biofungicide. Indian Journal of Microbiology 44: 91–94.

Stevenson, F.J. and Cole, M.A. 1999. Cycles of Soil : Carbon, Nitrogen, Phosphorus, Sulfur, Micronutrients, New York, Wiley.

Stewart, P.S. 2002. Mechanisms of antibiotic resistance in bacterial biofilms. International Journal of Medical Microbiology 292: 107–113.

Sylvia, D.M. 2005. Principles and Applications of Soil Microbiology. Upper Saddle River, N.J., Pearson Prentice Hall.

Tesfaye, M., Temple, S.J., Allan, D.L., Vance, C.P. and Samac, D.A. 2001. Overexpression of malate dehydrogenase in transgenic alfalfa enhances organic acid synthesis and confers tolerance to aluminum. Plant Physiol. 127: 1836–44.

Usta, C. 2013. Microorganisms in Biological Pest Control—A Review (Bacterial Toxin Application and Effect of Environmental Factors). Current Progress in Biological Research.

Uyeno, Y., Shigemori, S. and Shimosato, T. 2015. Effect of probiotics/prebiotics on cattle health and productivity. Microbes Environ. 30: 126–32.

Van Dommelen, A., Croonenborghs, A., Spaepen, S. and Vanderleyden, J. 2009. Wheat growth promotion through inoculation with an ammonium-excreting mutant of Azospirillum brasilense. Biology and Fertility of Soils 45: 549–553.

Van Rhijn, P., Goldberg, R.B. and Hirsch, A.M. 1998. Lotus corniculatus nodulation specificity is changed by the presence of a soybean lectin gene. Plant Cell 10: 1233–50.

Waksman, S.A. 1932. Principles of Soil Biology. Williams and Wilkins, Baltimore.

Wei, D., Yang, Q., Zhang, J.Z., Wang, S., Chen, X.L., Zhang, X.L. and Li, W.Q. 2008. Bacterial community structure and diversity in a black soil as affected by long-term fertilization. Pedosphere 18: 582–592.

Weller, D.M. 1988. Biological control of soilborne plant pathogens in the rhizosphere with bacteria. Annu. Rev. Phytopathol. 26: 379–407.

Weller, D.M. 2007. Pseudomonas biocontrol agents of soilborne pathogens: looking back over 30 years. Phytopathology 97: 250–6.

Wolinska, A., Kuzniar, A., Zielenkiewicz, U., Banach, A., Izak, D., Stepniewska, Z. and Blaszczyk, M. 2017. Metagenomic analysis of some potential nitrogen-fixing bacteria in arable soils at different formation processes. Microbial Ecology 73: 162–176.

Woodman, H.E. 2009. The mechanism of cellulose digestion in the ruminant organism. The Journal of Agricultural Science 17: 333–338.

Yang, J., Kloepper, J.W. and Ryu, C.M. 2009. Rhizosphere bacteria help plants tolerate abiotic stress. Trends Plant Sci. 14: 1–4.

Yue, C., Sun, M. and Yu, Z. 2005a. Broadening the insecticidal spectrum of Lepidoptera-specific *Bacillus thuringiensis* strains by chromosomal integration of cry3A. Biotechnol. Bioeng. 91: 296–303.

Yue, C., Sun, M. and Yu, Z. 2005b. Improved production of insecticidal proteins in *Bacillus thuringiensis* strains carrying an additional cry1C gene in its chromosome. Biotechnol. Bioeng. 92: 1–7.

Zhang, T., Yan, Y., He, S., Ping, S., Alam, K.M., Han, Y., Liu, X., Lu, W., Zhang, W., Chen, M., Xiang, W., Wang, X. and Lin, M. 2012. Involvement of the ammonium transporter AmtB in nitrogenase regulation and ammonium excretion in Pseudomonas stutzeri A1501. Res. Microbiol. 163: 332–9.

CHAPTER 6

Marine Algae as Green Agriculture

Simranjeet Singh,[1,#] *Vijay Kumar,*[2,#] *Shivika Datta,*[3,#] *Satyender Singh,*[4]
Daljeet Singh Dhanjal,[5] *Noyonika Kaul,*[6] *Praveen C. Ramamurthy*[1,*]
and *Joginder Singh*[6,*]

1. Introduction

At least 70% of the Earth's surface is covered with oceans representing various resources for potential therapeutic and clinical agents (Malve 2016). Over the past few decades, various novel compounds have been isolated from marine and coastal organisms that have interesting pharmacological activities, including biocidal activities (Gomes et al. 2016). Among various species, macroalgae or seaweeds are known to be found easily and are one of the major components of the marine ecosystem. It is classified into three variant classes Rhodophyta (red algae), Phaeophyta (brown algae), and Chlorophyta (green algae) (El Gamal 2010). The characteristic red color of Rhodophyta is due to the presence of pigment Phycoerythrin (Khalid et al. 2018). Fucoxanthin and xanthophylls constitute the major pigment system in Phaeophyta (Takaichi 2011). The green pigmentation of Chlorophyta is attributed to the dominance of chlorophyll 'a' and 'b' in higher plants (Grabski et al. 2016).

Marine algae are regarded as a potent source of bioactive compounds having a wide range of biocidal activities, including antifungal, antibacterial, and antibacterial properties (Bajpai 2016). The extracts of marine algae are also used as

[1] Interdisciplinary Centre for Water Research (ICWaR), Indian Institute of Sciences, Bangalore – 560012, India.
[2] Regional Ayurveda Research Institute for Drug Development, Gwalior – 474009, MP, India.
[3] Department of Zoology, Doaba College, Jalandhar, Punjab, India.
[4] Regional Advanced Water Testing Laboratory, Mohali – 160059, Punjab, India.
[5] Department of Biotechnology, Lovely Professional University, Phagwara – 144411, Punjab, India.
[6] Department of Microbiology, Lovely Professional University, Phagwara – 144411, Punjab, India.
# Equal contribution
* Corresponding authors: joginder.15005@lpu.co.in; onegroupb203@gmail.com

soil conditioners in the agricultural sector to enhance crop production (Calvo et al. 2014). The polysaccharide extract of marine algae can act as metal chelators. These metal chelators bind with insoluble microelements and convert them into soluble forms (Singh et al. 2019). Their extracts have also been used as bio-stimulants in agricultural practices (Battacharyya et al. 2015). These bio-stimulants act on plant physiology via various pathways to enhance crop vigour, quality, and yield (Khan et al. 2009). It also increases water holding capacity, nutrient availability, anti-oxidants, and chlorophyll concentration in plants (Ronga et al. 2019). The research on marine algal bodies is limited to field-scale only; however, no studies have been reported its uses on large scale usage (Barkia et al. 2019). This book chapter is considered the first of its kind that summarizes and highlights using marine algal products and biomass in biocidal control of agricultural diseases, improving plant growth, and integrated pest management for promising sustainable agricultural growth.

2. Morphological and Genetic Characterization of Marine Algae

About 71% of the earth's surface is covered by the oceans, which is the habitat for various marine organisms involving marine algae (Costello and Chaudhary 2017). Environmentally, marine algae are an important component in the aquatic food chain, aid as a habitat for various aquatic species, and it is essential for biogeochemical cycling (Beetul et al. 2016). These algae species can be either be eukaryotic or prokaryotic (Helliwell et al. 2016). Generally, these marine algal are inhabitants of shallow water and are classified into two key groups (Figure 1), i.e., microalgae and macroalgae (a.k.a seaweed) (El Gamal 2010). The microalgae are algal species of microscopic size and are classified into blue-green algae (Cyanophyta), green algae (Chlorophyta), dinoflagellates (Pyrrophyta) and diatoms and golden-brown algae (Chrysophyta), whereas macroalgae involve algal species of macroscopic size and are classified into three categories, i.e., brown macroalgae (Phaeophyta), red macroalgae (Rhodophyta), and green macroalgae (Chlorophyta) (Beetul et al. 2016). These marcoalgal are stated as non-vascular plants or thallophytes (Stanton and Reeb 2016). For taxonomical classification of these marine algae, various methods are used which are classified on the basis of genetic molecules, pigments, fatty acids and

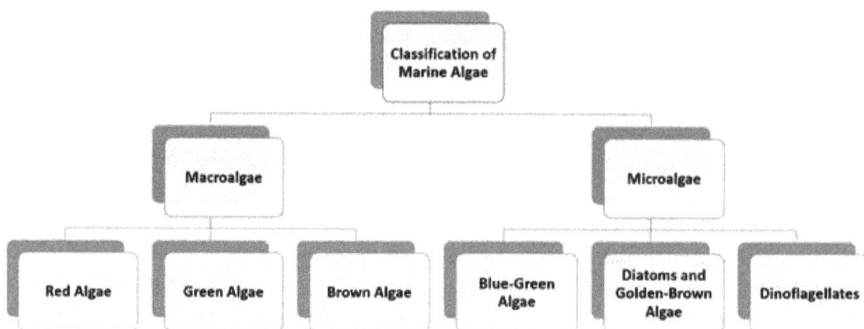

Figure 1. Marine algal classification.

secondary metabolite distribution as well as fluorescence parameters (measured by flow cytometry) (Wahby et al. 2014).

3. Benefits of Marine Algae

The cell walls of most marine algal bodies are rich in sulfated polysaccharides that exhibit various biological activities, such as anti-oxidative, anti-inflammation, anti-cancer, anti-viral anticoagulant, etc. (Xu et al. 2017). They are also regarded as an essential component for plant improvement due to the presence of high content of vitamins, amino acid, mineral substances and growth regulators, such as gibberellins, cytokinin, and auxins (Han et al. 2018). The extracts of these marine algae are particularly used in agriculture purposes to enhance the productivity of a wide range of agricultural plants, such as beet, tomato, citrus plants, grasses, potato, and leguminous plants (Hamed et al. 2018).

4. Improvement in Soil Health

Most of the marine microalgae and macroalgae are often regarded as biostimulators for plant growth and development as they have higher contents of plant growth regulators, vitamins, amino acids, mineral substances, etc. (Nabti et al. 2017). They secrete various substances which enhance nutrient availability (P and N) to the plants (Singh and Reddy 2014). Some of the species, such as Spirillum, Gallionella (iron bacteria), Nitrosospira, and Nitrosomonas, of order Nitrosomonadales, are reported to fix the atmospheric nitrogen by improving the soil fertility in irrigated and submerged rice cultivation areas (Sharma et al. 2012). They are used worldwide, including in Asia in rice fields by means to fix biological nitrogen to enhance soil fertility (Jochum et al. 2018). The species of Tolypothrix tenuis and Azolla are reported to directly add to the rice fields to fix the gaseous nitrogen (Vaishampayan et al. 2001). The organic matter formed from the decaying of marine algae is enriched in humus content and acts as a binding agent for nutrient availability and retains moisture for the growth and development of plants (Schoonover and Crim 2015). Marine algae also play an important role in soil reclamation by improving plant conditions under different stresses, such as salinity, temperature stress, etc. (Brodie et al. 2017). Marine algae also secrete some stimulants which act on plant physiology via various pathways to improve quality, yield, post-harvest, and crop vigour. The important role of marine algae and their byproducts are depicted in Table 1. They have also been shown to influence ion uptake, nucleic acid synthesis, respiration, and photosynthesis (Jadhao et al. 2015). These stimulants enhance water holding capacity, nutrient availability, increase in chlorophyll concentration, antioxidants, and metabolism in plants (Salim 2016).

Table 1. List of marine algae by-products commercially available for their usage in agriculture.

Macroalgae Name	Class	Family	Product Name	Company	Application			
					PGS	AF	BF	PB
Ascophyllum nodosum	Phaeophyceae	Fucaceae	Acadian®	Acadian Agritech	✓			
			Agri-Gro Ultra	Agri Gro Marketing Inc	✓			
			Alg-A-Mic	BioBizz Worldwide N.V.	✓			
			Bio-Genesis™	Green Air Products, Inc.	✓			
			Biovita	PI Industries Ltd	✓			
			Espoma	The Espoma Company	✓			
			Guarantee®	MaineStream Organics	✓			
			Kelp Meal	Acadian Seaplants Ltd	✓			
			Kelpro	Teniprocesos Biol., S.A. de C.V.	✓			
			Kelprosoil	Productos del Pacifico, S.A. de C.V.	✓			
			Maxicrop	Maxicrop USA, Inc.	✓			
			Nitrozime	Hydrodynamics International Inc.	✓			
			Soluble SW Extract	Technaflora Plant Products, LTD	✓			
			Stimplex®	Acadian Agritech	✓			
			Synergy	Green Air Products, Inc.	✓			
			Tasco®	Acadian Agritech		✓		
Durvillea Antarctica	Phaeophyceae	Durvillaeaceae	Profert®	BASF				✓
Durvillea potatorum	Phaeophyceae	Durvillaeaceae	Seasol®	Seasol International Pty Ltd	✓			
Ecklonia maxima	Phaeophyceae	Lessoniaceae	Kelpak	BASF	✓			
Lithothamnium calcareum	Florideophyceae	Lithothamniaceae	Acid Buf	Chance and Hunt Limited		✓		
Macrocystis pyrifera	Phaeophyceae	Laminariaceae	AgroKelp	Algas y Biod. Mar. S.A.	✓			
Unspecified	-	-	Sea Winner	China Ocean Univ. Product Dev. Co., Ltd				✓
Unspecified	-	-	Seanure	Farmura Ltd	✓			
Unspecified	-	-	Fartum®	Inversiones Patagonia S.A.			✓	

* PGS – Plant Growth Stimulant; AF – Animal Feed; BF- Bio-fertilizer; and PB – Plant Bio-stimulant

5. Role of Macroalgae in Bio-Stimulation of Plant Growth

Marine macroalgal are considered essential for plants as it enhances the mineral content, synthesizes essential amino acids, vitamins, and growth regulators, like auxin, gibberellins, and cytokinin (Kumar et al. 2015, Boghdady et al. 2016). Extracts of brown algae and algae themselves are extensively used in different agricultural practices. The involvement of the algal species has been known to increase the yield of plants, like beet, citrus plants, grasses, legumes, potato, and tomato (Singh et al. 2016, Mukherjee and Patel 2019). The use of marine algae in plant biotechnology has not improved the health of the plant but also increased the weight and number of fruits (Wells et al. 2017). Along with that, it also serves as an eco-friendly alternative in disease management (Wijesinghe and Jeon 2012). Additionally, it has been found that different aqueous extracts prepared via different methods have a positive impact on crop yield, growth, and health of various plants (Michalak et al. 2016).

There are a few points that separate plant regulators from fertilizers, i.e., (a) they can regulate and amend the cell division, (b) control the shoot and root elongation, and (c) triggers the flowering as well as other metabolic functions (Kumar et al. 2014). Whereas, fertilizers only provide the essential nutrient for plant growth (White and Brown 2010). For example, cytokinin is the essential growth regular in marine macroalgae (Stirk et al. 2003). On the other, the presence of trace elements in the extracts obtained from marine macroalgal plays a significant role in plant physiology by acting as an enzyme activator and providing essential nutrients (Latique et al. 2013). The use of *A. nodosum* extract on forage and turf grasses has been reported to increase the production of antioxidant molecules, like ascorbic acid, b-carotene and α-tocopherol along with antioxidant enzymes, like ascorbate peroxidase, GSH reductase, and superoxide dismutase (Allen et al. 2001).

Bio-stimulants are those organic compounds when used in very fewer amounts improve plant growth, which is not achieved through the conventional method (Yakhin et al. 2017). Now, various macroalgal extracts are used as agricultural bio-stimulants. The use of these bio-stimulants on plants has been reported to show various benefits, such as improved photosynthetic activity, high fruit as well as crop yield, enhanced rooting and development of resistance against bacteria, fungi, and viruses (Shukla et al. 2019). Agricultural bio-stimulants comprises diverse compounds and products, like microbes, enzymes, macroalgal extracts, plant growth regulators and trace elements, which are applied to plants to improve the physiological processes of the plant (Kocira et al. 2019). The benefits make them the target of interest (Buschmann et al. 2017). The role of these marine macroalgal in agricultural practices has been illustrated in Figure 2. These bio-stimulants are involved in various physiological pathways of the plant to enhance the quality and yield of the crop (Sharma et al. 2014). The macroalgal extracts have been reported to influence the ion uptake, respiration activity, nucleic acid synthesis, and photosynthesis processes in the plant. Additionally, these compounds also improve the water holding capacity, increase chlorophyll synthesis, improved nutrient availability and improve the metabolism of the plant (Cheng and Cheng 2015).

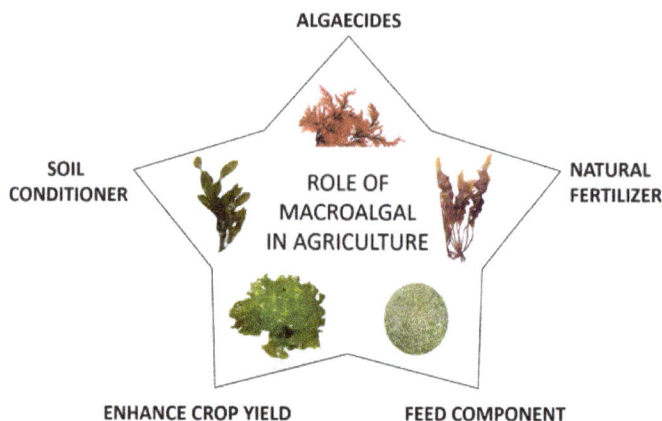

Figure 2. Application of macroalgal in agriculture practices.

6. Antimicrobial Potential of Marine Macroalgae

6.1 Antibacterial Activity

They also produce a wide range of chemically active metabolites, such as glycerols, lipids, alkaloids, cyclic peptides, polysaccharides, polyketides, phlorotannins, sterols, diterpenoids, etc., that have cytotoxic and antibacterial actions against various pathogens in the environment (Al-Saif et al. 2014). The methanolic extract of marine macroalgae, such as *S. swartzii* and *C. Agardh*, exhibits antibacterial activity against *Pseudomonas syringae* a causative agent for leaf spot disease in *Gymnema sylvestre* (Kumar et al. 2008). The acetonic extracts of *Sargassum polyceratium* also show a remarkable activity against *E. coli, Pectobacterium carotovora, Erwinia carotovora,* and *Staphylococcus aureus* using the disc diffusion method (Kumar et al. 2008). Other species of marine algae, such as *Gracilaria cervicornis, Caulerpa racemosa* and *S. polyceratium* and their ethanolic extracts, were found to be effective against *Staphylococcus aureus* (Borbón et al. 2012).

Moreover, the methanolic extract of *S. swartzii* has been reported to minimize the growth of *Xanthomonas oryzae* pv. *oryzae,* which is reported to cause blight disease in rice seedlings (Rajan et al. 2013). The extracts of *Fucus spiralis* and *Cystoseira myriophylloides* is found active against *Agrobacterium tumefaciens,* which causes crown gall diseases in tomato plants (Esserti et al. 2017). The antibacterial properties of marine algae are also attributed to various groups of biologically active fatty acids, such as palmitic acid (Pérez et al. 2016). Brown marine algae extract contributes antibacterial activity by producing both unsaturated and saturated fatty acids, such as eicosapentaenoic, palmitic, oleic, and myristic acids which plays a crucial defence against pathogenic gram-negative and gram-positive bacteria (Chowdhury et al. 2015). The brown algae contain phenolic compounds in their cell wall, which act as a chemical defender against pathogenic bacteria (Chauhan and Kasture 2014). In addition, carotenoids pigment also has antibacterial properties and could protect the cells of plants from oxidative stress (Sathasivam et al. 2018). Some marine algae such as *Fucus spiralis, Cystoseira, Laminaria digitata*, and myriophylloides are also reported to significantly increase the concentration of defense enzymes, such as peroxidases and polyphenol oxidases against various microbial populations (Esserti et al. 2017).

6.2 Anti-Viral Properties

Disease caused by viruses in plants is mainly responsible for great damage to the agricultural industry (Shalaby et al. 2011). Although various chemotherapeutic agents such as pesticides and insecticides are an effective and direct method for regulation of this agent, it also causes a chain of side effects including accumulation of pesticides residues in the food chain and in building up pathogen resistance to these chemicals (Singh et al. 2018, Nicolopoulou-Stamati et al. 2016). The framework of active substances in marine algae, includes oils, proteins, polysaccharides, polyphenols, and flavonoids which have efficient antiviral properties by inhibiting adsorption of viruses at the cell membrane of the plant cell (Ahmadi et al. 2015). Other compounds such as dicytol H, dicytol C, dictyodial, and betaines are also reported to be isolated from marine algae possessing antiviral and cytotoxic activities against some viruses (Wang et al. 2012). Beta/kappa carrageenans isolated from rhodophyte *Tichocarpus crinitus* suppress the activity of tobacco mosaic virus in leaves of Xanthi nc tobacco (Nagorskaia et al. 2008). Substantial quantities of proteins, omega 3 fatty acids, peptides, vitamin C, and amino acids are reported to present in red marine algae having noticeable activity against viruses (Hamed et al. 2015).

6.3 Antinematodal Activity

Many anti-nematotal compounds have been obtained from marine macroalgae (Dahms and Dobretsov 2017). In 2013, Baloch and his colleagues reported about treating the soil with powders of marine macroalgal (*Melanothamnus afaqhusainii, Polycladia indica,* and *Spatoglossum variabile*), which significantly reduced the *Meloidogyne incognita* (root-knot nematode) infection that majorly infected eggplants and watermelons (Baloch et al. 2013). These results were similar to what Sultana et al. (2012) found, wherein dry powders of marine macroalgal (*Halemida tuna, M. afaqhusainii,* and *S. variable*) were used to suppress the effect of *M. incognita*. Additionally, it was having a similar effect as carbofuran (toxic nematicide) in both field and greenhouse conditions (Sultana et al. 2012). Sunflower and tomato plants have also been assessed for reduction in nematode galls on roots as well as nematode penetration within the root system (Fuller et al. 2008). Another researcher recommended inoculating the agricultural macroalgal species synthesizing biostimulants as it will reduce the chances of *M. incognita* and *M. javanica* (root-knot nematodes) infection in tomato plants. A significant reduction in egg recovery from the root was also found in treated plants (Craigie 2011). In addition to that, macroalgae also synthesize precursors of ethylene biosynthesis, i.e., 1-aminocyclopropane-1-carboxylic acid and cytokinins which improve the susceptibility as well as resistance against root-knot nematodes in plants (Mukherjee and Patel 2019). Another researcher highlighted the potential of seaweed extracts obtained from *Ascophyllum nodosum* and *Ecklonia maxima* against *Meloidogyne chitwoodi* and *M. hapla in vivo* assays. Plus, they also reported that continuous exposure to *A. nodosum* extracts (alkaline) reduce the hatching percentage of *M. chitwoodi* egg (El-Deen and Issa 2016). Furthermore, pre-exposed juveniles of *M. chitwoodi* to *A. nodosum* or *E. maxima* grew on agar plant showed reduced attraction towards root diffusate of tomato and pre-exposure (24 hours) of *A. nodosum*, which also reduces the *M. chitwoodi* and *M. hapla* infectivity (Ngala et al. 2016).

6.4 Bioinsecticidal Activity

Marine algal is well known for showing insecticidal activity. Their extracts have been found efficient in integrated pest management. They are the natural reservoir of novel and eco-friendly insecticidal compounds (Asharaja and Sahayaraj 2013). Crude extracts from *Sargassum tenerrimum, Caulerpa scalpelliformis, Ulva fasciata, Padina pavonica*, and *Ulva lactuca* are potent insecticidal compounds against *Dysdercus* spp. (Cotton pest), which is responsible for severe crop loss (Karthikeyan et al. 2007, Cetin et al. 2010). Moreover, 96 hours of exposure to chloroform extract of *Padina pavoncia* and *Sargassum swartzii* exhibited *Dysdercus cingulatus* mortality (Sahayaraj and Jeeva 2012). Additionally, aqueous and chloroform extracts of *Sargassum swartzii* has been found responsible for the shortening female and male longevity of *Dysdercus cingulatus* (Asha et al. 2012). The methanolic and chloroform extracts of *Padina pavoncia* and *Sargassum swartzii* have been reported to reduce the hatchability and fecundity of *Dysdercus cingulatus* along with that mating period, which also increases on exposure to water extracts (Asharaja and Sahayaraj 2013). Later, hexane extracts have also been found to reduce the hatchability and fecundity of *Dysdercus cingulatus*. On extensive investigation of chloroform extract, the compounds stigmastan-6, 22-dien,3,5-dedihydro and hexadecanoic acid methyl ester, synthesized by *Padina pavoncia* and *Sargassum swartzii*, were responsible for the mortality of *Dysdercus cingulatus* nymphs (Asharaja and Sahayaraj 2013).

7. Conclusions and Future Perspectives

Algae are ubiquitous in world soil. In this book chapter, the beneficial roles of algae with respect to crop plants in agricultural systems have been analyzed. Algae forms to be an important component of arid and semi-arid ecosystems. They are also known as the biological indicator of the health of the environment. Algae have been ascertained for their usage as the 'biological conditioner' instead of artificial 'chemical conditioners'. The algae have the ability to reduce the resultant pollution for soil and plants and also play a vital role in improving soil and plant properties. The microalgae influence soil-plant systems by increasing the soil organic matter, formation of crust, fixation of nitrogen, and excretion of organic acids which increase phosphorus availability and uptake, PGPRs like activity, soil aggregation, and concentrate metal ions in the environment. There is enormous literature that discusses the antimicrobial activities of marine macroalgae against different plant pathogens. But most of these studies were undertaken in laboratory conditions. There may be many limitations *in vivo*. Thus, many deeper investigations of this trend of study need to be explored. The marine microalgae have the ability to prove themselves valuable for plant cultivation and can be considered as an important inoculant in organic farming for sustainable development.

References

Ahmadi, A., Zorofchian Moghadamtousi, S., Abubakar, S. and Zandi, K. 2015. Antiviral potential of algae polysaccharides isolated from marine sources: a review. Biomed. Res. Int.

Allen, V.G., Pond, K.R., Saker, K.E., Fontenot, J.P., Bagley, C.P., Ivy, R.L., Evans, R.R., Schmidt, R.E., Fike, J.H., Zhang, X. and Ayad, J.Y. 2001. Tasco: Influence of a brown seaweed on antioxidants in forages and livestock—A review. J. Anim. Sci. 79(suppl_E): E21–31.

Al-Saif, S.S., Abdel-Raouf, N., El-Wazanani, H.A. and Aref, I.A. 2014. Antibacterial substances from marine algae isolated from Jeddah coast of Red Sea, Saudi Arabia. Saudi J. Biol. Sci. 21(1): 57–64.

Asha, A., Rathi, J.M., Raja, D.P. and Sahayaraj, K. 2012. Biocidal activity of two marine green algal extracts against third instar nymph of *Dysdercus cingulatus* (Fab.) (Hemiptera: Pyrrhocoridae). J. Biopest. 5: 129–134.

Asharaja, A. and Sahayaraj, K. 2013. Screening of insecticidal activity of brown macroalgal extracts against *Dysdercus cingulatus* (Fab.) (Hemiptera: Pyrrhocoridae). J. Biopest. 6(2): 193–203.

Bajpai, V.K. 2016. Antimicrobial bioactive compounds from marine algae: A mini review. Indian J. Geo-Mar. Sci. 45(9): 1076–1085.

Baloch, G.N., Tariq, S., Ehteshamul-Haque, S., Athar, M., Sultana, V. and Ara, J. 2013. Management of root diseases of eggplant and watermelon with the application of asafoetida and seaweeds. J. Appl. Bot. Food Qual. 86(1): 138–142.

Barkia, I., Saari, N. and Manning, S.R. 2019. Microalgae for high-value products towards human health and nutrition. Mar. Drugs 17(5): 304.

Battacharyya, D., Babgohari, M.Z., Rathor, P. and Prithiviraj, B. 2015. Seaweed extracts as biostimulants in horticulture. Sci. Hort. 196: 39–48.

Beetul, K., Gopeechund, A., Kaullysing, D., Mattan-Moorgawa, S., Puchooa, D. and Bhagooli, R. 2016. Challenges and opportunities in the present era of marine algal applications. Thajuddin, N. and Dhanasekaran, D. (eds.). Algae-Organisms for Imminent Biotechnology. 29: 237–76.

Boghdady, M.S., Selim, D.A., Nassar, R.M. and Salama, A.M. 2016. Influence of foliar spray with seaweed extract on growth, yield and its quality, profile of protein pattern and anatomical structure of chickpea plant (*Cicer arietinum* L.). Middle East J. Appl. Sci. 6(1): 207–21.

Borbón, H., Herrera, J.M., Calvo, M., Sierra, H.T., Soto, R. and Vega, I. 2012. Antimicrobial activity of most abundant marine macroalgae of the Caribbean coast of Costa Rica. J. Asian Sci. Res. 2(5): 292.

Brodie, J., Chan, C.X., De Clerck, O., Cock, J.M., Coelho, S.M., Gachon, C., Grossman, A.R., Mock, T., Raven, J.A., Smith, A.G. and Yoon, H.S. 2017. The algal revolution. Trends Plant Sci. 22(8): 726–38.

Buschmann, A.H., Camus, C., Infante, J., Neori, A., Israel, Á., Hernández-González, M.C., Pereda, S.V., Gomez-Pinchetti, J.L., Golberg, A., Tadmor-Shalev, N. and Critchley, A.T. 2017. Seaweed production: overview of the global state of exploitation, farming and emerging research activity. Eur. J. Phycol. 52(4): 391–406.

Calvo, P., Nelson, L. and Kloepper, J.W. 2014. Agricultural uses of plant biostimulants. Plant Soil 383(1-2): 3–41.

Cetin, H., Gokoglu, M. and Oz, E. 2010. Larvicidal activity of the extract of seaweed, *Caulerpa scalpelliformis*, against *Culex pipiens*. J. Am. Mosq. Control Assoc. 26(4): 433–6.

Chauhan, J. and Kasture, A. 2014. Antimicrobial compounds of marine algae from Indian coast. Int. J. Curr. Microbiol. Appl. Sci. 7: 526–32.

Cheng, F. and Cheng, Z. 2015. Research progress on the use of plant allelopathy in agriculture and the physiological and ecological mechanisms of allelopathy. Front Plant Sci. 6: 1020.

Chowdhury, M.M., Kubra, K., Hossain, M.B., Mustafa, M.G., Jainab, T., Karim, M.R. and Mehedy, M.E. 2015. Screening of antibacterial and antifungal activity of freshwater and marine algae as a prominent natural antibiotic available in Bangladesh. Int. J. Pharmacol. 11: 828–33.

Costello, M.J. and Chaudhary, C. 2017. Marine biodiversity, biogeography, deep-sea gradients, and conservation. Curr. Biol. 27(11): R511–R527.

Craigie, J.S. 2011. Seaweed extract stimuli in plant science and agriculture. J. Appl. Psychol. 23(3): 371–93.

Dahms, H. and Dobretsov, S. 2017. Antifouling compounds from marine macroalgae. Mar. Drugs 15(9): 265.

El Gamal, A.A. 2010. Biological importance of marine algae. Saudi Pharm. J. 18(1): 1–25.

El-Deen, A.H. and Issa, A.A. 2016. Nematicidal properties of some algal aqueous extracts against root-knot nematode, *Meloidogyne incognita in vitro*. Egyptian Journal of Agronematology 15(1): 67–78.

Esserti, S., Smaili, A., Rifai, L.A., Koussa, T., Makroum, K., Belfaiza, M., Kabil, E.M., Faize, L., Burgos, L., Alburquerque, N. and Faize, M. 2017. Protective effect of three brown seaweed extracts against fungal and bacterial diseases of tomato. J. Appl. Psychol. 29(2): 1081–93.

Fuller, V.L., Lilley, C.J. and Urwin, P.E. 2008. Nematode resistance. New Phytol. 2008 Oct 180(1): 27–44.

Gomes, N., Dasari, R., Chandra, S., Kiss, R. and Kornienko, A. 2016. Marine invertebrate metabolites with anticancer activities: Solutions to the "supply problem". Mar. Drugs 14(5): 98.

Grabski, K., Baranowski, N., Skórko-Glonek, J. and Tukaj, Z. 2016. Chlorophyll catabolites in conditioned media of green microalga *Desmodesmus subspicatus*. J. Appl. Psychol. 28(2): 889–896.

Hamed, I., Özogul, F., Özogul, Y. and Regenstein, J.M. 2015. Marine bioactive compounds and their health benefits: a review. Compr. Rev. Food Sci. Food Saf. 2015 Jul 14(4): 446–65.

Hamed, S.M., El-Rhman, A.A.A., Abdel-Raouf, N. and Ibraheem, I.B. 2018. Role of marine macroalgae in plant protection and improvement for sustainable agriculture technology. Beni-Seuf Univ. J. Appl. Sci. 7(1): 104–110.

Han, X., Zeng, H., Bartocci, P., Fantozzi, F. and Yan, Y. 2018. Phytohormones and effects on growth and metabolites of microalgae: a review. Fermentation 4(2): 25.

Helliwell, K.E., Lawrence, A.D., Holzer, A., Kudahl, U.J., Sasso, S., Kräutler, B., Scanlan, D.J., Warren, M.J. and Smith, A.G. 2016. Cyanobacteria and eukaryotic algae use different chemical variants of vitamin B12. Curr. Biol. 26(8): 999–1008.

Jadhao, G.R., Chaudhary, D.R., Khadse, V.A. and Zodape, S.T. 2015. Utilization of seaweeds in enhancing productivity and quality of black gram [*Vigna mungo* (L.) Hepper] for sustainable agriculture. Indian J. Nat. Prod. Resour. 6(1): 16–22.

Jochum, M., Moncayo, L.P. and Jo, Y.K. 2018. Microalgal cultivation for biofertilization in rice plants using a vertical semi-closed airlift photobioreactor. PloS One 13(9): e0203456.

Karthikeyan, S., Balasubramanian, R. and Iyer, C.S. 2007. Evaluation of the marine algae *Ulva fasciata* and *Sargassum* sp. for the biosorption of Cu (II) from aqueous solutions. Bioresour. Technol. 98(2): 452–5.

Khalid, S., Abbas, M., Saeed, F., Bader-Ul-Ain, H. and Suleria, H.A.R. 2018. Therapeutic potential of seaweed bioactive compounds. *In*: Seaweed Biomaterials. IntechOpen.

Khan, W., Rayirath, U.P., Subramanian, S., Jithesh, M.N., Rayorath, P., Hodges, D.M., Critchley, A.T., Craigie, J.S., Norrie, J. and Prithiviraj, B. 2009. Seaweed extracts as biostimulants of plant growth and development. J. Plant Growth Regul. 28(4): 386–99.

Kocira, S., Szparaga, A., Kuboń, M., Czerwińska, E. and Piskier, T. 2019. Morphological and biochemical responses of *Glycine max* (L.) Merr. to the use of seaweed extract. Agronomy 9(2): 93.

Kumar, C.S., Sarada, D.V. and Rengasamy, R. 2008. Seaweed extracts control the leaf spot disease of the medicinal plant Gymnema sylvestre. Indian J. Sci. Technol. 3: 1–5.

Kumar, R., Khurana, A. and Sharma, A.K. 2014. Role of plant hormones and their interplay in development and ripening of fleshy fruits. J. Exp. Bot. 65(16): 4561–75.

Kumar, V., Singh, S., Singh, J. and Upadhyay, N. 2015. Potential of plant growth promoting traits by bacteria isolated from heavy metal contaminated soils. Bull. Environ. Contam. Toxicol. 94(6): 807–814.

Latique, S., Chernane, H., Mansori, M. and El Kaoua, M. 2013. Seaweed liquid fertilizer effect on physiological and biochemical parameters of bean plant (*Phaesolus vulgaris* variety *paulista*) under hydroponic system. Eur. Sci. J. 9(30): 174–191.

Malve, H. 2016. Exploring the ocean for new drug developments: Marine pharmacology. J. Pharm. Bioallied Sci. 8(2): 83.

Michalak, I., Chojnacka, K., Dmytryk, A., Wilk, R., Gramza, M. and Rój, E. 2016. Evaluation of supercritical extracts of algae as biostimulants of plant growth in field trials. Front Plant Sci. 7: 1591.

Mukherjee, A. and Patel, J.S. 2019. Seaweed extract: biostimulator of plant defense and plant productivity. Int. J. Environ. Sci. Technol. 1–6.

Nabti, E., Jha, B. and Hartmann, A. 2017. Impact of seaweeds on agricultural crop production as biofertilizer. Int. J. Environ. Sci. Technol. 14(5): 1119–1134.

Nagorskaia, V.P., Reunov, A.V., Lapshina, L.A., Ermak, I.M. and Barabanova, A.O. 2008. Influence of kappa/beta-carrageenan from red alga Tichocarpus crinitus on development of local infection induced by tobacco mosaic virus in Xanthi-nc tobacco leaves. Izv Akad Nauk Ser. Biol. (3): 360–4.

Ngala, B.M., Valdes, Y., Dos Santos, G., Perry, R.N. and Wesemael, W.M. 2016. Seaweed-based products from *Ecklonia maxima* and *Ascophyllum nodosum* as control agents for the root-knot nematodes *Meloidogyne chitwoodi* and *Meloidogyne hapla* on tomato plants. J. Appl. Psychol. 28(3): 2073–82.

Nicolopoulou-Stamati, P., Maipas, S., Kotampasi, C., Stamatis, P. and Hens, L. 2016. Chemical pesticides and human health: the urgent need for a new concept in agriculture. Front Public Health 4: 148.

Pérez, M., Falqué, E. and Domínguez, H. 2016. Antimicrobial action of compounds from marine seaweed. Mar Drugs 14(3): 52.

Rajan, D.S., Rajkumar, M., Srinivasan, R., Harikumar, R.P., Suresh, S. and Kumar, S. 2013. Antitumour activity of *Sargassum wightii* (greville) extracts against dalton's ascites lymphoma. Pak J. Biol. Sci. 16: 1336–41.

Ronga, D., Biazzi, E., Parati, K., Carminati, D., Carminati, E. and Tava, A. 2019. Microalgal biostimulants and biofertilisers in crop productions. Agronomy 9(4): 192.

Sahayaraj, K. and Jeeva, Y.M. 2012. Nymphicidal and ovipositional efficacy of seaweed *Sargassum tenerrimum* (J. Agardh) against *Dysdercus cingulatus* (Fab.) (Pyrrhocoridae). Chil J. Agr. Res. 72(1): 152–6.

Salim, B.B. 2016. Influence of biochar and seaweed extract applications on growth, yield and mineral composition of wheat (*Triticum aestivum* L.) under sandy soil conditions. Ann. Agric. Sci. 61(2): 257–65.

Sathasivam, R. and Ki, J.S. 2018. A review of the biological activities of microalgal carotenoids and their potential use in healthcare and cosmetic industries. Mar. Drugs 16(1): 26.

Schoonover, J.E. and Crim, J.F. 2015. An introduction to soil concepts and the role of soils in watershed management. J. Contemp Water Res. Educ. 154(1): 21–47.

Shalaby, E. 2011. Algae as promising organisms for environment and health. Plant Signal Behav. 6(9): 1338–50.

Sharma, H.S., Fleming, C., Selby, C., Rao, J.R. and Martin, T. 2014. Plant biostimulants: a review on the processing of macroalgae and use of extracts for crop management to reduce abiotic and biotic stresses. J. Appl. Psychol. 26(1): 465–90.

Sharma, R., Khokhar, M.K., Jat, R.L. and Khandelwal, S.K. 2012. Role of algae and cyanobacteria in sustainable agriculture system. Wudpecker J. Agric. Res. 1(9): 381–388.

Shukla, P.S., Mantin, E.G., Adil, M., Bajpai, S., Critchley, A.T. and Prithiviraj, B. 2019. *Ascophyllum nodosum*-based biostimulants: sustainable applications in agriculture for the stimulation of plant growth, stress tolerance, and disease management. Front Plant Sci. 10.

Singh, S., Singh, N., Kumar, V., Datta, S., Wani, A.B., Singh, D. and Singh, J. 2016. Toxicity, monitoring and biodegradation of the fungicide carbendazim. Environ. Chem. Lett. 14(3): 317–329.

Singh, S., Kumar, V., Chauhan, A., Datta, S., Wani, A.B., Singh, N. and Singh, J. 2018. Toxicity, degradation and analysis of the herbicide atrazine. Environ. Chem. Lett. 16(1): 211–237.

Singh, S., Kumar, V., Sidhu, G.K., Datta, S., Dhanjal, D.S., Koul, B. and Singh, J. 2019. Plant growth promoting rhizobacteria from heavy metal contaminated soil promote growth attributes of *Pisum sativum* L. Biocatal. Agric Biotechnol. 17: 665–671.

Singh, R.P. and Reddy, C.R.K. 2014. Seaweed–microbial interactions: key functions of seaweed-associated bacteria. FEMS Microbiol. Ecol. 88(2): 213–230.

Stanton, D.E. and Reeb, C. 2016. Morphogeometric approaches to non-vascular plants. Front Plant Sci. 7: 916.

Stirk, W.A., Novák, O., Strnad, M. and Van Staden, J. 2003. Cytokinins in macroalgae. Plant Growth Regul. 41(1): 13–24.

Sultana, V., Baloch, G.N., Ara, J., Ehteshamul-Haque, S., Tariq, R.M. and Athar, M. 2012. Seaweeds as an alternative to chemical pesticides for the management of root diseases of sunflower and tomato. J. Appl. Bot Food Qual. 2012 Mar 9; 84(2): 162–168.

Takaichi, S. 2011. Carotenoids in algae: distributions, biosyntheses and functions. Mar. Drugs 9(6): 1101–1118.

Vaishampayan, A., Sinha, R.P., Hader, D.P., Dey, T., Gupta, A.K., Bhan, U. and Rao, A.L. 2001. Cyanobacterial biofertilizers in rice agriculture. Bot Rev. 67(4): 453–516.

Wahby, I., Bennis, I., Tilsaghani, C. and Lubián, L.M. 2014. Potential use of flow cytometry in microalgae-based biodiesel project development. International J. Innov. Appl. Stud. 5(4): 333–43.

Wang, W., Wang, S.X. and Guan, H.S. 2012. The antiviral activities and mechanisms of marine polysaccharides: an overview. Mar. Drugs 10(12): 2795–816.

Wells, M.L., Potin, P., Craigie, J.S., Raven, J.A., Merchant, S.S., Helliwell, K.E., Smith, A.G., Camire, M.E. and Brawley, S.H. 2017. Algae as nutritional and functional food sources: revisiting our understanding. J. Appl. Psychol. 29(2): 949–82.

White, P.J. and Brown, P.H. 2010. Plant nutrition for sustainable development and global health. Ann. Bot. 105(7): 1073–80.

Wijesinghe, W.A. and Jeon, Y.J. 2012. Biological activities and potential industrial applications of fucose rich sulfated polysaccharides and fucoidans isolated from brown seaweeds: A review. Carbohydr. Polym. 88(1): 13–20.

Xu, S.Y., Huang, X. and Cheong, K.L. 2017. Recent advances in marine algae polysaccharides: Isolation, structure, and activities. Mar. Drugs 15(12): 388.

Yakhin, O.I., Lubyanov, A.A., Yakhin, I.A. and Brown, P.H. 2017. Biostimulants in plant science: a global perspective. Front Plant Sci. 7: 2049.

CHAPTER 7

Recovery of Value-Added Products and Biological Conversion of Coffee and Citrus Processing Waste Using Green Technologies

Erminta Tsouko, * *Sofia Maina,* * *Maria Alexandri,*
Harris Papapostolou and *Apostolis Koutinas*

1. Introduction

Increased waste generation and decrease of primary resources are the main problems of modern society closely interlinked with escalating raw material costs. Their efficient management is of utmost importance to meet sustainability. The alternative of a circular economy would turn goods that are at the end of their service life into resources, minimizing waste and replacing production with sufficiency. The transition to a low carbon economy must be based on the utilization of renewable resources and green manufacturing involving novel green technologies and bioprocessing to provide clean energy, green chemicals, and safe products of high quality and functionality (Clark 2019). Sectors of the EU bioeconomy have been reported to worth €2 trillion in an annual turnover, and they account for approximately 9% of the EU workforce (Act, Single Market 2011).

World energy-related CO_2 emissions are predicted to grow at an average of 0.6% annually within the period 2015–2040. Consequently, fossil resources depletion,

Institution – Agricultural University of Athens, Department of Food Science and Human Nutrition, Agricultural University of Athens, Iera Odos 75, 11855 Athens, Greece.
Emails: maryal1989@hotmail.gr; harris_papapostolou@yahoo.gr; akoutinas@aua.gr
* Corresponding author: eri.tsouko@gmail.com; sofiamaina@aua.gr

as a result of the growing demands for fuels and energy and severe environmental issues including global warming and toxic waste generation, have emerged seeking for renewable resources that are considered as a fast-growing source of energy for the period 2012–2040 (U.S. Energy Information Administration, IEO 2017). A study of seven European nations found that a shift towards a circular economy would reduce each nation's greenhouse-gas emissions by up to 70% (Stahel 2016) while simultaneously improving the economics and environmental sustainability of primary production and processing industries (Act, Single Market 2011).

Coffee and citrus fruits cultivation occupy million hectares of arable land globally. Agro-industrial and manufacturing activities for their processing generate huge amounts of by-product streams. Several waste fractions, i.e., citrus peel waste and spent coffee grounds (SCGs) possess high water content (around 70–90%) facilitating their decomposition process. Environmental burdens related to orange manufacturing facilities, are interlinked with high energy demands mostly for electricity and steam generation. Indicatively, 1.12 MJ are required for the production of 1 kg of orange juice (Pleissner et al. 2016). Considering coffee, it is estimated that 1 t of clean processed coffee has the same environmental impact as 273 m^3 of crude domestic sewage, while 8.4 m^3 of water are required per t of fruit being processed (Chanakya and Alwis 2004).

Waste generated from citrus and coffee manufacturing are underutilized streams with low nutritional value and limited applications. Conventional management strategies, including animal feed, composting, anaerobic digestion, disposal in landfills and incineration, create a high environmental footprint and/or are energy and cost-intensive (Murthy and Naidu 2012, Wei et al. 2017). In 2018, the European Parliament adopted the Circular Economy Package, setting legally binding targets for waste recycling and reduction of landfilling. It includes, among others, the amendment of the Landfilling Directive (1999/31/EC) and the Waste Framework Directive (2008/98/EC). Member States should ensure that, by 2030, waste that is suitable for recycling or recovery, will not be permitted to be disposed of. The use of landfills should remain exceptional rather than the norm. Directive 2018/851 requires member states to improve their waste management systems and ensure that waste is valued as a resource for the production of materials (EU, Circular Economy).

Conventional methods for the fractionation and bio-transformation of biomass, to products and materials employing toxic organic solvents are high energy, time demanding, and have a low extraction selectivity. The implementation of green technologies, i.e., microwave and ultrasonication, supercritical fluid extraction, and pulse electric field, is based on principles of green chemistry, which can change the course of the global economic growth toward sustainability, providing alternative socio-economic norms and industrial-scale transferability (Singh and Kumar 2017).

This chapter presents advancements made toward sustainable and green strategies to fractionate and bioconvert coffee and citrus processing by-product streams. A comprehensive insight into related studies is provided, with focus on the extraction of valuable compounds and fermentation products. The composition of all by-product streams generated from citrus and coffee processing stages is also included to highlight their potential for further valorization. The principal analytical techniques that have been entirely based on green technologies combined with green solvent systems are described. Biorefinery concepts based on principles of green chemistry are also discussed. This chapter may serve as an inclusive reference for the recovery and production of value-added products valorizing coffee and citrus processing waste via green and eco-friendly technologies.

2. Citrus and Coffee Waste Streams

2.1 Citrus Processing Waste

Citrus fruits belong to the family of *Rutaceae*, one of the largest fruit crops globally; the citrus industry ranks second in the fruit-processing sector for juice production. Citrus fruit consists of the epidermal layer of flavedo containing the essential oil and the layer of parenchymatous cells, namely albedo, where pectin formation occurs. The edible part of the fruit consists of vesicles rich in organic acids, sugars, vitamins, and amino acids, making the fruit a considerable candidate in health benefit issues (Chavan et al. 2018).

According to the US Department of Agriculture, global citrus production in 2018/19 was recorded to 91.5 million t with oranges dominating the market share at 57.9%, followed by tangerines and mandarins (34.6%), lemons and limes (8.6%), and grapefruits (7.5%) (USDA 2018). Global citrus production is dominated by Brazil, China, India, Mexico, Spain, and the USA, covering more than 67% of the whole production (Zema et al. 2018). Figure 1 presents the worldwide imports of various citrus fruits according to the latest FAOSTAT report (2017). Around 27% of citrus fruits are used for processing, while the majority is applied for domestic consumption. The citrus processing industry includes the sequential steps of harvesting, transportation to production plants, water-washing of fruits, and juice extraction via pressing and centrifugation (Figure 3). Citrus processing generates by-product streams, namely wastewater (5%) and solid residues (45%). The latter

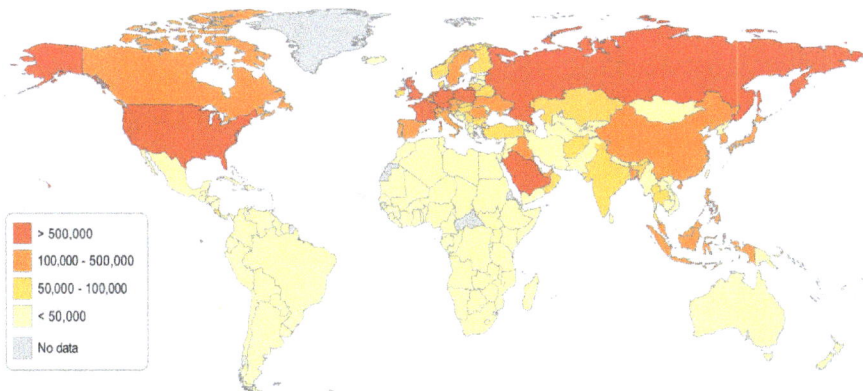

Figure 1. Worldwide imports (in tonnes) of citrus fruits, including grapefruit, lemons and limes, oranges, tangerines, mandarins, clementines, and satsumas (FAOSTAT 2017).

Table 1. Composition of major citrus fruits' by-products (% on dry basis).

Feedstock	Ash	Sugars	Essential oil	Protein	Antioxidants	Pectin	Lignin	Cellulose	Hemicellulose
Orange waste [1]	1.7–4.2	15.0–47.8	0.5–4.0	1.8–9.1	0.6–7.3	14.1–25.0	0.6–7.2	8.1–37.1	5.7–11.1
Mandarin waste [2]	3.2	31.6	-	5.8	-	22.6	0.6	10.1	4.3
Lemon waste [3]	2.5	6.5–9.0	1.5–3.1	7.0–8.7	4.5–12.5	13–22.5	7.6	23.1–36.2	8.1–11.1
Grapefruit waste [3]	8.1	8.0	0.5	12.5	3.0	8.5	11.6	26.6	5.6

[1] Negro et al. 2016, Pourbafrani et al. 2010, Khan et al. 2010, Pfaltzgraff et al. 2013, M'hiri et al. 2015.
[2] Oberoi et al. 2011.
[3] Marin et al. 2007.

comprises almost 50% of the total fruit mass, depending on the applied processing technology, fruit cultivar, and the harvesting period (Chavan et al. 2018).

Citrus by-product streams constitute highly promising renewable resources due to their wide availability and propensity to yield value-added chemicals and materials. Composition analysis of citrus waste, which is summarized in Table 1, presents high geographical diversity. Citrus waste are rich in soluble sugars (i.e., sucrose, glucose, and fructose), pectin, cellulose, several acids (mainly, citric acid and maleic acid), essential oil (mainly D-limonene), minerals (i.e., N, Ca, and K), antioxidants (i.e., flavonoids and polyphenols), limonoids, pigments, carotenoids, vitamins, and dietary fibers (Zema et al. 2018, Sharma et al. 2017). These compounds can prevent chronic diseases and act as antioxidants, anti-cancer, anti-tumor, anti-inflammatory, and anti-viral agents (Sharma et al. 2017).

2.2 Coffee Processing Waste

Coffee is one of the world's most prominent agricultural products, used as a beverage. *Coffea arabica* and *Coffea canephora* are the two main species cultivated for commercial production (Panusa et al. 2013). According to the International Coffee Organization, the world's coffee production reached 170.6 million bags in 2018 (ICO 2019). *Arabica* variety, derived from *C. arabica,* accounts for 75–80% of the world's production, while *C. canephora* is known as Robusta coffee. Figure 2 depicts the worldwide imports of green coffee and roasted coffee (FAOSTAT 2017).

Coffee fruit consists of skin, mesocarp, endocarp, and silverskin to obtain the coffee bean (Alves et al. 2017). Green coffee bean constitutes 50–55% of the ripe cherry, while the remaining part is generated as a by-product stream (Franca and Oliveira 2009). Coffee by-products, including those derived from the post-harvesting process, coffee roasting, and coffee brewing correspond to 80% of the total volume (Torres-Valenzuela et al. 2019).

Figure 4 depicts coffee processing steps and the by-products generated throughout coffee post-harvesting and brewing. During the primary coffee processing, coffee

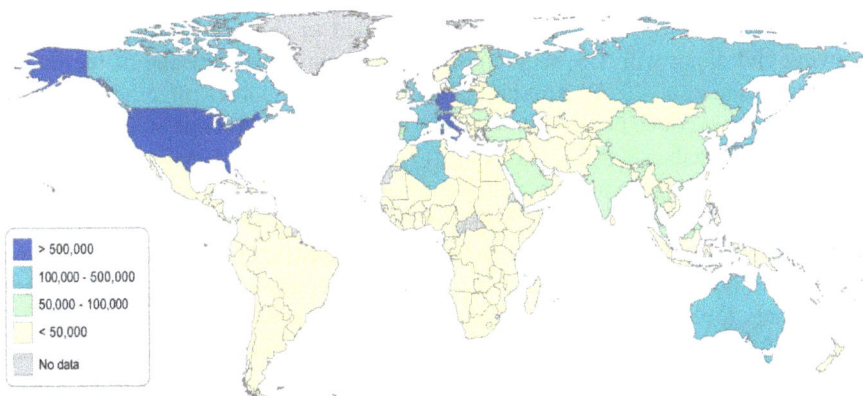

Figure 2. Worldwide imports (in tonnes) of green coffee and roasted coffee (FAOSTAT 2017).

```
                    ┌──────────────┐
                    │  Harvesting  │
                    └──────┬───────┘
                           ↓
                    ┌──────────────┐
                    │Transportation│
                    └──────┬───────┘
                           ↓
                    ┌──────────────┐
                    │   Sorting    │
                    │      &       │- - - - - - - - - - - - - ┐
                    │   Washing    │                          │
                    └──────┬───────┘                          ↓
                           │            Rotten fruits, stems, stalks
                           ↓
                    ┌──────────────┐
         ┌- - - - - │   Pressing   │
         │          │Juice extraction│
         ↓          └──────┬───────┘
  Seeds, pulp & peels      │
                           ↓
                    ┌──────────────┐
                    │ Clarification│
                    └──────┬───────┘
                           ↓
                    ┌──────────────┐
                    │Centrifugation│
                    └──────┬───────┘
                           ↓
                    ┌──────────────┐
                    │Concentration │
                    └──────┬───────┘
                           ↓
                    ┌──────────────┐
                    │Packing/storage│
                    └──────────────┘
```

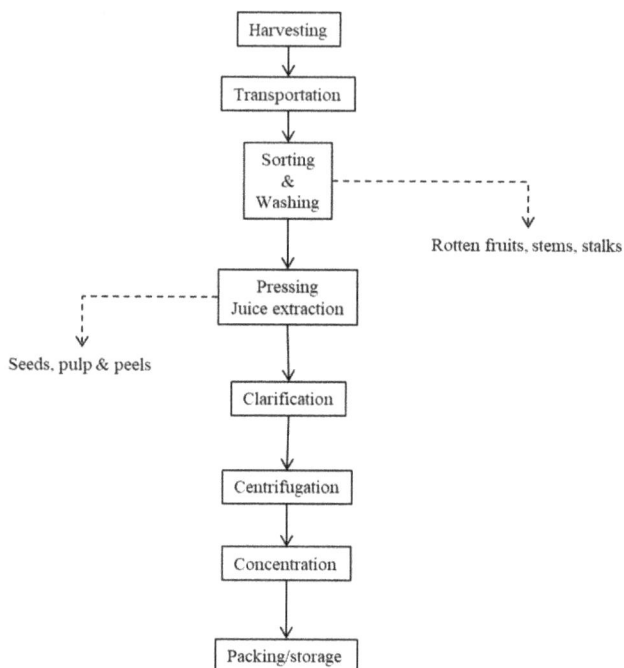

Figure 3. Citrus fruits processing and by-products generation during the juice production process.

berries obtained after the harvesting of the fruit, are processed using dry, wet, and semi-dry processing. Coffee husks (CHs) and coffee pulp (CP) along with defective coffee beans are obtained as a solid residue after de-hulling of the coffee cherries. CP is the by-product obtained during the wet processing of coffee, representing 29% of the whole coffee bean, while CHs are generated from the dry process accounting for around 12% of the coffee bean (Murthy and Naidu 2012). Coffee silverskin (CS) is the thin tegument of the coffee bean generated during coffee roasting (Mussatto et al. 2011). Approximately 30 kg of CS is generated from 4 t of industrial coffee production (Alves et al. 2017). SCGs are obtained as solid residues during coffee brewing and along with CS, they constitute the main waste streams of the coffee industry (Mata et al. 2018). Approximately 50% of the global coffee production is used for soluble coffee preparation resulting in about 6 million t of SCGs generated annually (Aristizábal-Marulanda et al. 2017). It is estimated that for 1 t of soluble coffee, 1.5 t of SCGs are produced (Echeverria and Nuti 2017). Furthermore, 650 kg of SCGs is produced from 1 t of green coffee beans (Karmee et al. 2018b).

Coffee by-product streams are rich in carbohydrates, proteins, lipids, caffeine, tannins and polyphenols with cellulose, and hemicellulose being the predominant components (Table 2). Their chemical composition depends on the geographical origin and variety of coffee. Due to the high organic matter along with functional compounds i.e., caffeine, tannins, and polyphenols that are contained in SCGs, they could be integrated into circular bioeconomy concepts to produce value added products.

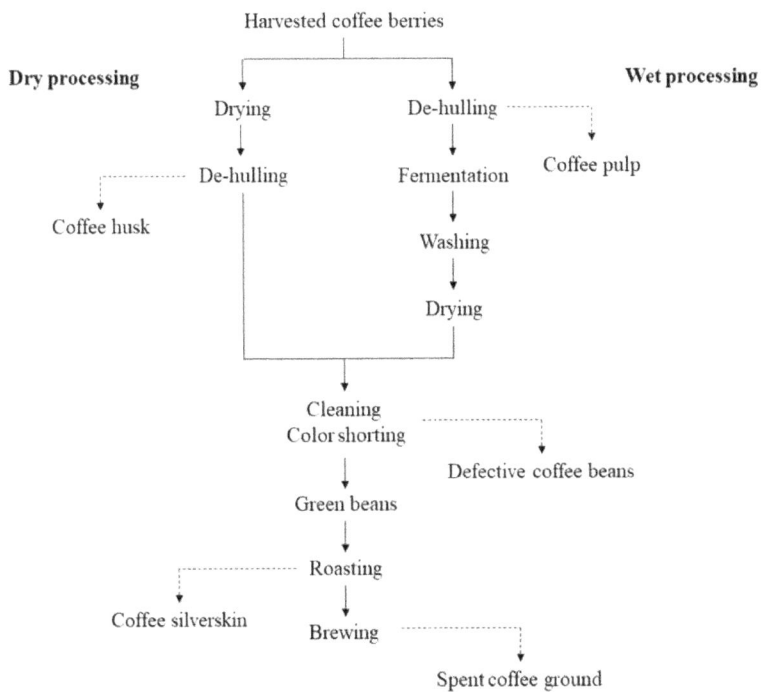

Figure 4. Coffee processing and by-products produced during the post-harvesting and brewing process (adapted from Alves et al. 2017 and Franca and Oliveira 2009).

Table 2. Composition composition of coffee by-products derived after post-harvesting and brewing process (% on dry basis).

	CH	CP	CS	SCG
Ash [1,6]	3.0–7.0	-	-	0.4–2.2
Protein [1,2,6]	8.0–11.0	4.0–12.0	18.0–19.0	6.7–13.7
Lipids [1,3,5,6]	0.5–3.0	1.0–2.0	1.6–3.3	10.0–15.0
Carbohydrates [1,3]	58.0–85.0	45.0–89.0	30.0–80.0	-
Cellulose [3,6]	19.0–26.0	63.0	18.0	8.6–15.3
Hemicellulose [3,4,6]	24.0–45.0	2.3	13.0	36.7
Lignin [3,6]	18.0–30.0	17.5		32.5–33.6
Minerals [3]	3.0–7.0	-	-	-
Caffeine [1,6]	1.0	1.0	0.8–1.4	0.02–0.45.0
Tannins [1,6]	5.0	1.0–9.0	-	0.02

[1] Narita et al. 2017, [2] Franca and Oliveira 2009, [3] Murthy and Naidu 2012, [4] Bekalo and Reinhardt 2010, [5] Bessada et al. 2018, [6] Mata et al. 2018

3. Green Technologies for the Recovery and Production of High Value-Added Products

Organic solvents are introduced systematically in industrial practice comprising more than 50% of industrial emissions and 30% of all volatile organic compound emissions globally (Cvjetko Bubalo et al. 2015). Their flammable, volatile, and toxic nature makes them hazardous to the environment and humans. EU environmental policy and legislation for the period 2010–2050 (Directive 8/2013), regarding the diminution of petrochemical solvents and volatile organic compounds, has narrowed the margins to fossil-based production, leading the manufacturers to the unavoidable implementation of green technologies that employ solvent-free processes or 'green solvents' considered as 'GRAS' (EU, Industrial Emissions).

Conventional petrochemical solvents are mainly used for the extraction of bioactive compounds from natural sources. Existing extraction technologies require up to 50% of the total capital investment, and they demand high energy and prolonged time. Renewable agro-solvents have gained increasing attention in scientific and industrial terms due to high purity and low price (ethanol and glycerol), low polarity, high solvent power (terpenes), good technical performance (fatty acid methyl esters), and high chemical and thermal stability (ionic liquids) (Cvjetko Bubalo et al. 2015).

The development of integrated green processes could lead to a reduction of equipment size, time, energy and processing steps (Klemeš et al. 2019). Ultrasound-assisted extraction (UAE), microwave-assisted extraction (MAE), subcritical water extraction (SWE), supercritical fluid extraction (SFE), pulsed electric fields (PEF), and instant controlled pressure drop are highly promising and innovative green extraction techniques, which are gradually becoming more attractive. Especially, in cases where the aforementioned novel technologies are combined, hybrid systems can be developed for maximum efficiency of processes enabling commercialization (Chemat et al. 2017)

The UAE is a sustainable alternative that is employed for the extraction of bioactive compounds from plant material due to the high extraction efficiencies and good potential for pilot-scale implementation (Boonkird et al. 2008). The UAE is based on the principle of acoustic cavitation that enhances the release of bioactive compounds with frequencies of ultrasonic waves within the range of 20 kHz to 100 MHz. The mechanisms involved in ultrasound extraction are fragmentation, erosion, sono-capillarity, detexturation, and sonoporation (Medina-Torres 2017). The Optimization of UAE includes solvent and temperature conditions, ultrasound frequency, sonication power and time as well as ultrasonic wave distribution. The UAE is easy to handle, safe, economical and reproducible technology that allows its implementation under atmospheric pressure and at ambient temperature (Khan and Dangles 2014).

The MAE is a novel method to extract components into a fluid, from a wide range of materials. Electromagnetic energy with frequencies between 300 MHz–300 GHz is generated and converted to thermal energy, which is transferred to the biological matrix via dipole rotation and ionic conduction (Kumari and Singh 2018). The physical properties of the tissue are modified, improving the extraction of the targeted compounds. MAE reduces the extraction time and the amount of solvent required, leading to a high extraction rate at a low cost (Conde et al. 2011).

SWE is an eco-friendly technique for the extraction of various bioactive compounds from plants and food materials (Xu et al. 2015). SWE uses water in the temperature range of 100 to 374°C under pressure to maintain water in a liquid state. Under subcritical conditions, the intermolecular hydrogen bond of water break down and the dielectric constant of water decreases leading to the extraction of polar materials with high solubility in water under ambient temperatures (Asl and Khajenoori 2013, Rodríguez-Meizoso 2010). SWE promotes phenomena of oxidation, hydrolysis, and decomposition of compounds and therefore its application is rather restricted. SWE has been efficiently applied in the extraction of antioxidants, lactones, pesticides, and herbicides mostly from plant material (Cvjetko Bubalo et al. 2015).

SFE employs supercritical fluids as an alternative to conventional organic solvents. SFE is based on the change of physicochemical properties of fluids (density, dielectric constant, and viscosity) by altering pressure or temperature. Supercritical fluids have a high diffusion coefficient and low surface tension leading to the effective penetration of the supercritical solvent into the solid material to release the targeted compound. CO_2 is the most used supercritical fluid as it is GRAS. CO_2 is non-toxic, non-flammable, inexpensive, and it is widely applied in food components to prevent degradation (Shilpi et al. 2013).

Auto-hydrolysis employing hot water has been proposed as an eco-friendly method for selective degradation of biomass and extraction of high value-added compounds. Autohydrolysis operates in mild conditions, eliminate corrosive issues, and reduces the operational cost (Ballesteros).

PEF is a technique mostly applied for the extraction of intracellular components from non-wood plant material. The electric pulses in combination with a moderate electric field increase mass transfer phenomena due to the permeabilization of cell membranes resulting in favorable yields, low energy requirements, and eco-friendly conditions (Brabosa-Pereira 2018). PEF gives high-quality end products, and it can operate under continuous flow which is a key factor for its transferability at an industrial level (Luengo et al. 2013). Optimization parameters of the technique are the strength of the electric field, shape of waves, number and duration of pulses, and temperature. Most PEF systems operate under a square wave and alternate directional pulses. Studies have demonstrated that low electric field strength in combination with longer extraction time is effective in pre-treating and obtaining valuable products from biological materials (Rocha et al. 2018).

3.1 Antioxidants

Antioxidants are bioactive compounds widely distributed in agro-industrial wastes and residues. Natural antioxidants from plants materials mainly include polyphenols, carotenoids and vitamins exhibiting anti-inflammatory, anti-bacterial, anti-viral, anti-aging, and anti-cancer properties (Xu et al. 2017). Polyphenols are classified into flavonoids and non-flavonoids. Non-flavonoids include mainly phenolic acids, stilbenes, and tannins (Khan and Dangles 2014). The chemical variability of antioxidants depends on the source, the pre-treatment and the extraction method leading to variation in their composition and antioxidant capacity (Kim et al. 2016).

Natural antioxidants could substitute synthetic antioxidant agents, such as butylated hydroxyanisole and butylated hydroxytoluene, that have been reported to damage the liver or be cancer-causing (Khan and Dangles 2014). Their wide application in food, nutraceuticals, and pharmaceuticals, creates an imperative necessity for their recovery through green and non-conventional processes to avoid prolonged extraction times, high temperatures, organic solvents, rationally avoiding their degradation (M'hiri et al. 2015). Figure 5 presents the main phenolic compounds found in coffee and citrus by-products along with their chemical structure. Several innovative methods have been widely investigated for the efficient extraction of antioxidants from citrus and coffee waste biomass.

Most studies evaluating the antioxidant capacity of citrus renewable biomass have employed response surface methodology and a central composite design approach to optimize the extraction process in terms of solvent, solid to liquid ratio, and process operative parameters. Phenolic compounds and flavonoids are among the most studied molecules. Trabelsi et al. (2016) reported the efficient production of peel extracts containing a range of compounds, including fatty acid esters, phenols, coumarin derivates, and terpenes from *Citrus aurantium* peels using CO_2 extraction with pure ethanol as co-solvent. The highest recovery of osthole (coumarin derivative) was obtained under 40°C, extraction time of 120 minutes, 170 bar and flow rate of 2.7 kg/h. The same extraction system was employed for phenolic compounds recovered from orange pomace after industrial juice processing. It was demonstrated that high-pressure values enhanced global extraction yield, while antioxidant activity was reduced. The total phenolic content (TPC) of extracts was maximized at 21.8 mg gallic acid equivalents (GAE)/g extract (40°C and 35 MPa pressure). The global extraction yield was 50% lower compared to ethanol Soxhlet extraction, while SFE required 78% less time and 10 folds lower ethanol consumption (Espinosa-Pardo et al. 2017). SFE employing CO_2 was used for flavonoids recovery from pomelo peels (*Citrus grandis*) and their antioxidant activity was investigated. Under the optimal conditions (80°C, 39 MPa, extraction time of 49 minutes, and 85% ethanol) the experimental yield was 2.4%, coinciding with the value predicted by the model. The

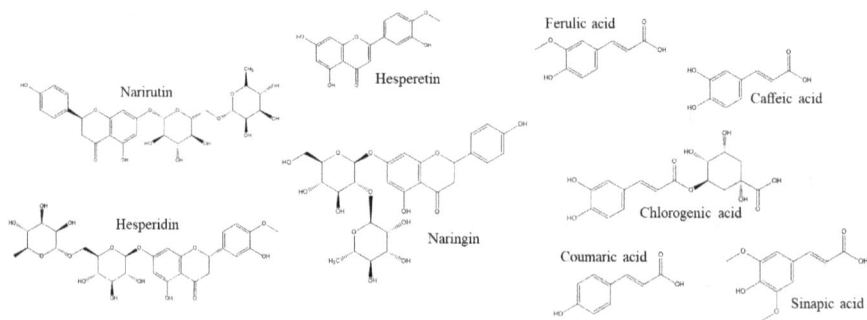

Figure 5. Chemical structure of predominant phenolic compounds found in citrus and coffee processing by-products.

extraction yield of flavonoids and their antioxidant activity were higher in the case of SFE compared to the conventional solvent extraction (He et al. 2012). Lachos-Perez et al. (2018) optimized flavanones (hesperidin and narirutin) extraction from defatted orange peels using SWE. The maximum yields of hesperidin (188.7 mg/g extract) and narirutin (22 mg/g extract) were achieved at 150°C and 10 mL/min water flow rate. These yields accounted for around 21% of the total amount of flavanones in the extracts. SWE was superior to conventional low-pressure extraction in terms of process efficiency and activity/purity of bioactive compounds.

UAE and MAE employing aqueous ethanol (50–65%) were assessed for the recovery of phenolic compounds from *Citrus limon* peels. Optimization of MAE at 2 minutes, 400 W and initial solid concentration of 35.7 g/L resulted in similar TPC (15.2–15.8 mg GAE/g) with UAE at 15.1 minutes, 77.8% amplitude, and initial solid concentration of 25 g/L. MAE appeared to be superior to UAE as well as to conventional solvent extraction in terms of higher yield, reduced operation time, and ethanol utilization as well as improved antioxidant activity of the extracts (Dahmoune et al. 2013). Another scientific team applied UAE of polyphenols from orange peels, focusing on flavanones (hesperidin and naringin). After process optimization (40°C, 150 W, 75% ethanol), a TPC of 2.8 mg GAE/g, 0.7 mg of naringin, and 2.1 mg of hesperidin/g were obtained. Extraction yield (10.9%) of UAE was higher when compared to conventional solvent extraction, while energy input was decreased (Khan et al. 2010).

A sustainable separation process for the recovery of polyphenols from orange peel employing deep eutectic solvents (DES) was proposed by Ozturk et al. (2018). Choline chloride-based DES paired with glycerol and ethylene glycol were more efficient than conventional ethanol extraction concerning TPC and antioxidant activity of the extracts. Additionally, their high selectivity in certain compounds was demonstrated. Under the optimized conditions (10% water, 60°C, 1:10 solid/liquid ratio and 100 minutes), choline chloride-based DES paired with ethylene glycol (1:4) showed the highest TPC (3.61 mg GAE/g) and antioxidant activity (30.6 μg/mL). Luengo et al. (2013) evaluated the efficacy of PEF extraction of polyphenols and flavonoids from orange peel. After 30 minutes of pressurization followed by PEF treatment of orange peel (60 μs, 5 kV/cm), polyphenol extraction yield and their antioxidant activity increased to 153% (0.3 mg GAE/g) and 192%, respectively, compared to the untreated biomass. Naringin and hesperidin increased 3 folds (0.03 mg/g) and 3.5 folds (0.05 mg/g), respectively.

Coffee contains several classes of antioxidants including hydroxycinnamic acids (caffeic, chlorogenic, coumaric, ferulic, and sinapic), melanoidins, and caffeine (Komes and Bušić 2014). The content of antioxidants in coffee is influenced by the origin, the species, and the brewing process. Table 3 includes several green extraction methods that have been evaluated for the efficient recovery of phenolic compounds from various waste streams of the coffee manufacturing process. Extracts from coffee by-products have been identified as important antioxidants possessing anti-inflammatory, anti-tumor, and anti-allergic activity (Kim et al. 2016). Specifically, chlorogenic acid is an important phenolic compound showing antioxidant, anti-bacterial, anti-viral, and anti-microbial activity (Naveed et al. 2018), while

Table 3. Green extraction techniques of phenolic compounds from several coffee by-product streams.

Waste streams	TPC[1]	Method	Solvent	Conditions	Reference
SCG	33.8	UAE	ethanol	[2]S/L 1:17, 244W, 40°C, 40 min	Al-Dhabi et al. 2017
SCG	46.0	MAE	ethanol 54%	S/L 1:20, 150°C for 90 min	Pettinato et al. 2019
CS	9.9	UAE	ethanol 60%	S/L 1:35, 40 kHz, 300 W, 29.5 min, 80°C	Guglielmetti et al. 2017
	7.3	MAE	ethanol 60%	S/L 1:35, 280 W, 32 min, 51.5°C	
SCG	28.3	Conventional	ethanol 60%	S/L 2:100, 60°C, 30 min	Panusa et al. 2013
SCG	88.3	SWE	water	14.1 g/L, 179°C, 36 min	Xu et al. 2015
SCG	587.7	UAE	ethanol	S/L 7:210, 55 kHz, 2 h	Andrade et al. 2012
SCG	57.0	SFE	CO₂ and ethanol 4%	200 bar and 50°C	
CH	151.0	UAE	ethanol	S/L 7:210, 55 kHz, 2 h	
CH	36.0	SFE	CO₂ and ethanol 8%	200 bar and 50°C	
SCG	40.4	LHW	water	S/L 1:15, 200°C, 50 min	Ballesteros et al. 2017

[1] TPC: total phenolic content expressed as mg GAE/g solid residue; [2] S/L: solid to liquid ratio

caffeine has pharmacological activity in stroke, heart arrhythmias, and heart attacks (Turnbull et al. 2017).

UAE has been optimized for the extraction of antioxidants from SCGs. A TPC of 33.8 mg GAE/g SCGs has been reported by Al-Dhabi et al. (2017) under the optimum extraction conditions of 244 W, 40°C and 34 minutes. Pettinato et al. (2019) demonstrated the efficient extraction of caffeine (32 mg/L) and chlorogenic acid (9 mg/L) using MAE and aqueous ethanol as a solvent (150°C and 9 minutes). Optimized MAE parameters (550 W and 3 minutes) by Ranic et al. (2014) led to an extraction yield of total phenolics equal to 31.2 mg/g SCGs. The highest TPC (68.6–79.8% in dry extract) was achieved at extraction time in the range of 0.7–1.8 min and microwave power of 240–400 W. Guglielmetti et al. (2017) studied the effect of SC particle size on the extraction of phenolic compounds using UAE and MAE. UAE resulted in higher TPC (10.0 g GAE/kg), while no significant differences were observed among the applied particle sizes of SC.

The employment of SWE for phenolic compounds recovery from SCG has been investigated. The effect of SWE parameters on the yield of TPC and antioxidant activity was optimized by Xu et al. (2015). TPC was increased at temperatures varying within 110–170°C, while further temperature increase resulted in lower TPC values. A TPC of 88.34 mg GAE/g with antioxidant activity of 32.28 mmol Trolox equivalents/100 g was obtained at 179°C, 36 minutes, and a solid to liquid ratio of 14.1 g/L. Mayanga-Torres et al. (2017) reported a TPC of 55.7 mg GAE/g coffee powder using SWE at 175°C and a pressure of 22.5 MPa.

Andrade et al. (2012) evaluated the extraction of antioxidants from SCGs and CHs using low-pressure extraction (Soxhlet and UAE) and supercritical CO_2 extraction. UAE resulted in a higher extraction yield compared to CO_2 extraction with a TPC of 133 mg GAE/g for CHs extracts and 587.7 mg GAE/g for SCGs extracts. Ballesteros et al. (2017) applied liquid hot water extraction to recover phenolic compounds from SCGs. A TPC of 40.36 mg GAE/g with a significant antioxidant activity was achieved under 200°C and 50 minutes of extraction.

3.2 Oils and Essential Oils

SCGs contain significant lipids amounts (10–15%). The chemical composition of SCG oil depends on the source and the brewing process with up to 80–90% being glycerides and free fatty acids, while the rest include terpenes, sterols, and tocopherols. The main fatty acids found in SCG oil are palmitic acid (39–45%), linoleic acid (35–40%), stearic acid (5–7%), and oleic acid (7–10%). SCG oil can serve as the raw material for biodiesel production via transesterification/esterification reactions or for commodity chemicals production, such as lubricants, surfactants, and solvents.

Citrus essential oil mainly contains D-limonene (94–96%). Emerging applications of D-limonene include a bio-solvent, a natural flavouring and detoxifying agent in the food sector, an anti-asthmatic and anti-cancer agent in medicine, and a pest control agent. Due to its volatile nature, it is also applied in perfumery. The price of D-limonene that is highly dependent on the extraction technology, its production scale and the source of the biowaste, shows great fluctuations varying between $ 3.3–11/kg from 2011 to 2014 (Ciriminna et al. 2014).

The long-term implementation of non-green organic solvents for the extraction of oils from solid matrices has raised health, safety, and environmental concerns that limit industrial application. Recovery of lipids from SCGs has been extensively carried out using Soxhlet extraction with organic solvents, i.e., hexane, octane, pentane, toluene, isopropanol, chloroform, methanol, or a mixture of solvents resulting in high extraction yields up to 26.5%, while extended extraction time renders such processes non-viable (Al-Hamamre et al. 2012). Essential oil contained in waste streams of citrus fruit juice processing is so far obtained by Soxhlet extraction, cold pressing, steam, and hydrodistillation (Marathu et al. 2016). Prolonged duration of extraction and elevated temperatures of such techniques lead to degradation of the essential oil quality.

Alternative and consistent with green concepts have been developed to eliminate solvents toxicity, increase process yield and profitability, deteriorate degradation of thermostable compounds in the oil, and provide high-quality end products (Putnik et al. 2017). Optimization parameters for increased oil extraction efficiency include the moisture content of the solid matrix, particle size, solid to solvent ratio, type of solvent, and extraction duration (Efthymiopoulos et al. 2018). Green solvent extraction (ethanol, CO_2, and terpenes) integrated with novel technologies, i.e., ionic liquids, MAE, and UAE (Abdullah and Koc 2013, Barbosa-Pereira et al. 2018) have been recently developed for enhanced extraction.

D-limonene is a key factor that act as an inhibitor to microorganisms (Negro et al. 2016). Its removal and recovery prior to any treatment of citrus biomass is a prerequisite. Ndayishimiye et al. (2017) demonstrated that oil recovery and its oxidative stability from a mixture of citrus seeds and peels with hexane (70°C) were higher than supercritical CO_2 extraction (200/250 bar, 45/60°C), while the antioxidant and anti-microbial properties were weaker.

MAE applied in several citrus wastes have been reported as superior to other methods, such as hydro-distillation (HD) and cold pressing in terms of energy requirements, environmental footprint, the yield of essential oil, purity, and higher value of the final product (Bousbia et al. 2009, Bustamante et al. 2016, Golmakani and Moayyedi 2015). Bousbia et al. (2009) demonstrated that microwave hydro-diffusion used for essential oil extraction from lime peels resulted in a 15-fold reduction of energy requirements and environmental footprint, while the yield of essential oil (1.0%) was higher than the HD. In another study, microwave-assisted HD and solvent-free microwave extraction (SFME) were proposed as efficient green technologies for essential oil extraction from *Citrus limon*. Extraction duration was about eight folds lower in comparison to conventional HD, while essential oil varied between 1.2–1.4% (Golmakani and Moayyedi 2015). Microwave-assisted HD has also been evaluated to obtain essential oil from wet citrus peel waste. The highest oil yield (1.8%) was achieved under optimal conditions (20 min, 300 mbar), and it was quite comparable to that obtained by conventional HD (1.7%) (Bustamante et al. 2016).

The effect of solvent pressure and temperature of CO_2 extraction on oil recovery from SCGs has been studied by Couto et al. (2009). An extraction yield of 15.4 g oil/100 g SCGs was obtained at 300 bar and 55°C. Supercritical CO_2 extraction has been applied for oil extraction from SCG yielding up to 11.4% of oil (Akgün et al. 2014). Ahangari and Sargolzaei (2013) reported that CO_2 extraction of oil from SCGs led to enhanced oil yields at increased pressure values and solvent volume, while increased temperatures resulted in decreased extraction yield. A maximum oil yield of 15.4% was obtained with pure CO_2, at a pressure of 250–300 bars and a temperature of 50–55°C.

The economic viability of SCGs oil extraction using supercritical extraction has been assessed by de Melo et al. (2014) showing high efficiency. Operation conditions of 2 hours extraction, 300 bar, 50°C, and 30 kg CO_2/kg SCGs/h resulted in a manufacturing cost of €2.4 M and process net income of €56.6 M for the production of 454 t oil/year.

Oil extracted from coffee waste streams and mostly SCGs has been utilized as the raw material for the production of biodiesel. Thus, none of the reported studies has been entirely based on green processes either regarding oil extraction or its subsequent use for biodiesel production. Although biodiesel production has been carried out employing enzymatic catalysis with ethanol and a co-solvent (Caetano et al. 2017), the oil has been extracted from SCGs using hexane. Other studies evaluated biodiesel production, employing transesterification procedures, that were carried out via lipases catalyzed methanolysis (Karmee et al. 2018a, Ferrario et al. 2013) or ultrasound-assisted oil extraction with hexane as a solvent and subsequent methanolysis (Rocha et al. 2014).

An interesting approach for the valorization of oil contained in SCGs and green coffee defective beans has been presented by Marto et al. (2016). Sunscreen formulations in the form of water-in-oil emulsions with oil extracted via CO_2 from the aforementioned waste streams were evaluated. The cold emulsification process employed for sunscreens production showed that emulsions containing 35% of SCGs oil presented improved SPF when compared to emulsions containing 35% of green coffee oil. A promising potential of SCG oil for the cosmetic industry has been studied by Ribeiro et al. (2013). The formulation of oil in water creams containing 10% SCGs oil extracted with supercritical CO_2 improved skin hydration and sebum capacity.

3.3 Pectin

Pectin comprises the main component of the primary cell walls of most vegetables and fruits (Ciriminna et al. 2015). Pectin is a natural bioactive hydrocolloid with a complex internal structural organization and growing applications in the cosmetics and food industry as a gelling agent, stabilizer, thickener, emulsifier and active food packaging (Hosseini et al. 2016, Colodel et al. 2018). Pectin is also used as a novel biomaterial in the pharmaceutical and biomedical sectors for the formulation of polymer hydrogels, nanoparticles, and scaffolds (Ciriminna et al. 2015).

The chemical characterization of pectin is directly related to extraction conditions and the sourcing plant. Pectins suitable for food applications are recommended to have a galacturonic acid content of more than 65%. Galacturonic acid is an indicator of the overall quality of the food matrix where pectin is incorporated affecting its chemical and sensorial properties. The esterification degree (DE) of pectins define their potent applications and classify them into high methoxyl pectins (DE > 50%) and low methoxyl pectins (DE < 50%). High methoxyl pectins form gels at high sugar concentration or low pH and do not react with calcium ions, while low methoxyl pectins form gels with low concentrations of calcium ions and at high pH values introducing a vast spectrum of end-uses in the field of and dairy products (Jiang et al. 2020, Rahmani et al. 2020).

Industrial pectin extraction from apple pomace and citrus peels is commonly carried out with several acids, i.e., oxalic acid, HCl, HNO_3, and H_2SO_4. Similar strategies are followed by the majority of experimental studies, which focus on the optimization of conditions in terms of solid to liquid ratio, extraction time, pH, and acid concentration. Despite the fact that acidic extraction methods offer satisfying extraction yields and process economics, they are associated with the formation of furfural derivatives, equipment corrosion, generation of considerable amounts of toxic waste increasing environmental impact and defiance of legislative norms (Liu et al. 2017). Even in cases where aqueous extraction conditions are applied, the extraction times are prolonged up to 180 minutes combined with elevated temperatures around 90°C. Eco-friendly innovative methods to obtain efficient yields of pectins with desirable physicochemical properties need to be developed.

Table 4 depicts an overview of methods and techniques based on principles of green chemistry that have been so far employed for pectin extraction from several citrus fruits waste. Citric acid has been widely applied as a green solvent for pectin

Table 4. Effect of different green extraction methods on pectin extraction from various citrus sources.

Source of waste	Extraction	Temperature (°C)	Time (min)	Solid:liquid (w/v)	Yield (%)	DE (%)	References
Lemon peel	MAE, 700 W		3	1:15	25.3	5.8	Rahmani et al. 2020
Orange peel	[1]MS-AE, 400 W		7	1:21.5	32.8	69.8	Su et al. 2019
Lime peel	MAE, 700 W		5	1:40	9.7	88.3	Rodsamran and Sothornvit 2019
Orange peel	UAE, 150 W	< 30	10	1:20	28.1	6.8	Hosseini et al. 2019
Pomelo	UAE followed by MAE (643 W)		27.5 min sonication/6.4 min irradiation	1:30	38.0	56.9	Liew et al. 2016
Pomelo	SWE, 30 bars	120			19.6	38.2	Liew et al. 2018b
Pomelo	Citric acid	88	141	1:29	39.7	57.6	Liew et al. 2018a
Pomelo	[2]Deep eutectic solvents	50	60		23.0	79.2	Liew et al. 2018a
Citrange	Electromagnetic induction		30	1:50	24.0	61.0	Zouambia et al. 2017

[1] Microwave-surfactant assisted extraction; [2] Lactic acid:Glucose:Water (molar ratio of 6:1:6)

extraction from various citrus sources in conventional heating extraction or MAE and UAE. In several studies, other agents including surfactants have been used to enhance extraction.

The MAE using citric acid as pH regulator was carried out to recover pectin from sweet lemon peel and optimized using the Box-Behnken design. The optimal extraction conditions (700 W, 3 minutes of irradiation, and 1.5 pH) resulted in a pectin yield of 25.3% with a galacturonic acid content of 60%, low DE of 5.8%, and satisfying emulsifying properties and antioxidant activity (Rahmani et al. 2020). Su et al. (2019) developed a microwave-surfactant assisted extraction of pectin from orange peel using Tween-80. Optimized conditions of 400 W, 1.2 pH, 7 minutes of irradiation, and 8 g/L Tween-80 led to a pectin yield of 32.8%, galacturonic acid content of 78.1%, and DE of 69.8%. MAE using citric acid as solvent was compared with a conventional heating method to extract pectin from lime peel waste. Conventional citric acid extraction and MAE using citric acid were applied for pectin extraction from lime peel waste. Pectins recovered with both techniques were characterized as high methoxyl pectins with DE varying between 85.6–88.3% and galacturonic acid content in the range of 85.2–91.1%. Conventional citric acid heating resulted in higher pectin yield (19.6%) compared to MAE (9.7%), thus extraction time was 12 fold extended (Rodsamran and Sothornvit 2019). UAE combined with citric acid was optimized via Box-Behnken design for efficient pectin recovery from sour orange peels. A maximum pectin yield of 28.1% with satisfying emulsifying and antioxidant properties but low DE (6.8%) were obtained under the optimum conditions (150 W, 10 minutes irradiation, and pH of 1.5) (Hosseini et al. 2019).

Electromagnetic induction was applied as an innovative method for pectins extraction from citrange (*Citrus sinensis* and *Poncirus trifoliata*) albedo. After 30 minutes of electromagnetic induction, pectin yield was similar (24%) to acidic heating after 90 minutes of extraction, while galacturonic acid content and DE varied between 29.1–29.4% and 61.0–62.5%, respectively (Zouambia et al. 2017).

Liew et al. (2016, 2018a, 2018b) have applied several techniques and strategies for pectin extraction from pomelo (*Citrus grandis*). UAE, followed by MAE with citric acid as solvent, has been reported as an attractive combination, avoiding product degradation. Pectin yield and galacturonic acid content of the extracted pectin were enhanced in comparison to MAE or UAE techniques used alone. The same group conducted a face-centered central composite design for pectin extraction using SWE. Extraction at 120°C and 30 bars resulted in an experimental pectin yield of 19.6% with a low DE (38.2%) and a galacturonic acid content of 76.6% (Liew et al. 2018b). Other methods evaluated by Liew et al. (2018a) included pectin recovery using citric acid and DES as extractant solvents. The optimized conditions for citric acid assisted extraction based on the Box-Behnken design led to a pectin yield of 39.7% with a DE of 57.6% and galacturonic acid content of 68.5%. Regarding DES-assisted extraction, lactic acid-glucose-water DES with a molar ratio of 6:1:6 gave the highest pectin yield of 23.0%. Overall evaluation of the applied methods demonstrated that all pectins showed similar viscosity values and pseudoplastic behavior. Higher pectin yields with higher DE were obtained in the case of citric acid assisted extraction. Low methoxyl pectins could be obtained by SWE, while UAE followed by MAE gives pectins with high methoxyl content. SWE was the most energy-intensive technique, while UAE followed by MAE required the lowest energy (Liew et al. 2019).

3.4 Other Products

Coffee beans and SCGs contain substantial amounts of lignocellulosic derived carbohydrates, making them challenging for the production of bioenergy, i.e., biochar and bio-oil as well as for the recovery of value-added carbon sources. Fast pyrolysis of SCGs was evaluated using a screw conveyor reactor to produce bio-oil ad char. The model which was developed using a central composite surface response predicted a maximum yield of bio-oil (61.8%) and char (17.1%) at 505°C and 70 rpm (Kelkar et al. 2015). SCGs were reported as a promising adsorbent for selected xenobiotics removal, such as dyes via magnetic fluid treatment (Safarik et al. 2012). Galactomannans, which can be applied in the medical sector, were efficiently obtained from SCGs using sequential microwave superheated water extraction (200°C for 3 minutes) (Passos et al. 2014). The recovery of a stream rich in polysaccharides (29.3%) and mostly galactose and mannose with high antioxidant activity was achieved using autohydrolysis of SCGs at 160°C for 10 minutes. The high thermostability and antioxidant activity of the obtained polysaccharides demonstrated their potential use as food encapsulated additives or as prebiotics (Ballesteros et al. 2017). SCGs after UAE oil extraction were evaluated as reinforcement agents to formulate polypropylene composites. The mechanical properties of the produced composites demonstrated the good compatibility between filler (SCG) and the polypropylene matrix as well as their good resistance properties (Wu et al. 2016).

4. Biorefinery Concepts Employing Principles of Green Chemistry

The development of biorefineries via the sustainable processing of biomass into a spectrum of marketable products (i.e., chemicals, biopolymers, and high-added value products) and energy could pave the way toward innovative green technologies as well as sustainable and economically viable fermentative processes. Figure 6 illustrates conventional and green valorization approaches of citrus and coffee residues. Stand-alone processes normally are less efficient than biorefinery concepts (Aristizábal-Marulanda et al. 2017). Biorefineries are dynamic systems and their successful design presupposes the evaluation and analysis of economic, environmental, and social factors of processes developed and carried out. Kookos et al. (2018) performed a techno-economic and environmental evaluation of biodiesel production using oil derived from SCGs, including glycerol production and electricity generation in the proposed process. The author demonstrated that the economic feasibility of the process could be achieved only in large scale production,

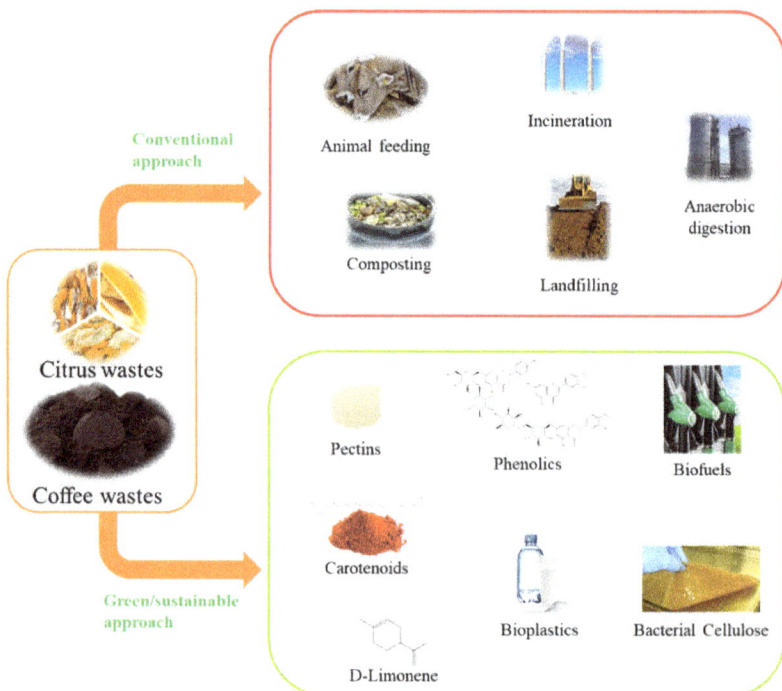

Figure 6. Schematic representation of the conventional and the green valorization approaches of citrus and coffee residues.

while environmental performance in terms of GHG emissions and non-renewable energy use was comparable to the currently best available technologies for biodiesel production. The scale level of processes is critical for the successful designation of biorefineries taking into consideration that the Net Present Value and production costs are interrelated to the process. This is attributed to the fact that scale energy consumption and equipment size does not increase at the same rate as the production scale. Consequently, the co-production of high-value-added compounds is a prerequisite for process viability (Moncada and Aristizábal 2016).

Boukroufa et al. (2017) developed a biorefinery based on orange peel waste applying green and solvent-free extraction techniques to recover essential oil, polyphenols, and pectin. The microwave hydro-diffusion resulted in an essential oil yield of 4.2%, the UAE led to a TPC of 0.05%, and MAE gave a pectin yield equal to 24.2%. Pfaltzgraff et al. (2013) reported the successful use of hydrothermal low-temperature microwave process at a large scale to obtain D-limonene, pectin, and flavonoids from waste orange peels. The economic viability of the proposed biorefinery for the processing of 50,000 t of orange waste was demonstrated taking into consideration an extraction yield of 1.5% for D-limonene and 10.8% for pectin. Innovative microwave and ultrasound configurations, namely coaxial solventless MAE (SMAE) and simultaneous ultrasound coaxial microwave-assisted hydrodistillation (UC-MA-HD) were developed for D-limonene extraction from waste orange peels and compared to microwave assisted HD and HD. The oil-free residues were subsequently treated via MAE for pectin extraction. The yield of D-limonene and pectin varied between 1.2–1.6% and 14.2–17.4, respectively, in all the applied green processes. SMAE was the most promising technology considering energy requirements followed by microwave-assisted HD and UC-MA-HD, while UC-MA-HD was less time consuming (González-Rivera et al. 2016). Clark et al. (2016) developed and patented a novel citrus waste-based biorefinery using a microwave-assisted hydrothermal treatment applying low temperature for the simultaneous recovery of pectin (9.2%), D-limonene (1.1%), flavonoids, monosaccharides, and mesoporous cellulose.

Vardon et al. (2013) evaluated the concept of entire utilization of SCGs. Lipids extracted from SCGs were transesterified to biodiesel via alkali catalysis. Biodiesel blends B5 and B20 met ASTM specifications. The defatted fraction was further treated via green slow pyrolysis to produce bio-oil and biochar. The produced biochar mixed with fertilizer resulted in a two-fold increase in sorghum-sudangrass cultivation yield compared to control preparations containing only fertilizer. The potential of biochar as a soil improver was demonstrated. Ktori et al. (2018) also used renewable SCGs for the recovery of bio-oil, char, and biogas via pyrolysis. A maximum yield of bio-oil (36%) and char (29%) was achieved at 540°C, while gas yield was maximized at 700°C. The low heating value (13.3 MJ/m^3) of the biogas classifies it as medium level fuels, suitable for application in engines, turbines, and boilers for power production.

5. Biological Conversion

The recovery of valuable compounds from coffee and citrus by-products is essential prior to their subsequent valorization as inexpensive and abundant feedstock in solid-state or submerged fermentations. The efficient conversion to bio-products of both coffee and citrus residues is highly dependent on the pre-treatment method that would allow the efficient hydrolysis of hemicellulose and cellulose into fermentable sugars. In the case of coffee waste streams that are rich in lignin, more intensive methods are required to make the complex susceptible to enzymatic hydrolysis.

In most cases, biomass pre-treatment is inevitable, even though costly, as it solubilises lignin and hemicellulose, alters the structure, and crystallinity of cellulose and thus improves enzymes accessibility (Capolupo and Faraco 2016). Favorable pretreatment methods that have low impacts to the environment, should provide low cellulose damage and inhibitors formation combined with low cost, energy and chemicals requirements (Adsul et al. 2011). Effective and environmentally friendly methods, including extrusion, liquid hot water, steam explosion, ammonia fiber explosion, and supercritical CO_2 explosion have been evaluated on different feedstocks on both pilot and industrial scales (Capolupo and Faraco 2016).

Dilute acid pretreatment using sulfuric acid is the most commonly applied strategy for the solubilization of hemicellulose, present in lignocellulosic biomass (Kumari and Singh 2018). Despite its high effectiveness, the generation of inhibitors, i.e., furfural and hydroxymethylfurfural, toxicity, and waste production have led research to alternative and more environmentally friendly processes including organosolv, ozone, ionic liquids (ILs), and popping pretreatment. Organosolv has been so far tested for the pretreatment of coffee and citrus wastes prior to their biotechnological conversion (Capolupo and Faraco 2016). An advantage of this method is that lignin can be obtained in high purities and further utilized for the production of value-added chemicals (Zhao et al. 2009). Popping pre-treatment was applied in mandarin peel wastes to increase the enzymatic hydrolysis of polysaccharides. A scanning electron microscope showed that particle size was decreased, while the surface area was increased, facilitating the action of cell-wall degrading enzymes (Choi et al. 2013).

DES has been also investigated as an appealing green pretreatment method to enhance the saccharification of lignocellulosic biomass. They are cost-effective and act under milder conditions in comparison to the aforementioned processes, hence presenting lower energy requirements and inhibitory compounds formation (Procentese and Rehmann 2018). DES technology has been applied in CS prior to its enzymatic hydrolysis (Procentese and Rehmann 2018). Under the optimized conditions of choline-chloride with glycerol DES (150°C and solid to liquid ratio 1:32 w/v), a maximum hydrolysis yield of 0.73 g/g was obtained with minor concentrations of HMF and furfural.

Following pretreatment, enzymatic hydrolysis is employed mainly targeting cellulose hydrolysis to glucose. The produced hydrolysates can be subsequently converted to various biotechnological products, which are described in detail in the following sections.

5.1 Enzyme Production

Solid-state fermentation (SSF) with various fungal strains has been widely applied to citrus biomass focusing mainly on the production of pectinolytic enzymes since citrus peels are rich in pectins. Mahmoodi et al. (2019) valorized orange pomace as a substrate for polygalacturonases production with *Aspergillus niger*. High pectinase activities (2181 U/L) were also achieved using the same fungal strain and lemon peels as substrate (Ruiz et al. 2012). *Penicillium oxalicum* PJO2 was cultivated on orange peels for the production of pectinases, carboxymethylcellulases (CMC), and xylanases. Exo-pectinases and endo-pectinases activity was maximized at 36.5°C with the addition of 1.1 g/L NH_4Cl with respective values of 36.88 U/mL and 0.62 U/mL (Li et al. 2015). Ahmed et al. (2016) mixed citrus waste peel with Czapeck medium and carried out submerged fermentations with *A. niger*. The maximum pectinase yield was 117.1 µM/mL/min after five days of fermentation. In another study, a pilot-scale packed bed bioreactor was employed for pectinases production growing *A. oryzae* on a mixture of citrus pulp and sugarcane bagasse. Pectinase activities varying between 33–41 U/g were achieved (Biz et al. 2016). Ni et al. (2019) evaluated the effect of sterilization techniques, namely irradiation and autoclave on SSF of citrus waste for the production of enzymes. Sterilization via irradiation favored *A. aculeatus* growth while enzyme production was enhanced due to the preservation of flavonoids, which serve as enzyme inducers.

Citrus by-products can be used as substrates for multi-enzyme complexes, i.e., cellulases and hemicellulases (Mamma and Christakopoulos 2008). Being used industrially in several applications, such as cotton processing, animal feed production, juice extraction, paper recycling, and detergent manufacturing, cellulases rank third in terms of market value (Kuhad et al. 2011). Díaz et al. (2013) investigated the production of hydrolytic enzymes with *A. awamori* grown on a mixture of grape pomace and orange peels. SSF in a packed bed and a tray bioreactor resulted in similar xylanases and cellulases activities of 42.6 IU/g and 2.2 IU/g, respectively. The produced enzymes were subsequently tested in orange juice clarification leading in up to 95% juice clarity. Untreated orange peel waste was used for cellulases production, employing several fungal strains. *Emericella variecolor* NS3 gave the highest cellulase activity (31 IU/g) amongst all the studied strains. Tao et al. (2011) demonstrated the effective SSF of multi-enzymes using cultures of *Eupenicillium javanicum* on sweet orange processing waste. The maximum enzyme activity was achieved in the case of xylanases (106.42 U/g), while activities of endoglucanase, β-glucosidase, and pectinase varied between 46.8–51.9 U/g.

CHs have been assessed as a potent lignocellulosic substrate for SSF targeting cellulases production (10 FPU/g) (Cerda et al. 2017a). The same group evaluated the co-production of cellulases and hemicellulases. Increased temperature (71°C) during cultivation on a 50 L bioreactor led to almost 50% cellulase reduction. The overall enzymatic activities achieved were 3.1 FPU/g for cellulases and 48 U/g for xylanases (Cerda et al. 2017b). The environmental impact of cellulases production under SSF on CHs has been recently evaluated (Catalán et al. 2019). Electricity requirements for their downstream showed the highest environmental impact, accounting for 98.8% of total emissions. The study demonstrated that when the lyophilization after filtration and ultrafiltration of enzymes was omitted, the environmental impact could be decreased by 94%.

Enzyme production from SCGs has been investigated using wild fungi isolated from SCG and coffee beans. Out of 51 screened strains, 2 fungal strains belonging to *Aspergillus* ssp. and *Penicillium* ssp. were found to be promising for mannanase production. However, the highest hydrolysis yield (17%) was achieved when a recombinant *A. niger* was used (Jooste et al. 2013). Kandasamy et al. (2016) reported a protease activity of 920 U/mL when cultivating *Bacillus* sp. on CP mixed with corncobs at a ratio of 3:2 after 60 hours of fermentation.

5.2 Bioethanol Production

Ethanol has been arisen as a promising alternative to gasoline, owing to its chemical properties, i.e., higher octane number (Zabed et al. 2017). The fact that ethanol is oxygenated leads to a 15% higher combustion efficiency with lower emissions. The valorization of various waste and byproduct streams derived from citrus and coffee could contribute to the minimization of ethanol costs combined with sustainable production.

Bioethanol fermentation with the whole ground and aqueous extract from ground CHs was investigated by Gouvea et al. (2009). The highest bioethanol concentration of 13.6 g/L was observed when whole CHs were used as a substrate for *S. cerevisiae*. Acid, alkali, or steam explosion pretreatment was applied on CHs cherries to enhance enzymes secreted by *Sphingobacterium* sp. The steam explosion was the most efficient pretreatment leading to the production of a hydrolysate containing 49 g/L fermentable sugars. Co-culture of *S. cerevisiae* and *Candida tropicalis* resulted in 18.2 g/L bioethanol production when they were cultivated in CHs enzymatic hydrolysates (Shankar et al. 2019).

Coffee residue waste (CRW) was investigated for bioethanol production. After being subjected to popping pretreatment at 1.47 MPa, enzymatic hydrolysis of CRW was carried out using commercial cellulases. A hydrolysis efficiency of 85.6% was achieved (40.2 g/L fermentable sugars). The hydrolysate was subsequently utilised for bioethanol production with *S. cerevisiae*, leading to a 15.2 g/L final concentration (Choi et al. 2012). Another study evaluated CRW as a potent substrate for the co-production of D-mannose and ethanol. Pre-treatment of CRW using 60% ethanol in a high-pressure vessel (150°C for 2 hours) led to extractives and lignin removal while eliminating monosaccharide degradation. Hydrolysates derived from enzymatic hydrolysis of pre-treated CRW were sequentially used for bioethanol production with *S. cerevisiae* KCTC 7906. From the overall process, 15.7 g D-mannose and 11.3 g bioethanol could be obtained from 150 g of pre-treated CRW (Nguyen et al. 2017).

Bioethanol production from citrus by-products, especially orange peels, has been a topic of research since the 1990s. Simultaneous saccharification and fermentation present several advantages when compared to sequential saccharification and fermentation for bioethanol production, including simplicity of operation and low capital investment (Oberoi et al. 2011). Factor optimization, i.e., cellulase and pectinase concentration, fermentation time and temperature. Mandarin peel waste subjected to hydrothermal or popping pretreatment prior to enzymatic hydrolysis, have been reported very promising for the production of fermentable sugars and the subsequent ethanol production (42.0–46.2 g/L) (Oberoi et al. 2011, Choi et al. 2013, Boluda-Aguilar et al. 2010). Mandarin peels used as the sole substrate to produce bio-ethanol (29.5 g/L) were more efficient compared to mixtures with banana peels, apple pomace and pear waste (Choi and Song 2015).

Boluda-Aguilar et al. (2010) utilized mandarin citrus peel wastes (MCPW) as a substrate for bioethanol fermentation. The steam explosion was applied in MCPW, to separate D-limonene and facilitate their enzymatic saccharification. Sugars production reached 46.9 g/100 g dry matter. Simultaneous saccharification and fermentation in 5 L bioreactors, could lead to the production of 50–60 L of ethanol from 1,000 kg of fresh MCPW. D-limonene, galacturonic acid and citrus pulp pellets were obtained as co-products. Steam-exploded lemon peels have been reported to yield 60 L of bio-ethanol from 1,000 kg of fresh material (Boluda-Aguilar and López-Gómez 2013).

5.3 Biohydrogen and Biogas Production

Biohydrogen is a "green" alternative to conventional petroleum-based fuels. Biohydrogen contains a high energy yield, almost three times higher than gasoline (Tandon et al. 2018). The co-digestion of various coffee waste streams (coffee pulp, husk, and processing wastewater) for hydrogen production was investigated in a hydrothermal reactor with microbial consortia indigenous to the waste streams. The digestion of 2 g/L pulp and CHs and 30 g COD/L coffee processing wastewater resulted in the production of 82 mL of hydrogen, including organic acids, and ethanol (Villa Montoya et al. 2019). Effluents from the citrus-processing industry were individually used or mixed with vinasse for hydrogen production under the dark fermentation process. Citrus wastewater led to the production of up to 85.3 mmol H_2/L and up to 13.4 mmol H_2/L when citrus wastewater was mixed with vinasse (Torquato et al. 2017). Citrus peels were also evaluated for the production of hydrogen-rich syngas via an air-steam gasification process. Gasification under 750°C and a biomass ratio of 1.25 led to hydrogen yields of 0.65 to 0.69 Nm^3/kg (Prestipino et al. 2017).

Biogas production from coffee and citrus residues constitutes another environmentally friendly process for energy generation. The steam explosion was implemented for the pretreatment of CHs prior to methane production. The highest methane yield (144.96 NmL CH_4/g COD) and electricity production (0.59 kWh/kg CH) was observed when mild conditions were chosen (120°C for 60 minutes) (Baêta et al. 2017). Biomethane production was also studied by Girotto et al. (2018) using SCGs. A final methane yield of 0.36 m^3 CH_4/kg volatile solids was obtained when a substrate to inoculum ratio of 2:1 was employed. The co-digestion of coffee residues (i.e., green coffee powder, parchment and deffated cake) with sugarcane vinasse have also been employed for the production of both bio-hydrogen and bio-methane (Pinto et al. 2018).

5.4 Production of Phenolic-Rich Extracts and Their Biotransformation

D-limonene is an important commodity chemical and its extraction using green methods have been extensively discussed in Section 3.2. This terpene can be also utilized as a precursor for the production of other fine chemicals, such as carveol or perrillyl alcohol (Badee et al. 2011). Badee et al. (2011) reported the biotransformation of D-limonene derived from orange peels into α-terpineol. D-limonene was extracted via cold pressure and fed to *Penicillium digitatum* NRRL 1202 as an ethanolic solution (20% v/v). The fungus was able to fully convert D-limonene into α-terpineol after approximately 2 days of cultivation. The biotransformation of citrus flavonoids has been also reported. Lee et al. (2012) reported the potential to produce hesperetin-7-*O*-glucoside from hesperidin via enzymatic bioconversion using an *A. sojae* naringinase. Hesperidin was extracted from orange juice and peels and its conversion to hesperetin-7-*O*-glucoside was carried out under a repetitive batch reaction to increase naringinase's productivity. A production yield of up to 90% on a molar basis was achieved. Hesperetin-7-*O*-glucoside solubility presented higher solubility than hesperidin demonstrating its higher bioavailability.

Phenolic compounds can be bound to the plant cell wall, mainly with ester bonds. During SSF, the ester bonds break and increase the release of phenolic compounds in the extract (Arellano-González et al. 2011, Palmieri et al. 2018). CP was subjected to SSF for the production of a phenolic-rich extract in a laboratory (0.4 kg), semi-pilot (12 kg), and pilot scales (90 kg) (Santos da Silveira et al. 2019). The yeast *S. cerevisiae* was employed for the fermentation of CP, aiming to enhance the extraction of chlorogenic acid. Compared to the control sample, the concentration of chlorogenic acid increased almost four-fold. Even though the composition of CP presented significant variations, the process could be conducted both on semi-pilot and pilot scales leading to the production of almost 600 mg chlorogenic acid/kg of CP (Santos da Silveira et al. 2019).

5.5 Other Products

Xanthan gum is a hetero-polysaccharide produced by *Xanthomonas campestris* with high commercial value since it is industrially used as a viscosity enhancer and stabilizer in the food industry (Mohsin et al. 2019). Studies carried out in the early nineties indicated that efficient xanthan gum production could be achieved using *X. campestris* grown on whole citrus wastes (Mamma and Christakopoulos 2008). Mohsin et al. (2018) utilized orange peels for the production of sugar-rich hydrolysate via dilute acid hydrolysis. Even though inorganic acids are not in compliance with green concepts, it is worth noting that 30.2 g/L of xanthan gum were produced.

Butanol presents various advantages as a fuel in comparison to ethanol such as higher flash point and energy content combined with lower volatility and hygroscopicity. Butanol can be mixed with gasoline, so the same pipelines could be still used (Hijosa-Valsero et al. 2018). CS were subjected to autohydrolysis followed by sequential enzymatic hydrolysis with Cellic CTec2 (Hijosa-Valsero et al. 2018). The produced hydrolysate was rich in monosaccharides (34.4 g/L) and highly efficient for butanol production with *C. beijerinckii* (Table 5).

Bacterial cellulose, produced mainly via fermentative procedures of *Acetobacter* species, is a highly functional biopolymer that can be produced from numerous carbon sources. Possessing properties, such as high mechanical strength, high crystallinity, high water-binding capacity, and an ultra-fine and highly pure fiber network, bacterial cellulose can be applied in numerous sectors including pharmaceutical, broadcasting, food industry, paper manufacture, and mining (Tsouko et al. 2015). The production of bacterial cellulose (5.7 g/L) with the strain *Komagataeibacter xylinus* using citrus peel and pomace as fermentation substrate has been reported by Fan et al. (2016). Andritsou et al. (2018) evaluated waste oranges, grapefruits, and lemons from open markets for BC production with the same bacterial strain. Bacterial cellulose concentrations higher than 6 g/L were achieved when the bacterial strain *Komagataeibacter sucrofermentans* DSM 15973 was cultivated on orange and grapefruit juices (Table 6). The properties of bacterial cellulose were enhanced compared to nano-cellulose isolated from orange peels.

Several publications have dealt with polyhydroxyalkanoates (PHAs) production from SCGs either from the extracted oil or from carbon sources derived from the hydrolysis of the lignocellulosic fraction (Kourmentza et al. 2018, Obruca et al. 2015). Even though in most of these studies, oil is extracted using organic solvents and the lignocellulosic fraction is hydrolyzed via chemical methods, it is worth noting that PHAs production has been reported to be feasible leading to high production yields. Indicatively, Obruca et al. (2014) reported an intracellular PHAs content of 89.1% (49.4 g/L) in fed-batch cultures of *Cupriavidus necator* H16 using oil fraction. Cruz et al. (2014) evaluated coffee oil extracted via supercritical CO_2 for PHA production by *C. negator* resulting in 0.77 kg of PHA per kg of SCG oil.

Carotenoids production has also been reported utilizing SCGs, thus pretreatment techniques were not in line with green chemistry principles (Petrik et al. 2014). SCGs, after oil removal (with hexane) and diluted acid pretreatment, were hydrolyzed to glucose using commercial enzymatic preparations. Fed-batch fermentation with the yeast *Sporobolomyces roseus* using SCG hydrolysates resulted in 29.9 mg/L of total carotenoids with approximately 50% corresponding to β-carotene.

6. Conclusions and Future Directions

The growing awareness of the escalating consequences of greenhouse gases for climate change has turned economic dynamics towards the opening-up of new green markets having as an ally the reformation of the existing regulatory and policy structure. A greener and safer industry could mitigate environmental change, provide better health outcomes and a stronger foundation for the long-run world sustainability combined with technological innovations and better-performing markets for consumers and businesses.

The implementation of efficient and green processes is a key challenge for the valorization of by-products of the industrial manufacturing process. The selection of suitable technologies for the recovery of value-added components and the subsequent

Table 5. Bioconversion of coffee processing by-products to value added and biobased products.

Substrate	Treatment	Strain	Product	Production (g/L)	Yield (g/g)	Productivity (g/L/h)	References
CRW	Popping (150°C, 1.47 MPa, 10 min); enzymatic hydrolysis with Celluclast	*S. cerevisiae*	bioethanol	15.3	0.87	0.16	Choi et al. 2012
CRW	60% EtOH at solid to liquid ratio 1:5, 150°C, 2 h; enzymatic hydrolysis with cellulase and pectinase	*S. cerevisiae* KCTC 7906		16.4	0.51	1.82	Nguyen et al. 2017
CHs				13.6	0.92	1.2	Gouvea et al. 2009
CHs	Steam (121°C, 10 min, 15 psi); enzymatic hydrolysis with cocktail by *Sphingobacterium* sp.	*S. cerevisiae* & *C. tropicalis*		18.21	-	-	Shankar et al. 2019
CHs	Steam explosion (120°C, 60 min)	-	biogas	-	145.0[1]	-	Lobo Baêta et al. 2017
SCGs		Mixed cultures	biomethane		0.36[2]		Girotto et al. 2018
Coffee waste fractions		Mixed cultures	biohythane		0.14[3]	-	Maciel Pinto et al. 2018
Coffee waste		Indigenous microbes in coffee	bio-H_2	82[4]	-	-	Villa-Montoya et al. 2019
CS	Autohydrolysis (170°C, 20 min); enzymatic hydrolysis with Cellic CTec2	*Clostridium beijerinckii* CECT	butanol	7.02	0.27		Hijosa-Valsero et al. 2018
CHs	Boiling water, 30 min	*Gluconacetobacter hansenii* UAC09	bacterial cellulose	6.24	-	0.45[5]	Rani et al. 2011
				5.6-8.2	-	-	Rani and Appaiah 2013
Coffee oil	Supercritical CO_2	*C. necator*	PHAs		0.77	4.7[5]	Cruz et al. 2014

[1] NmL CH_4/g COD; [2] $m^3 CH_4$/kg VS; [3] mL CH_4/g VS$_{added}$; [4] mL; [5] g/L/day

Table 6. Bioconversion of citrus processing by-products to value added and biobased products.

Substrate	Pretreatment	Strain	Product	Production (g/L)	Yield (%)	Productivity (g/L/h)	References
Orange peel	Enzymatic hydrolysis with enzymes from *A. citrisporus* & *T. longibrachiatum*	*S. cerevisiae* KTC 7906	bioethanol	27.1	92.4	3.01	Choi et al. 2015
Mandarin peel				29.5	93.1	3.28	
Grapefruit peel				21.6	90.7	2.40	
Lemon peel				20.4	90.2	2.27	
Lime peel				20.3	91.8	2.26	
Mixed citrus peel				19.6	91.1	2.18	
Mixed fruit wastes				14.4	90.8	1.60	
Mandarin peel waste	Popping (150 °C, 10 min); enzymatic hydrolysis with pectinase, xylanase & β-glucosidase	*S. cerevisiae* KTC 7906		46.1	90.6	3.84	Choi et al. 2013
Mandarin peel wastes (simultaneous saccharification and fermentation)	Steam explosion (160 °C for 5 min); enzymatic hydrolysis with pectinase, β-glucosidase & cellulase	*S. cerevisiae* CECT 1329			60		Boluda-Aguilar et al. 2010
Mandarin waste (simultaneous saccharification and fermentation)	Autoclave (121 °C, 15 min); Enzymatic hydrolysis with cellulase & pectinase	*S. cerevisiae*		42		3.50	Oberoi et al. 2011
Citrus peel and pomace		*K. xylinus*	bacterial cellulose	5.7			Fan et al. 2016
Grapefruit juice	Mixed peels with water, boiling for 1 h	*K. xylinus* DSM 15973		6.7	0.36[2]	0.61[3]	Andritsou et al. 2018
Orange juice				6.1			
Lemon peel aqueous extract				5.2			
Grapefruit peel aqueous extract							
Orange peel aqueous extract				5.0			

Table 6 contd. ....

...*Table 6 contd.*

Substrate	Pretreatment	Strain	Product	Production (g/L)	Yield (%)	Productivity (g/L/h)	References
Citrus waste	Filtration with PVDF membranes	Mixed cultures	methane	0.33[4]			Wikandari et al. 2014, Forgács et al. 2012, Su et al. 2016
	Steam explosion			0.537[4]			
	Biodegradation with *P. chrysosporium & A. niger*			176.05[5]			
Citrus wastewater		Mixed cultures	bio-H_2	85.3[6]			Torquato et al. 2017
Citrus peels				0.65–0.69[4]			Prestipino et al. 2017
Citrus peel waste		*Alcaligenes faecalis*	bioflocculants	3.49	0.09[2]		Qi et al. 2020
Orange juice & peel	EtOH extraction with ultrasounds	*A. sojae* naringinase	Hes-7-G	1.2[7]	90		Lee et al. 2012
Orange peel oil	Cold-pressing	*P. digirarum* NRRL 1202	α-terpineol		100		Badee et al. 2011

[1] L/1000 kg peel; [2] g/g; [3] g/L/day; [4] Nm^3/kg VS; [5] mL/gVS; [6] mmolH_2/L; [7] mM

biological conversion of renewable feedstock should be initially based on principles of green chemistry and sustainability. Especially in the case that recovered or produced products re-enter the food chain or personal care sector, the application of safe solvents and green methods, including SWE, SLE, MAE, UAE, PEF and combined with green solvents should be employed. The potential applicability of methods from laboratory scale to pilot and industrial scale should be also taken into consideration. Technologies that fail to embed the aforementioned criteria will lose their edge in future applications.

The constant increase in industrial applications of clean and safe materials (i.e., pectins, D-limonene), bio-based products (i.e., enzymes), biofuels (i.e., bioethanol, biobutanol, and biohydrogen), biopolymers (i.e., PHAs), organic acids, and bacterial cellulose that are obtained from by-products and waste streams of citrus and coffee industry could eliminate the landfill practice supplying the corresponding biorefineries. Concepts of integrated biorefineries located at source feedstock, localized, or centrally located receiving waste streams from producers could introduce the bio-economy era in which waste is considered a valued resource, eventually boosting the citrus and coffee industries.

Future research should focus on the development of novel methods and strategies for the cascade valorization of citrus and coffee wastes. Coffee and citrus processing waste should be considered important and highly promising renewable resources for high-value materials, chemicals, pharmaceutical products, and bioenergy. Their fractionation and the application of biotechnological processes is a dominant choice for entire valorization. Processing efficiencies, clean production, and profitability should be the drivers of related scientific investigation to obtain functional products suitable for application in food, pharmaceuticals, and cosmetic industries (Chemat et al. 2017). Synergies between industry, researchers, and entrepreneurs are critical for the successful development and implementation of novel technologies.

Acknowledgments

We acknowledge the support of this work by the project "Research Infrastructure on Food Bioprocessing Development and Innovation Exploitation – Food Innovation RI" (MIS 5027222), which is implemented under the Action "Reinforcement of the Research and Innovation Infrastructure", funded by the Operational Programme "Competitiveness, Entrepreneurship and Innovation" (NSRF 2014–2020) and co-financed by Greece and the European Union (European Regional Development Fund).

Abbreviations

CP	Coffee Pulp
CH	Coffee Husks
CS	Coffee Silverskin
CRW	Coffee Residue Waste
SCG	Spent Coffee Ground
MCPW	Mandarin Citrus Peel Wastes

UAE	Ultrasound-Assisted Extraction
MAE	Microwave-Assisted Extraction
SWE	Subcritical Water Extraction
SFE	Supercritical Fluid Extraction
PEF	Pulsed Electric Fields
DES	Deep Eutectic Solvents
MAD	Microwave-Accelerated Distillation
SFME	Solvent Free Microwave Extraction
UC-MA-HD	Ultrasound Coaxial Microwave-Assisted Hydro-Distillation
SMAE	Solventless Microwave-Assisted Extraction
HD	Hydro-Distillation
ILs	Ionic Liquids
FPU	Filter Paper Units
GAE	Gallic Acid Equivalents
TPC	Total Phenolic Content
DE	Degree of Esterification
PHAs	Polyhydroxyalkanoates

References

Abdullah, M. and Koc, A.B. 2013. Oil removal from waste coffee grounds using two-phase solvent extraction enhanced with ultrasonication. Renewable Energy 50: 965–970.

Act, Single Market. 2011. Communication from the Commission to the European Parliament, the Council, the Economic and Social Committee and the Committee of the Regions.

Adsul, M.G., Singhvi, M.S., Gaikaiwari, S.A. and Gokhale, D.V. 2011. Development of biocatalysts for production of commodity chemicals from lignocellulosic biomass. Bioresource Technology 102: 4304–4312.

Ahangari, B. and Sargolzaei, J. 2013. Extraction of lipids from spent coffee grounds using organic solvents and supercritical carbon dioxide. Journal of Food Processing and Preservation 37(5): 1014–1021.

Ahmed, I., Zia, M.A., Hussain, M.A., Akram, Z., Naveed, M.T. and Nowrouzi, A. 2016. Bioprocessing of citrus waste peel for induced pectinase production by *Aspergillus niger*; its purification and characterization. Journal of Radiation Research and Applied Sciences 9: 148–154.

Akgün, N.A., Bulut, H., Kikic, I. and Solinas, D. 2014. Extraction behavior of lipids obtained from spent coffee grounds using supercritical carbon dioxide. Chemical Engineering & Technology 37(11): 1975–1981.

Al-Dhabi, N.A., Ponmurugan, K. and Jeganathan, P.M. 2017. Development and validation of ultrasound-assisted solid-liquid extraction of phenolic compounds from waste spent coffee grounds. Ultrasonics Sonochemistry 34: 206–213.

Al-Hamamre, Z., Foerster, S., Hartmann, F., Kröger, M. and Kaltschmitt, M. 2012. Oil extracted from spent coffee grounds as a renewable source for fatty acid methyl ester manufacturing. Fuel 96: 70–76.

Alves, R.C., Rodrigues, F., Nunes, M.A., Vinha, A.F. and Oliveira, M.B.P. 2017. State of the art in coffee processing by-products. pp. 1–26. *In*: Handbook of Coffee Processing By-products. Academic Press.

Andrade, K.S., Gonçalvez, R.T., Maraschin, M., Ribeiro-do-Valle, R.M., Martínez, J. and Ferreira, S.R. 2012. Supercritical fluid extraction from spent coffee grounds and coffee husks: antioxidant activity and effect of operational variables on extract composition. Talanta 88: 544–552.

Andritsou, V., De Melo, E.M., Tsouko, E., Ladakis, D., Maragkoudaki, S., Koutinas, A.A. and Matharu, A.S. 2018. Synthesis and characterization of bacterial cellulose from citrus-based sustainable resources. ACS Omega 3: 10365–10373.

Arellano-González, M.A., Ramírez-Coronel, M.A., Torres-Mancera, M.T., Pérez-Morales, G.G. and Saucedo-Castañeda, G. 2011. Antioxidant activity of fermented and nonfermented coffee (*Coffea arabica*) pulp extracts. Food Technology and Biotechnology 49: 374–378.

Aristizábal-Marulanda, V., Chacón-Perez, Y. and Alzate, C.A.C. 2017. The biorefinery concept for the industrial valorization of coffee processing by-products. pp. 63–92. *In*: Handbook of Coffee Processing By-Products. Academic Press.

Asl, A.H. and Khajenoori, M. 2013. Subcritical water extraction. Mass Transfer-Advances in Sustainable Energy and Environment Oriented Numerical Modelling, 459–487.

Badee, A.Z.M., Helmy, S.A. and Morsy, N.F.S. 2011. Utilisation of orange peel in the production of α-terpineol by *Penicillium digitatum* (NRRL 1202). Food Chemistry 126: 849–854.

Baêta, B.E.L., Cordeiro, P.H. de M., Passos, F., Gurgel, L.V.A., de Aquino, S.F. and Fdz-Polanco, F. 2017. Steam explosion pretreatment improved the biomethanization of coffee husks. Bioresource Technology 245: 66–72.

Ballesteros, L.F., Teixeira, J.A. and Mussatto, S.I. 2017. Extraction of polysaccharides by autohydrolysis of spent coffee grounds and evaluation of their antioxidant activity. Carbohydrate Polymers 157: 258–266.

Barbosa-Pereira, L., Guglielmetti, A. and Zeppa, G. 2018. Pulsed electric field assisted extraction of bioactive compounds from cocoa bean shell and coffee silverskin. Food and Bioprocess Technology 11(4): 818–835.

Bekalo, S.A. and Reinhardt, H.W. 2010. Fibers of coffee husk and hulls for the production of particleboard. Materials and Structures 43(8): 1049–1060.

Bessada, S.M., C Alves, R., Oliveira, P.P. and Beatriz, M. 2018. Coffee silverskin: A review on potential cosmetic applications. Cosmetics 5(1): 5.

Bharathiraja, S., Suriya, J., Krishnan, M., Manivasagan, P. and Kim, S.K. 2017. Production of enzymes from agricultural wastes and their potential industrial applications, 1st ed, Advances in Food and Nutrition Research. Elsevier Inc.

Biz, A., Finkler, A.T.J., Pitol, L.O., Medina, B.S., Krieger, N. and Mitchell, D.A. 2016. Production of pectinases by solid-state fermentation of a mixture of citrus waste and sugarcane bagasse in a pilot-scale packed-bed bioreactor. Biochemical Engineering Journal 111: 54–62.

Boluda-Aguilar, M., García-Vidal, L., González-Castañeda, F. del P. and López-Gómez, A. 2010. Mandarin peel wastes pretreatment with steam explosion for bioethanol production. Bioresource Technology 101: 3506–3513.

Boluda-Aguilar, M. and López-Gómez, A. 2013. Production of bioethanol by fermentation of lemon (*Citrus limon* L.) peel wastes pretreated with steam explosion. Industrial Crops and Products 41: 188–197.

Boonkird, S., Phisalaphong, C. and Phisalaphong, M. 2008. Ultrasound-assisted extraction of capsaicinoids from *Capsicum frutescens* on a lab-and pilot-plant scale. Ultrasonics Sonochemistry 15(6): 1075–1079.

Boukroufa, M., Boutekedjiret, C. and Chemat, F. 2017. Development of a green procedure of citrus fruits waste processing to recover carotenoids. Resource-Efficient Technologies 3(3): 252–262.

Bousbia, N., Vian, M.A., Ferhat, M.A., Meklati, B.Y. and Chemat, F. 2009. A new process for extraction of essential oil from Citrus peels: Microwave hydrodiffusion and gravity. Journal of Food Engineering 90(3): 409–413.

Bustamante, J., van Stempvoort, S., García-Gallarreta, M., Houghton, J.A., Briers, H.K., Budarin, V.L., Matharu, A.S. and Clark, J.H. 2016. Microwave assisted hydro-distillation of essential oils from wet citrus peel waste. Journal of Cleaner Production 137: 598–605.

Caetano, N.S., Caldeira, D., Martins, A.A. and Mata, T.M. 2017. Valorisation of spent coffee grounds: production of biodiesel via enzymatic catalysis with ethanol and a co-solvent. Waste and Biomass Valorization 8(6): 1981–1994.

Capolupo, L. and Faraco, V. 2016. Green methods of lignocellulose pretreatment for biorefinery development. Applied Microbiology and Biotechnology 100: 9451–9467.

Catalán, E., Komilis, D. and Sánchez, A. 2019. Environmental impact of cellulase production from coffee husks by solid-state fermentation: A life-cycle assessment. Journal of Cleaner Production 233: 954–962.

Cerda, A., Gea, T., Vargas-García, M.C. and Sánchez, A. 2017a. Towards a competitive solid state fermentation: Cellulases production from coffee husk by sequential batch operation and role of microbial diversity. The Science of the Total Environment 589: 56–65.

Cerda, A., Mejías, L., Gea, T. and Sánchez, A. 2017b. Cellulase and xylanase production at pilot scale by solid-state fermentation from coffee husk using specialized consortia: The consistency of the process and the microbial communities involved. Bioresource Technology 243: 1059–1068.

Chanakya, H.N. and De Alwis, A.A.P. 2004. Environmental issues and management in primary coffee processing. Process Safety and Environmental Protection 82(4): 291–300.

Chavan, P., Singh, A.K. and Kaur, G. 2018. Recent progress in the utilization of industrial waste and by-products of citrus fruits: A review. Journal of Food Process Engineering 41(8): e12895.

Chemat, F., Rombaut, N., Sicaire, A.G., Meullemiestre, A., Fabiano-Tixier, A.S. and Abert-Vian, M. 2017. Ultrasound assisted extraction of food and natural products. Mechanisms, techniques, combinations, protocols and applications. A review. Ultrasonics Sonochemistry 34: 540–560.

Chiyanzy, I., Brienzo, M., García-Aparicio, M., Agudelo, R. and Görgens, J. 2015. Spent coffee ground mass solubilisation by steam explosion and enzymatic hydrolysis. Journal of Chemical Technology and Biotechnology 90: 449–458.

Choi, I.S., Wi, S.G., Kim, S.B. and Bae, H.J. 2012. Conversion of coffee residue waste into bioethanol with using popping pretreatment. Bioresource Technology 125: 132–137.

Choi, I.S., Kim, J.H., Wi, S.G., Kim, K.H. and Bae, H.J. 2013. Bioethanol production from mandarin (*Citrus unshiu*) peel waste using popping pretreatment. Applied Energy 102: 204–210.

Choi, S. and Song, C.W. 2015. Biorefineries for the production of top building block chemicals and their derivatives. Metabolic Engineering 28: 223–239.

Ciriminna, R., Lomeli-Rodriguez, M., Cara, P.D., Lopez-Sanchez, J.A. and Pagliaro, M. 2014. Limonene: a versatile chemical of the bioeconomy. Chemical Communications 50(97): 15288–15296.

Ciriminna, R., Chavarría-Hernández, N., Inés Rodríguez Hernández, A. and Pagliaro, M. 2015. Pectin: A new perspective from the biorefinery standpoint. Biofuels, Bioproducts and Biorefining 9(4): 368–377.

Clark, J.H., Pfaltzgraff, L.A., Budarin, V.L. and De Bruyn, M. 2013. Microwave assisted citrus waste biorefinery. Patent WO 2013150262, A1.

Clark, J.H., Pfaltzgraff, L.A., Budarin, V.L. and De Bruyn, M. 2016. U.S. Patent No. 9,382,339. Washington, DC: U.S. Patent and Trademark Office.

Clark, J.H. 2019. Green biorefinery technologies based on waste biomass. Green Chemistry 21(6): 1168–1170.

Colodel, C., Vriesmann, L.C., Teófilo, R.F. and de Oliveira Petkowicz, C.L. 2018. Extraction of pectin from ponkan (*Citrus reticulata blanco* cv. *ponkan*) peel: Optimization and structural characterization. International Journal of Biological Macromolecules 117: 385–391.

Conde, E., Moure, A., Domínguez, H. and Parajó, J.C. 2011. Production of antioxidants by non-isothermal autohydrolysis of lignocellulosic wastes. LWT-Food Science and Technology 44(2): 436–442.

Couto, R.M., Fernandes, J., Da Silva, M.G. and Simões, P.C. 2009. Supercritical fluid extraction of lipids from spent coffee grounds. The Journal of Supercritical Fluids 51(2): 159–166.

Cruz, M.V., Paiva, A., Lisboa, P., Freitas, F., Alves, V.D., Simões, P., Barreiros, S. and Reis, M.A.M. 2014. Production of polyhydroxyalkanoates from spent coffee grounds oil obtained by supercritical fluid extraction technology. Bioresource Technology 157: 360–363.

Cvjetko Bubalo, M., Vidović, S., Radojčić Redovniković, I. and Jokić, S. 2015. Green solvents for green technologies. Journal of Chemical Technology & Biotechnology 90(9): 1631–1639.

Dahmoune, F., Boulekbache, L., Moussi, K., Aoun, O., Spigno, G. and Madani, K. 2013. Valorization of Citrus limon residues for the recovery of antioxidants: evaluation and optimization of microwave and ultrasound application to solvent extraction. Industrial Crops and Products 50: 77–87.

Díaz, A.B., Alvarado, O., De Ory, I., Caro, I. and Blandino, A. 2013. Valorization of grape pomace and orange peels: Improved production of hydrolytic enzymes for the clarification of orange juice. Food and Bioproducts Processing 91: 580–586.

Echeverria, M.C. and Nuti, M. 2017. Valorisation of the residues of coffee agro-industry: perspectives and limitations. The Open Waste Management Journal 10(1): 13–22.

Efthymiopoulos, I., Hellier, P., Ladommatos, N., Russo-Profili, A., Eveleigh, A., Aliev, A., Kay, A. and Mills-Lamptey, B. 2018. Influence of solvent selection and extraction temperature on yield and composition of lipids extracted from spent coffee grounds. Industrial Crops and Products 119: 49–56.

Espinosa-Pardo, F.A., Nakajima, V.M., Macedo, G.A., Macedo, J.A. and Martínez, J. 2017. Extraction of phenolic compounds from dry and fermented orange pomace using supercritical CO_2 and co-solvents. Food and Bioproducts Processing 101: 1–10.

Fan, X., Gao, Y., He, W., Hu, H., Tian, M., Wang, K. and Pan, S. 2016. Production of nano bacterial cellulose from beverage industrial waste of citrus peel and pomace using *Komagataeibacter xylinus*. Carbohydrate Polymers 151: 1068–1072.

Ferrario, V., Veny, H., De Angelis, E., Navarini, L., Ebert, C. and Gardossi, L. 2013. Lipases immobilization for effective synthesis of biodiesel starting from coffee waste oils. Biomolecules 3(3): 514–534.

Flores-Gómez, C.A., Escamilla Silva, E.M., Zhong, C., Dale, B.E., Da Costa Sousa, L. and Balan, V. 2018. Conversion of lignocellulosic agave residues into liquid biofuels using an AFEXTM-based biorefinery. Biotechnology for Biofuels 11: 1–18.

Forgács, G., Pourbafrani, M., Niklasson, C., Taherzadeh, M.J. and Hováth, I.S. 2012. Methane production from citrus wastes: Process development and cost estimation. Journal of Chemical Technology and Biotechnology 87: 250–255.

Franca, A.S. and Oliveira, L.S. 2009. Coffee processing solid wastes: current uses and future perspectives. Agricultural Wastes 9: 155–189.

Girotto, F., Pivato, A., Cossu, R., Nkeng, G.E. and Lavagnolo, M.C. 2018. The broad spectrum of possibilities for spent coffee grounds valorisation. Journal of Material Cycles and Waste Management 20: 695–701.

Golmakani, M.T. and Moayyedi, M. 2015. Comparison of heat and mass transfer of different microwave-assisted extraction methods of essential oil from *Citrus limon* (Lisbon variety) peel. Food Science & Nutrition 3(6): 506–518.

González-Rivera, J., Spepi, A., Ferrari, C., Duce, C., Longo, I., Falconieri, D., Piras, A. and Tinè, M.R. 2016. Novel configurations for a citrus waste based biorefinery: from solventless to simultaneous ultrasound and microwave assisted extraction. Green Chemistry 18(24): 6482–6492.

Gouvea, B.M., Torres, C., Franca, A.S., Oliveira, L.S. and Oliveira, E.S. 2009. Feasibility of ethanol production from coffee husks. Biotechnology Letters 31: 1315–1319.

Guglielmetti, A., D'ignoti, V., Ghirardello, D., Belviso, S. and Zeppa, G. 2017. Optimisation of ultrasound and microwave-assisted extraction of caffeoylquinic acids and caffeine from coffee silverskin using response surface methodology. Italian Journal of Food Science 29(3): 409–423.

Hijosa-Valsero, M., Garita-Cambronero, J., Paniagua-García, A.I. and Díez-Antolínez, R. 2018. Biobutanol production from coffee silverskin. Microbial Cell Factories 17: 1–9.

Hosseini, S.S., Khodaiyan, F. and Yarmand, M.S. 2016. Optimization of microwave assisted extraction of pectin from sour orange peel and its physicochemical properties. Carbohydrate Polymers 140: 59–65.

Hosseini, S.S., Khodaiyan, F., Kazemi, M. and Najari, Z. 2019. Optimization and characterization of pectin extracted from sour orange peel by ultrasound assisted method. International Journal of Biological Macromolecules 125: 621–629.

EU, Industrial Emissions, https://ec.europa.eu/environment/archives/air/stationary/solvents/index.htm.

EU, Circular Economy, https://ec.europa.eu/environment/circular-economy/.

ICO. 2019. International coffee organization accessed 3/2/2019 www.ico.org.

Jiang, W.X., Qi, J.R., Huang, Y.X., Zhang, Y. and Yang, X.Q. 2020. Emulsifying properties of high methoxyl pectins in binary systems of water-ethanol. Carbohydrate Polymers 229: 115420.

Jooste, T., García-Aparicio, M.P., Brienzo, M., Van Zyl, W.H. and Görgens, J.F. 2013. Enzymatic hydrolysis of spent coffee ground. Applied Biochemistry and Biotechnology 169: 2248–2262.

Kandasamy, S., Muthusamy, G., Balakrishnan, S., Duraisamy, S., Thangasamy, S., Seralathan, K.K. and Chinnappan, S. 2016. Optimization of protease production from surface-modified coffee pulp waste and corncobs using *Bacillus* sp. by SSF. 3 Biotech 6: 1–11.

Karmee, S.K., Swanepoel, W. and Marx, S. 2018a. Biofuel production from spent coffee grounds via lipase catalysis. Energy Sources, Part A: Recovery, Utilization, and Environmental Effects 40(3): 294–300.

Karmee, S.K. 2018b. A spent coffee grounds based biorefinery for the production of biofuels, biopolymers, antioxidants and biocomposites. Waste Management 72: 240–254.

Kelkar, S., Saffron, C.M., Chai, L., Bovee, J., Stuecken, T.R., Garedew, M., Li, Z. and Kriegel, R.M. 2015. Pyrolysis of spent coffee grounds using a screw-conveyor reactor. Fuel Processing Technology 137: 170–178.

Khan, M.K., Abert-Vian, M., Fabiano-Tixier, A.S., Dangles, O. and Chemat, F. 2010. Ultrasound-assisted extraction of polyphenols (flavanone glycosides) from orange (*Citrus sinensis* L.) peel. Food Chemistry 119(2): 851–858.

Khan, M.K. and Dangles, O. 2014. A comprehensive review on flavanones, the major citrus polyphenols. Journal of Food Composition and Analysis 33(1): 85–104.

Kim, J.H., Ahn, D., Eun, J. and Moon, S. 2016. Antioxidant effect of extracts from the coffee residue in raw and cooked meat. Antioxidants 5(3): 21.

Klemeš, J.J., Varbanov, P.S., Ocłoń, P. and Chin, H.H. 2019. Towards efficient and clean process integration: utilisation of renewable resources and energy-saving technologies. Energies 12(21): 4092.

Komes, D. and Bušić, A. 2014. Antioxidants in coffee. pp. 25–32. *In*: Processing and Impact on Antioxidants in Beverages. Academic Press.

Kookos, I.K. 2018. Technoeconomic and environmental assessment of a process for biodiesel production from spent coffee grounds (SCGs). Resources, Conservation and Recycling 134: 156–164.

Kourmentza, C., Economou, C.N., Tsafrakidou, P. and Kornaros, M. 2018. Spent coffee grounds make much more than waste: Exploring recent advances and future exploitation strategies for the valorization of an emerging food waste stream. Journal of Cleaner Production 172: 980–992.

Ktori, R., Kamaterou, P. and Zabaniotou, A. 2018. Spent coffee grounds valorization through pyrolysis for energy and materials production in the concept of circular economy. Materials Today: Proceedings 5(14): 27582–27588.

Kuhad, R.C., Gupta, R. and Singh, A. 2011. Microbial cellulases and their industrial applications. Enzyme Research, 2011.

Kumar, A.K., Parikh, B.S. and Pravakar, M. 2016. Natural deep eutectic solvent mediated pretreatment of rice straw: bioanalytical characterization of lignin extract and enzymatic hydrolysis of pretreated biomass residue. Environmental Science and Pollution Research 23: 9265–9275.

Kumari, D. and Singh, R. 2018. Pretreatment of lignocellulosic wastes for biofuel production: A critical review. Renewable and Sustainable Energy Reviews 90: 877–891.

Lachos-Perez, D., Baseggio, A.M., Mayanga-Torres, P.C., Junior, M.R.M., Rostagno, M.A., Martínez, J. and Forster-Carneiro, T. 2018. Subcritical water extraction of flavanones from defatted orange peel. The Journal of Supercritical Fluids 138: 7–16.

Lee, Y.S., Huh, J.Y., Nam, S.H., Moon, S.K. and Lee, S.B. 2012. Enzymatic bioconversion of citrus hesperidin by *Aspergillus sojae* naringinase: Enhanced solubility of hesperetin-7-O-glucoside with in vitro inhibition of human intestinal maltase, HMG-CoA reductase, and growth of *Helicobacter pylori*. Food Chemistry 135: 2253–2259. doi:10.1016/j.foodchem.2012.07.007.

Li, P.J., Xia, J.L., Shan, Y., Nie, Z.Y., Su, D.L., Gao, Q.R., Zhang, C. and Ma, Y.L. 2015. Optimizing production of pectinase from orange peel by *Penicillium oxalicum* PJ02 using response surface methodology. Waste and Biomass Valorization 6: 13–22.

Liew, S.Q., Ngoh, G.C., Yusoff, R. and Teoh, W.H. 2016. Sequential ultrasound-microwave assisted acid extraction (UMAE) of pectin from pomelo peels. International Journal of Biological Macromolecules 93: 426–435.

Liew, S.Q., Ngoh, G.C., Yusoff, R. and Teoh, W.H. 2018a. Acid and deep eutectic solvent (DES) extraction of pectin from pomelo (*Citrus grandis* (L.) *Osbeck*) peels. Biocatalysis and Agricultural Biotechnology 13: 1–11.

Liew, S.Q., Teoh, W.H., Tan, C.K., Yusoff, R. and Ngoh, G.C. (2018b). Subcritical water extraction of low methoxyl pectin from pomelo (*Citrus grandis* (L.) *Osbeck*) peels. International Journal of Biological Macromolecules 116: 128–135.

Liew, S.Q., Teoh, W.H., Yusoff, R. and Ngoh, G.C. 2019. Comparisons of process intensifying methods in the extraction of pectin from pomelo peel. Chemical Engineering and Processing-Process Intensification 143: 107586.

Liu, Z., Qiao, L., Gu, H., Yang, F. and Yang, L. 2017. Development of Brönsted acidic ionic liquid-based microwave assisted method for simultaneous extraction of pectin and naringin from pomelo peels. Separation and Purification Technology 172: 326–337.

M'hiri, N., Ioannou, I., Boudhrioua, N.M. and Ghoul, M. 2015. Effect of different operating conditions on the extraction of phenolic compounds in orange peel. Food and Bioproducts Processing 96: 161–170.

Mahmoodi, M., Najafpour, G.D. and Mohammadi, M. 2019. Bioconversion of agroindustrial wastes to pectinases enzyme via solid state fermentation in trays and rotating drum bioreactors. Biocatalysis and Agricultural Biotechnology 21: 101280.

Mamma, D. and Christakopoulos, P. 2008. Citrus peels: an excellent raw material for the bioconversion into value-added products. Tree and Forestry Science and Biotechnology 2: 83–97.

Marto, J., Gouveia, L.F., Chiari, B.G., Paiva, A., Isaac, V., Pinto, P., Simões, P., Almeida, A.J. and Ribeiro, H.M. 2016. The green generation of sunscreens: Using coffee industrial sub-products. Industrial Crops and Products 80: 93–100.

Mata, T.M., Martins, A.A. and Caetano, N.S. 2018. Bio-refinery approach for spent coffee grounds valorization. Bioresource Technology 247: 1077–1084.

Matharu, A.S., de Melo, E.M. and Houghton, J.A. 2016. Opportunity for high value-added chemicals from food supply chain wastes. Bioresource Technology 215: 123–130.

Mayanga-Torres, P.C., Lachos-Perez, D., Rezende, C.A., Prado, J.M., Ma, Z., Tompsett, G.T., Timko, M.T. and Forster-Carneiro, T. 2017. Valorization of coffee industry residues by subcritical water hydrolysis: recovery of sugars and phenolic compounds. The Journal of Supercritical Fluids 120: 75–85.

Medina-Torres, N., Ayora-Talavera, T., Espinosa-Andrews, H., Sánchez-Contreras, A. and Pacheco, N. 2017. Ultrasound assisted extraction for the recovery of phenolic compounds from vegetable sources. Agronomy 7(3): 47.

Min, B., Lim, J., Ko, S., Lee, K.G., Lee, S.H. and Lee, S. 2011. Environmentally friendly preparation of pectins from agricultural byproducts and their structural/rheological characterization. Bioresource Technology 102(4): 3855–3860.

Mohsin, A., Zhang, K., Hu, J., Salim-ur-Rehman, Tariq, M., Zaman, W.Q., Khan, I.M., Zhuang, Y. and Guo, M. 2018. Optimized biosynthesis of xanthan via effective valorization of orange peels using response surface methodology: A kinetic model approach. Carbohydrate Polymers 181: 793–800.

Mohsin, A., Ni, H., Luo, Y., Wei, Y., Tian, X., Guan, W., Ali, M., Khan, I.M., Niazi, S., Rehman, S. ur, Zhuang, Y. and Guo, M. 2019. Qualitative improvement of camel milk date yoghurt by addition of biosynthesized xanthan from orange waste. Lwt 108: 61–68.

Moncada, J. and Aristizábal, V. 2016. Design strategies for sustainable biorefineries. Biochemical Engineering Journal 116: 122–134.

Murthy, P.S. and Naidu, M.M. 2012. Sustainable management of coffee industry by-products and value addition—A review. Resources, Conservation and Recycling 66: 45–58.

Mussatto, S.I., Carneiro, L.M., Silva, J.P., Roberto, I.C. and Teixeira, J.A. 2011. A study on chemical constituents and sugars extraction from spent coffee grounds. Carbohydrate Polymers 83(2): 368–374.

Narita, Y. and Inouye, K. 2014. Review on utilization and composition of coffee silverskin. Food Research International 61: 16–22.

Naveed, M., Hejazi, V., Abbas, M., Kamboh, A.A., Khan, G.J., Shumzaid, M., Ahmad, F., Babazadeh, D., FangFang, X., Modarresi-Ghazani, F., WenHua, L. and XiaoHui, Z. 2018. Chlorogenic acid (CGA): A pharmacological review and call for further research. Biomedicine & Pharmacotherapy 97: 67–74.

Ndayishimiye, J., Getachew, A.T. and Chun, B.S. 2017. Comparison of characteristics of oils extracted from a mixture of citrus seeds and peels using hexane and supercritical carbon dioxide. Waste and Biomass Valorization 8(4): 1205–1217.

Negro, V., Mancini, G., Ruggeri, B. and Fino, D. 2016. Citrus waste as feedstock for bio-based products recovery: Review on limonene case study and energy valorization. Bioresource Technology 214: 806–815.

Nguyen, Q.A., Cho, E., Trinh, L.T.P., Jeong, J. su and Bae, H.J. 2017. Development of an integrated process to produce D-mannose and bioethanol from coffee residue waste. Bioresource Technology 244: 1039–1048.

Ni, H., Zhang, T., Guo, X., Hu, Y., Xiao, A., Jiang, Z., Li, L. and Li, Q. 2019. Comparison between irradiating and autoclaving citrus wastes as substrate for solid-state fermentation by *Aspergillus aculeatus*. Letters in Applied Microbiology 69: 71–78.

Oberoi, H.S., Vadlani, P.V., Nanjundaswamy, A., Bansal, S., Singh, S., Kaur, S. and Babbar, N. 2011. Enhanced ethanol production from Kinnow mandarin (*Citrus reticulata*) waste via a statistically optimized simultaneous saccharification and fermentation process. Bioresource Technology 102: 1593–1601.

Obruca, S., Petrik, S., Benesova, P., Svoboda, Z., Eremka, L. and Marova, I. 2014. Utilization of oil extracted from spent coffee grounds for sustainable production of polyhydroxyalkanoates. Applied Microbiology and Biotechnology 98: 5883–5890.

Obruca, S., Benesova, P., Kucera, D., Petrik, S. and Marova, I. 2015. Biotechnological conversion of spent coffee grounds into polyhydroxyalkanoates and carotenoids. New Biotechnology 32: 569–574.

Ozturk, B., Parkinson, C. and Gonzalez-Miquel, M. 2018. Extraction of polyphenolic antioxidants from orange peel waste using deep eutectic solvents. Separation and Purification Technology 206: 1–13.

Palmieri, M.G.S., Cruz, L.T., Bertges, F.S., Húngaro, H.M., Batista, L.R., da Silva, S.S., Fonseca, M.J.V., Rodarte, M.P., Vilela, F.M.P. and Amaral, M.D.P.H. 2018. Enhancement of antioxidant properties from green coffee as promising ingredient for food and cosmetic industries. Biocatalysis and Agricultural Biotechnology 16: 43–48.

Panusa, A., Zuorro, A., Lavecchia, R., Marrosu, G. and Petrucci, R. 2013. Recovery of natural antioxidants from spent coffee grounds. Journal of Agricultural and Food Chemistry 61(17): 4162–4168.

Passos, C.P., Moreira, A.S., Domingues, M.R.M., Evtuguin, D.V. and Coimbra, M.A. 2014. Sequential microwave superheated water extraction of mannans from spent coffee grounds. Carbohydrate Polymers 103: 333–338.

Petrik, S., Obruča, S., Benešová, P. and Márová, I. 2014. Bioconversion of spent coffee grounds into carotenoids and other valuable metabolites by selected red yeast strains. Biochemical Engineering Journal 90: 307–315.

Pettinato, M., Casazza, A.A., Ferrari, P.F., Palombo, D. and Perego, P. 2019. Eco-sustainable recovery of antioxidants from spent coffee grounds by microwave-assisted extraction: Process optimization, kinetic modelling and biological validation. Food and Bioproducts Processing 114: 31–42.

Pfaltzgraff, L.A., Cooper, E.C., Budarin, V. and Clark, J.H. 2013. Food waste biomass: a resource for high-value chemicals. Green Chemistry 15(2): 307–314.

Pinto, M.P.M., Mudhoo, A., de Alencar Neves, T., Berni, M.D. and Forster-Carneiro, T. 2018. Co-digestion of coffee residues and sugarcane vinasse for biohythane generation. Journal of Environmental Chemical Engineering 6: 146–155.

Pleissner, D., Qi, Q., Gao, C., Rivero, C.P., Webb, C., Lin, C.S.K. and Venus, J. 2016. Valorization of organic residues for the production of added value chemicals: a contribution to the bio-based economy. Biochemical Engineering Journal 116: 3–16.

Pourbafrani, M., Forgács, G., Horváth, I.S., Niklasson, C. and Taherzadeh, M.J. 2010. Production of biofuels, limonene and pectin from citrus wastes. Bioresource Technology 101(11): 4246–4250.

Prestipino, M., Chiodo, V., Maisano, S., Zafarana, G., Urbani, F. and Galvagno, A. 2017. Hydrogen rich syngas production by air-steam gasification of citrus peel residues from citrus juice manufacturing: Experimental and simulation activities. International Journal of Hydrogen Energy 42: 26816–26827.

Procentese, A., Johnson, E., Orr, V., Garruto Campanile, A., Wood, J.A., Marzocchella, A. and Rehmann, L. 2015. Deep eutectic solvent pretreatment and subsequent saccharification of corncob. Bioresource Technology 192: 31–36.

Procentese, A., Raganati, F., Olivieri, G., Russo, M.E., Rehmann, L. and Marzocchella, A. 2017. Low-energy biomass pretreatment with deep eutectic solvents for bio-butanol production. Bioresource Technology 243: 464–473.

Procentese, A. and Rehmann, L. 2018. Fermentable sugar production from a coffee processing by-product after deep eutectic solvent pretreatment. Bioresource Technology Reports 4: 174–180.

Putnik, P., Bursać Kovačević, D., Režek Jambrak, A., Barba, F.J., Cravotto, G., Binello, A. and Shpigelman, A. 2017. Innovative "green" and novel strategies for the extraction of bioactive added value compounds from citrus wastes—A review. Molecules 22(5): 680.

Qi, X., Zheng, Y., Tang, N., Zhou, J. and Sun, S. 2020. Bioconversion of citrus peel wastes into bioflocculants and their application in the removal of microcystins. Science of Total Environment 715: 136885.

Rahmani, Z., Khodaiyan, F., Kazemi, M. and Sharifan, A. 2020. Optimization of microwave-assisted extraction and structural characterization of pectin from sweet lemon peel. International Journal of Biological Macromolecules 147: 1107–1115.

Rani, M.U., Rastogi, N.K. and Anu Appaiah, K.A. 2011. Statistical optimization of medium composition for bacterial cellulose production by *Gluconacetobacter hansenii* UAC09 using coffee cherry husk extract—an agro-industry waste. Journal of Microbiology and Biotechnology 21: 739–745.

Rani, M.U. and Appaiah, K.A.A. 2013. Production of bacterial cellulose by *Gluconacetobacter hansenii* UAC09 using coffee cherry husk. Journal of Food Science and Technology 50: 755–762.

Ravindran, R., Jaiswal, S., Abu-Ghannam, N. and Jaiswal, A.K. 2017. Two-step sequential pretreatment for the enhanced enzymatic hydrolysis of coffee spent waste. Bioresource Technology 239: 276–284.

Ribeiro, H., Marto, J., Raposo, S., Agapito, M., Isaac, V., Chiari, B.G. and Simões, P. 2013. From coffee industry waste materials to skin-friendly products with improved skin fat levels. European Journal of Lipid Science and Technology 115: 330–336.

Rocha, C.M., Genisheva, Z., Ferreira-Santos, P., Rodrigues, R., Vicente, A.A., Teixeira, J.A. and Pereira, R.N. 2018. Electric field-based technologies for valorization of bioresources. Bioresource Technology 254: 325–339.

Rocha, M.V.P., de Matos, L.J.B.L., de Lima, L.P., da Silva Figueiredo, P.M., Lucena, I.L., Fernandes, F.A.N. and Gonçalves, L.R.B. 2014. Ultrasound-assisted production of biodiesel and ethanol from spent coffee grounds. Bioresource Technology 167: 343–348.

Rodríguez-Meizoso, I., Jaime, L., Santoyo, S., Señoráns, F.J., Cifuentes, A. and Ibáñez, E. 2010. Subcritical water extraction and characterization of bioactive compounds from *Haematococcus pluvialis* microalga. Journal of Pharmaceutical and Biomedical Analysis 51(2): 456–463.

Rodsamran, P. and Sothornvit, R. 2019. Microwave heating extraction of pectin from lime peel: Characterization and properties compared with the conventional heating method. Food Chemistry 278: 364–372.

Ruiz, H.A., Rodríguez-Jasso, R.M., Rodríguez, R., Contreras-Esquivel, J.C. and Aguilar, C.N. 2012. Pectinase production from lemon peel pomace as support and carbon source in solid-state fermentation column-tray bioreactor. Biochemical Engineering Journal 65: 90–95.

Safarik, I., Horska, K., Svobodova, B. and Safarikova, M. 2012. Magnetically modified spent coffee grounds for dyes removal. European Food Research and Technology 234(2): 345–350.

Santos da Silveira, J., Durand, N., Lacour, S., Belleville, M.P., Perez, A., Loiseau, G. and Dornier, M. 2019. Solid-state fermentation as a sustainable method for coffee pulp treatment and production of an extract rich in chlorogenic acids. Food and Bioproducts Processing 115: 175–184.

Seifollahi, M. and Amiri, H. 2018. Phosphoric acid-acetone process for cleaner production of acetone, butanol, and ethanol from waste cotton fibers. Journal of Cleaner Production 193: 459–470.

Severini, C., Derossi, A. and Fiore, A.G. 2017. Ultrasound-assisted extraction to improve the recovery of phenols and antioxidants from spent espresso coffee ground: a study by response surface methodology and desirability approach. European Food Research and Technology 243(5): 835–847.

Shankar, K., Kulkarni, N.S., Jayalakshmi, S.K. and Sreeramulu, K. 2019. Saccharification of the pretreated husks of corn, peanut and coffee cherry by the lignocellulolytic enzymes secreted by *Sphingobacterium* sp. ksn for the production of bioethanol. Biomass and Bioenergy 127: 105298.

Sharma, K., Mahato, N., Cho, M.H. and Lee, Y.R. 2017. Converting citrus wastes into value-added products: Economic and environmentally friendly approaches. Nutrition 34: 29–46.

Shilpi, A., Shivhare, U.S. and Basu, S. 2013. Supercritical CO_2 extraction of compounds with antioxidant activity from fruits and vegetables waste—a review. Focusing on Modern Food Industry 2(1): 43–62.

Silva, I.M., Gonzaga, L.V., Amante, E.R., Teófilo, R.F., Ferreira, M.M. and Amboni, R.D. 2008. Optimization of extraction of high-ester pectin from passion fruit peel (*Passiflora edulis flavicarpa*) with citric acid by using response surface methodology. Bioresource Technology 99(13): 5561–5566.

Singh, R. and Kumar, S. (eds.). 2017. Green Technologies and Environmental Sustainability. Springer.

Srivastava, N., Srivastava, M., Manikanta, A., Singh, P., Ramteke, P.W., Mishra, P.K. and Malhotra, B.D. 2017. Production and optimization of physicochemical parameters of cellulase using untreated orange waste by newly isolated *Emericella variecolor* NS3. Applied Biochemistry and Biotechnology 183: 601–612.

Stahel, W.R. 2016. The circular economy. Nature News 531(7595): 435.

Su, D.L., Li, P.J., Quek, S.Y., Huang, Z.Q., Yuan, Y.J., Li, G.Y. and Shan, Y. 2019. Efficient extraction and characterization of pectin from orange peel by a combined surfactant and microwave assisted process. Food Chemistry 286: 1–7.

Su, H., Tan, F. and Xu, Y. 2016. Enhancement of biogas and methanization of citrus waste via biodegradation pretreatment and subsequent optimized fermentation. Fuel 181: 843–851.

Tandon, M., Thakur, V., Tiwari, K.L. and Jadhav, S.K. 2018. *Enterobacter ludwigii* strain IF2SW-B4 isolated for bio-hydrogen production from rice bran and de-oiled rice bran. Environmental Technology and Innovation 10: 345–354.

Tao, N., Shi, W., Liu, Y. and Huang, S. 2011. Production of feed enzymes from citrus processing waste by solid-state fermentation with *Eupenicillium javanicum*. International Journal of Food Science and Technology 46: 1073–1079.

Torquato, L.D.M., Pachiega, R., Crespi, M.S., Nespeca, M.G., de Oliveira, J.E. and Maintinguer, S.I. 2017. Potential of biohydrogen production from effluents of citrus processing industry using anaerobic bacteria from sewage sludge. Waste Management 59: 181–193.

Torres-Valenzuela, L.S., Serna-Jiménez, J.A. and Martínez, K. 2019. Coffee by-Products: Nowadays and Perspectives. In Coffee. IntechOpen.

Trabelsi, D., Aydi, A., Zibetti, A.W., Della Porta, G., Scognamiglio, M., Cricchio, V., Langa, E., Abderrabba, M. and Mainar, A.M. 2016. Supercritical extraction from *Citrus aurantium amara* peels using CO_2 with ethanol as co-solvent. The Journal of Supercritical Fluids 117: 33–39.

Tsouko, E., Kourmentza, C., Ladakis, D., Kopsahelis, N., Mandala, I., Papanikolaou, S., Paloukis, F., Alves, V. and Koutinas, A. 2015. Bacterial cellulose production from industrial waste and by-product streams. International Journal of Molecular Sciences 16(7): 14832–14849.

Turnbull, D., Rodricks, J.V., Mariano, G.F. and Chowdhury, F. 2017. Caffeine and cardiovascular health. Regulatory Toxicology and Pharmacology 89: 165–185.

USDA. 2018/2019. https://apps.fas.usda.gov/psdonline/circulars/citrus.pdf.

U.S. Energy Information Administration: International Energy Outlook 2017 (IEO, 2017) www.eia.gov/ieo.

Vardon, D.R., Moser, B.R., Zheng, W., Witkin, K., Evangelista, R.L., Strathmann, T.J., Rajagopalan, K. and Sharma, B.K. 2013. Complete utilization of spent coffee grounds to produce biodiesel, bio-oil, and biochar. ACS Sustainable Chemistry & Engineering 1(10): 1286–1294.

Villa Montoya, A.C., Cristina da Silva Mazareli, R., Delforno, T.P., Centurion, V.B., Sakamoto, I.K., Maia de Oliveira, V., Silva, E.L. and Amâncio Varesche, M.B. 2019. Hydrogen, alcohols and volatile fatty acids from the co-digestion of coffee waste (coffee pulp, husk, and processing wastewater) by applying autochthonous microorganisms. International Journal of Hydrogen Energy 44: 21434–21450.

Wei, Y., Li, J., Shi, D., Liu, G., Zhao, Y. and Shimaoka, T. 2017. Environmental challenges impeding the composting of biodegradable municipal solid waste: A critical review. Resources, Conservation and Recycling 122: 51–65.

Wikandari, R., Millati, R., Cahyanto, M.N. and Taherzadeh, M.J. 2014. Biogas production from citrus waste by membrane bioreactor. Membranes (Basel) 4: 596–607.

Wu, H., Hu, W., Zhang, Y., Huang, L., Zhang, J., Tan, S., Cai, X. and Liao, X. 2016. Effect of oil extraction on properties of spent coffee ground-plastic composites. Journal of Materials Science 51(22): 10205–10214.

Xu, H., Wang, W., Liu, X., Yuan, F. and Gao, Y. 2015. Antioxidative phenolics obtained from spent coffee grounds (*Coffea arabica* L.) by subcritical water extraction. Industrial Crops and Products 76: 946–954.

Xu, D.P., Li, Y., Meng, X., Zhou, T., Zhou, Y., Zheng, J. and Li, H.B. 2017. Natural antioxidants in foods and medicinal plants: Extraction, assessment and resources. International Journal of Molecular Sciences 18(1): 96.

Zabed, H., Sahu, J.N., Suely, A., Boyce, A.N. and Faruq, G. 2017. Bioethanol production from renewable sources: Current perspectives and technological progress. Renewable and Sustainable Energy Reviews 71: 475–501.

Zema, D.A., Calabrò, P.S., Folino, A., Tamburino, V., Zappia, G. and Zimbone, S.M. 2018. Valorisation of citrus processing waste: A review. Waste Management 80: 252–273

Zhao, X., Cheng, K. and Liu, D. 2009. Organosolv pretreatment of lignocellulosic biomass for enzymatic hydrolysis. Applied Microbiology and Biotechnology 82: 815–827.

Zouambia, Y., Ettoumi, K.Y., Krea, M. and Moulai-Mostefa, N. 2017. A new approach for pectin extraction: Electromagnetic induction heating. Arabian Journal of Chemistry 10(4): 480–487.

CHAPTER 8

Green Technologies in Food Processing

Sofia Chanioti, Paraskevi Siamandoura and *Constantina Tzia**

1. Introduction

Plant materials include fruits, vegetables, medicinal and aromatic herbs, and their wastes and by-products (Putnik et al. 2018). The antioxidant activity of these plant sources has been attributed to their bioactive compounds' content, including vitamins, sterols, squalene, various phenolic compounds, flavonoids, and others (Žlabur et al. 2016). In the human body, phenolic compounds exhibit their positive effect through several mechanisms: (i) free radicals neutralization (antioxidant effect); (ii) protection and regeneration of other antioxidants (vitamin E); and (iii) chelating properties of the oxidizing metal ions (bind themselves to metal ions and thus inhibit its absorption and utilization) (Garcia-Salas et al. 2010). It should be noted that the antioxidant activity of the phenolic compounds primarily depends on the number of hydroxyl groups (—OH), i.e., their position in the molecule, as well as on the type of substitution on the aromatic ring (Vo et al. 2018).

The mentioned compounds are widely used as nutritional supplements in food products by means of enrichment processes. Synthetic antioxidants are widely used because of their stability and their widespread availability; however, they are related to mutagenic and carcinogenic effects and this has led to the search for natural antioxidant compounds extracted from plant materials (Soquetta et al. 2018). The idea of sustainability has become increasingly prominent worldwide, and it is applied and discussed in the academic and industrial sectors. In this way, the food industry has to become conscious of sustainable green food production and alternative food processing techniques. There is an increasing public awareness of health, environment, and safety hazards associated with the organic solvents' use

Laboratory of Food Chemistry and Technology, School of Chemical Engineering, National Technical University of Athens, 5 Iroon Polytechniou St., 15780, Zografou, Greece.
* Corresponding author: tzia@chemeng.ntua.gr

in food processes, including extraction for recovery/delivery of food components, i.e., vegetable oils or bioactive substances. The high cost of organic solvents and the increasingly stringent environmental regulations have pointed out the need for the development of new and clean technologies for food product processing. The new and environmentally friendly techniques and their combinations become more than important. These innovative techniques have been focused to eliminate energy consumption, solvent volume, and processing time by applying the regulations on food safety and control (Uzel 2018).

The extraction of bioactive compounds depends on several factors, such as the plant source, the extraction technique, and the extraction solvent (Tiwari 2015). The techniques can be classified into conventional or non-conventional. Conventional techniques require the use of organic solvents, temperature, and agitation, including Soxhlet and maceration. Several new assisted extractions methods, such as homogenization (HAE), microwave heating (MAE), ultrasound (UAE), high hydrostatic pressure (HHPAE), pulsed electric field (PEF), high-voltage electrical discharge (HVED), infrared radiation (IR), and accelerated conditions of temperature and pressure (ASE) are currently investigated in the extraction of bioactive compounds from various plant resources (Chanioti and Tzia 2018, Cheaib et al. 2018, Jerman and Mozetic 2010, Yang et al. 2010, Liu et al. 2013, Zhang et al. 2011). Moreover, a new generation of green solvents, named deep eutectic solvents can have the potential to act as effective solvents for the extraction of a wide range of non-polar and polar compounds and thus have been proposed as alternatives to several conventional and toxic organic solvents. There is an increasing number of studies on the extraction of bioactive plant compounds, including flavonoids, catechin, and phenolic acids by applying NADESs as extraction media (Bi et al. 2013, Bubalo et al. 2016, Huang et al. 2016, Paradiso et al. 2016, Qi et al. 2015, Ruesgas-Ramon et al. 2017, Wei et al. 2015, Yao et al. 2015).

The aim of this chapter was to discuss the mechanisms, principles, and applications of green solvents emerging and technologies in the extraction of bioactive compounds, including sterols, squalene, phenolic compounds, and flavonoids from plant sources offering an updated glimpse of the huge technological effort which is being developed at educational and industrial scale.

2. Green Solvents

The most common solvents used in the extraction processes are organic, including ethanol, diethyl ether, *N,N*dimethylformamide, hexane, toluene, and their aqueous solutions (Byrne et al. 2016). Many of the extraction methods, especially the conventional ones, require large solvent volume and produce a high amount of toxic generated waste that possesses a negative impact on the environment as well as human health and safety (Vian et al. 2017). Nowadays, current regulations have a widely progressive and direct impact on diminishing consumption of petrochemical solvents and Volatile Organic Compounds (VOCs). Therefore, manufacturers using organic solvents should certify the absence of risk during the extraction process and demonstrate the safety of the received target compounds in regard to solvent traces.

For these reasons, the reduction of organic solvent consumption and the development of greener solvents has been strongly encouraged (Cvjetko Bubalo et al. 2018).

The solvents can be characterized as green if they are non-toxic, non-volatile, recyclable, biodegradable, and do not involve a high energy cost of synthesis (Das et al. 2017). Nowadays, four main categories include green solvents, supercritical fluids, neoteric, and bio-based and supramolecular solvents. It should be noted that novel challenges and limitations can arise with the replacement of an organic solvent with a green one because of the different physicochemical properties of the latter solvents.

2.1 Supercritical Fluids

The supercritical fluid is in an intermediate stage between a gas and a liquid with a liquid-like density and gas-like viscosity (Sihvonen et al. 1999, Wang et al. 2008). Supercritical fluids have improved mass transfer properties with respect to liquids because their density is regulated by varying the pressure and temperature. In addition, supercritical fluids have a relatively low surface tension, thereby avoiding the degradation of susceptible compounds.

Many fluids (ethylene, methane, nitrogen, or fluorocarbons) can be used as supercritical fluids, but most separation systems use carbon dioxide and water. Supercritical carbon dioxide extraction takes place in the supercritical CO_2 area at a temperature and pressure above the respective critical CO_2 gas (31.1°C and 7.4 MPa) and is an alternative to the use of organic solvents (Herrero et al. 2006). The advantages are that CO_2 is a non-explosive, non-toxic gas, non-carcinogenic, non-mutagenic, non-flammable, thermodynamically stable low cost, solubilizes lipophilic compounds and can be readily recovered by providing solvent-free extracts (Wang et al. 2008, Knez et al. 2014). In addition, this gas is environmentally friendly and "generally recognized as safe" (GRAS) by the Food and Drug Administration (FDA) and EFSA (European Food Safety Authority).

Because of its low polarity, CO_2 is not effective in extracting highly polarized components. The polar components dissolve barely in the supercritical CO_2 and are practically not extracted. For this reason, co-solvents (hexane, methanol, ethanol, isopropanol, acetonitrile, and dichloromethane) are used in small amounts to increase the solubility of the components and the selectivity of the process (Sihvonen et al. 1999). Among the mentioned co-solvents, ethanol is usually preferred because of its low toxicity and good miscibility with CO_2 (Liza et al. 2010).

Supercritical water exists at temperatures above 374°C and pressures above 22.1 MPa and is considered the cleanest solvent. Under these extreme conditions, the hydrogen bonding is lost, so the supercritical water behaves as a nonpolar solvent (DeSimone 2002). SFE using water has been applied in the extraction of bioactive compounds from plant materials, such as quercetin from onion waste. However, the use of supercritical water is connected with corrosion problems limiting its application on an industrial scale. An alternative is considered the use of pressurized hot water extraction (PHWE) using water at temperatures above its boiling point and atmospheric pressure (Plaza and Turner 2015) (Section 3.8).

The advantages of using supercritical liquids for the extraction of bioactive compounds can be understood by considering the following:

1) The supercritical fluid has a higher diffusion coefficient, lower viscosity, and less surface tension than a liquid solvent, leading to greater penetration into the solid matrix and enhancing the mass transfer. The extraction time is significantly reduced by the application of SFE compared to conventional methods.

2) The selectivity of a supercritical fluid is higher than that of a liquid solvent since it is possible to adjust its solubilization by changing the temperature and/ or pressure.

2.2 *Neoteric Solvents*

Neoteric solvents are novel solvents in terms of their structure and they are characterized by physical and chemical properties (Gutiérrez-Arnillas et al. 2016). Neoteric solvents are classified into two groups: ionic liquids and eutectic solvents.

2.2.1 *Ionic Liquids*

Ionic liquids (ILs) are a class of salts composed of discrete cations and anions with melting points below 100°C (Henderson et al. 2011). They have unique physicochemical properties, pre-organized and tunable solvent structures, negligible vapor pressure, excellent thermal and chemical stability, wide electrochemical potential window and outstanding solubility for organic, and inorganic and organometallic substances. Due to the above, ionic liquids are considered interesting materials for extraction (Henderson et al. 2011, Passos et al. 2014, Ventura et al. 2017). Until now, the use of ILs in food methods is not regulated by the Federal Drug Administration, FDA (Martins et al. 2017). 1-alkyl-3-methylimidazolium-based ILs are the most studied ILs and are usually combined with [BF4], Cl–, and Br– counterions. The application of "greener" ILs, e.g., ammonium-based cations, such as cholinium, is still rare (Ventura et al. 2017).

In Table 1, some applications of ILs for the extraction of bioactive compounds from plant materials are presented. For the extraction using ILs, high temperature (up to 140°C) and long extraction time (2 hours) are applied due to their high viscosity as well as to enhance the mass transfer process and fluid flow (Khan et al. 2018) and to promote the dissolution of biomass (Hou et al. 2015). The most investigated parameters for ILs extraction methods are the concentration and the composition as they consist of various components. A special advantage of ILs for the extraction of bioactive compounds is their ability to permeate and modify the cell walls and tissues and facilitate the release of the target compounds. Protic ILs may facilitate the hydrolysis of polysaccharides and other components for cell lysis via strong hydrogen bonding. This has been exploited for the extraction of asthaxanthin for algae and levulinic acid from lignocellulosic biomass (Shankar et al. 2017, Khan et al. 2018). Acidic ionic liquids have been proposed for improving the hydrolysis of lignocellulosic materials (Li et al. 2008). The extraction of levunilic acid also involved a catalytic method favored by acidic ILs (Khan et al. 2018). Recently, Kiyonga et al. (2021) examined the back-extraction to recover target components, such as coumarins oxypeucedanin

Table 1. Extraction of bioactive compounds from plant materials using ILs.

Plant material	Bioactive compound	Type of ILs	Extraction conditions	Extraction efficiency	Reference
Corn stalk	Total reducing sugars	C4mimBr, C4mimCl, C4mimHSO₄, C6mimCl, 1-Allyl-3-methylimidazolium chloride, **C4mimCl**	100°C, 60 minutes, HCl/sample ratio 7%	71%	(Li et al. 2008)
Soybean hulls	Reducing sugars	[C4(Mim)2] hydrogen sulfate + pretreatment with 1-allyl-3-imidazolium chloride [AMIM]Cl	95°C, 1 hour; ultrasonic-assisted extraction; water/sample 20:1	275.4 mg/g	(Hu et al. 2014)
Rice husk	Levulinic acid	[C4(Mim)2][(2HSO₄)(H₂SO₄)0], **[C4(Mim)2][(2HSO₄)(H₂SO₄)₂]** C4(Mim)2][(2HSO₄)(H₂SO₄)₄]	110°C, 60 minutes IL:water 10:1	47.52%	(Khan et al. 2018)
Olive tree leaves	Oleanolic acid	[C6mim]Cl, [C8mim]Cl, [C10mim]Cl, **[C12mim]Cl,** [C12mim]Br, [C12mim]I, **[C14mim]Cl,** [C16mim]Cl and [C18mim]Cl	80°C for 2 hours or microwave assisted extraction for 30 minutes; IL in water (500 mM)	2.5%	(Cláudio et al. 2018)
Olive mill wastewater	Tyrosol	**[P4441][Tf2N],** [N4441][[Tf2N], and [N8881] [Tf2N]	30°C, 2 hours	78%	(Larriba et al. 2016)
Coconut husk	Cellulose, lignin	[N2220][HSO₄]	120°C, 2 hours	56.5% (cellulose) 12.8% (lignin)	(Zahari et al. 2018)
Sugarcane bagasse	Lignin	C3mim acetate	140°C, 120 minutes	90.1%	(Saha et al. 2017)
Angelica dahurica	Oxypeucedamin Hydrate, Byakangelicin	**[Bmin]Tf₂N** [Bmim]PF6	Solvent/solid ratio 8:1, 60°C 180 minutes	98.06% (Oxypeucedamin Hydrate) 99.52% (Byakangelicin)	(Kiyonga et al. 2021)

Optimal ILs shown in bold; Cmim: 1-alkyl-3-methylimidazolium cation; Cnmim: 1-alkyl-3-methylimidazolium cation; [Tf2N]: bis(tri-fuoromethylsulfonyl)imide anion; [N2220]: Triethylammonium cation; [N4441]: tributyl(methyl)phosphonium cation; [N8881]: tricaprylmethylammonium; [P441]: tributylmethylphosphonium cation; [Bmin]: 1-butyl-3-methylimidazolium; PF6: Hexafluorophosphate anion hydrate and byakangelicin from *Angelica dahurica* by using ILs. They demonstrated that the imidazolium-based hydrophobic IL [Bmim]Tf2N resulted in an outstanding performance for extracting.

However, their further practical use has been limited mainly due to their high costs and potential toxicity. The development of more environment-friendly ILs for the extraction process is still under investigation (Passos et al. 2014). In order to reduce the operation costs, some researchers used co-solvents such as methanol or reused solvents (Cooney and Benjamin 2016, Khan et al. 2018).

2.2.2 Deep Eutectic Solvents

Deep Eutectic Solvents (DESs) were developed to overcome the environmental issues of ILs. They have physical and chemical properties comparable to ionic liquids, but they are easier to be synthesized and more stable, environmental-friendly and low-cost (Zhang et al. 2012, Satlewal et al. 2018). In the green extraction process, DESs have shown great potential.

A DES is usually composed of a mixture of two inexpensive and safe components; a hydrogen bond acceptor, usually choline chloride, and a hydrogen bond donor, such as amino acids, carboxylic acids, sugars, etc., in a solid-state. Such components form a eutectic mixture with a lower melting point than that of each component (Wei et al. 2015). In particular, in the case of the combination of certain components which are present in nature and play a role in metabolic processes of living cells, such as choline chloride, urea, organic acids and sugars, the DES mixture is called "natural deep eutectic solvent" (NADES) (Dai et al. 2013, García et al. 2016). NADESs have unique physicochemical properties, i.e., adjustable viscosity, low volatility, solubility in water, and a high degree of solubilization strength for various compounds, while they also have many advantages, such as easy preparation, non-flammability, and pharmaceutically acceptable toxicity (Dai et al. 2013, García et al. 2016, Huang et al. 2016, Qi et al. 2015). However, the preparation and application of NADESs should always be accompanied by toxicity studies to enable scientists and technologists to develop a truly environmentally friendly process rather than simply hypothesizing that non-toxic raw materials used in their preparation could lead to the formation of non-toxic solvents (Huang et al. 2017, Radošević et al. 2016). Therefore, they can have the potential to act as effective solvents for the extraction of a wide range of non-polar and polar compounds and thus have been proposed as alternatives to several conventional and toxic organic solvents. There is an increasing number of studies on the extraction of bioactive plant compounds, including flavonoids, catechin and phenolic acids by applying NADESs as extraction media (Table 2).

The most important physicochemical properties of DESs that significantly influenced the extraction are polarity, viscosity, and surface tension. The high viscosity and surface tension of DES is major disadvantages since they reduce the mass transfer rate of bioactive compounds (Chanioti and Tzia 2018). Viscosity and surface tension can be reduced by the use of different hydrogen bond donors (i.e., sugars, polyhydric alcohols and organic acids), by increasing the extraction temperature and by mixing DES with water. In addition, water addition increases the polarity of DES (Huang et al. 2017). Sugar-based DESs have the greatest viscosity and higher polarity

Table 2. Extraction of bioactive compounds from plant materials using DES.

Plant Material	Bioactive Compound	Type of DES	Extraction Conditions	Extraction Efficiency	Reference
Olive, almond, sesame, cinnamon	Phenolic compounds	ChCl: glycol, ChCl: glycerol	Ultrasound-assisted extraction, 5 minutes	27–83 µg/L	(Khezeli et al. 2016)
Crude palm oil	Tocopherols	ChCl: acetic acid, ChCl: malonic acid, ChCl: citric acid	3 hours IL diluted in methanol, sample diluted in hexane	14,689–18,525 mg/Kg	(Hadi et al. 2015)
Wine lees (*Merlot grapes*)	Total anthocyanins and related compounds	ChCl: citric acid, ChCl: malic acid, ChCl: oxalic acid, ChCl: glucose, ChCl: fructose, ChCl: xylose, ChCl: glycerol	30 minutes and UAE Water in NADES 35.4 w/w	5.2–6.5 mg/g (total anthocyanins)	(Bosiljkov et al. 2017)
Catharanthus roseus	anthocyanins	1,2–propanediol–ChCl **LA–glucose** proline–malic acid malic acid–ChCl glucose–ChCl glucose–fructose–sucrose	Stirring and UAE at 40°C for 30 minutes		(Dai et al. 2016)
Pigeon pea roots	Genistin, genistein apigenin	ChCl: sucrose ChCl:1,2-propanediol, ChCl: glucose, ChCl: sorbitol, ChCl: glycol, ChCl: glycerol, ChCl:1,3-Butanediol, ChCl: 1,4-Butanediol, **ChCl: 1,6- Hexanediol,** glucose: L-proline, glucose: lactid acid	11 minutes, 80°C, microwave assisted extraction	0.449 mg genistin/g, 0.617 mg genistein/g and 0.221 mg apigenin/g	(Cui et al. 2015)
Kappaphycus alvarezii	k-carrageenan	ChCl.: Urea, ChCl: Glycol ChCl-Glycerol	autoclaved at 107°C for 1.5 hours	30.9–60.3%	(Das et al. 2016)

Rice straw	Lignin	**LA: betaine; LA: ChCl**	12 h, 60°C DES with 5% water v/v	68 mg/g	(Kumar et al. 2016)
Lonicerae japonicae Flos	Phenolic compounds	ChCl–1,2-propanediol, ChCl–glycerol, ChCl–glycol, choline **ChCl–1,3 butanediol,** ChCl–1,4 butanediol, ChCl–urea, ChCl–propanedioic acid, ChCl–glucose, ChCl-sorbitol, oxalic acid–sucrose, LA–sucrose LA–glucose	Microwave, hot reflux, ultrasonic irradiation solid/liquid ratio (5/1–15/1), temperature (40–80°C) and time (10–30 minutes)		(Peng et al. 2016)
Lemon peels, olive leaves, onion solid wastes, red grape pomace, spent filter coffee and wheat bran	Polyphenols	**ChCl: glycerol: sodium acetate,** glycerol: sodium-potassium, tartrate: water	90 minutes, 80°C, ultrasound-assisted extraction DES and 10% water	88.03 mg GAE/g in onion solid wastes (with glycerol: sodium–potassium, tartrate:water), 53.76 mg GAE/g in lemon waste peels, 36.75 mg GAE/g in olive leaves, 53.63 mg GAE/g in red grape pomace, 22.59 mg GAE/g in spent coffee grounds and 17.78 mg GAE/g in wheat bran	(Mouratoglou et al. 2016)
Orange Peel Waste	Phenolic compounds	**ChCl: glycol,** ChCl: glycerol	Conventional extraction 100 minutes at 40°C	3.61 mg GAE/g	(Ozturk et al. 2018)

Table 2 contd. ...

...Table 2 contd.

Plant Material	Bioactive Compound	Type of DES	Extraction Conditions	Extraction Efficiency	Reference
Corncob	Fermentable sugar	**ChCl: glycerol, ChCl:imidazole,** ChCl: urea	15 hours, 80°C (ChCl: imidazole) 15 hours, 180°C (ChCl:glicerol) Washing and evaporation to remove DES + enzymatic Treatment	Glucose 91.5–92.3% Xylose 59.5–95.5%	(Procentese et al. 2015)
Grape skins	Total phenolic content and total anthocyanins	ChCl: glucose, ChCl: fructose, ChCl: xylose, ChCl: glycerol, **ChCl: malic acid**	50 minutes, 65°C, ultrasound-assisted extraction	91 mg/g polyphenols and 24 mg/g anthocyanins	(Radošević et al. 2016)
Cajanus cajan leaves	Phenolic compounds (*n* = 14)	ChCl: glycerol, ChCl: 1,4-butanediol, ChCl: ethylene glycol, ChCl: glucose ChCl: sucrose, **ChCl: maltose,** ChCl: sorbitol, ChCl: citric acid, ChCl: malic acid, ChCl: lactic acid, citric acid: glucose, citric acid: sucrose, lactic acid: glucose, lactic acid: sucrose	12 minutes, 60°C, microwave-assisted extraction DES and 20% water	Stilbenes cajaninstilbene acid 6.9 mg/g; longistyline C 4.4 mg/g	(Wei et al. 2015)

Corncob	Lignin removal and glucose yield	ChCl: lactic acid, ChCl: glycolic acid, ChCl: levulinic acid, ChCl: malonic acid, ChCl: glutaric acid, ChCl: oxalic acid, ChCl: malic acid, ChCl: ethylene glycol, **ChCl: glycerol**	24 hours, 90°C Enzymatic treatment	71.3% (lignin); 96.4% (glucose)	(Zhang et al. 2016)
Grape skins	Flavonoids	ChCl: glycerol ChCl: oxalic acid ChCl: malic acid ChCl: sorbose **ChCl: proline: malic acid**	50 minutes, 65°C, ultrasound-assisted extraction DES and 25% water	~ 25 mg/g (sum of anthocyanins and cynidine-3-O-glucosides)	(Cvjetko Bubalo et al. 2016)
Morus alba L. leaves	Phenolic compounds	ChCl: Urea, ChCl: Ethylene glycol, ChCl: Glycerol, **ChCl: Citric acid,** ChCl: malic acid, Betaine: levulinic acid, betaine: lactic acid, betaine: glycerol, proline: malic acid, proline: Glycerol, L-proline: levulinic acid, L-proline: lactic acid	30 minutes, 40°C, ultrasonic-assisted extraction DES:water 3:1 v/v	22.66 mg/g	(Zhou et al. 2018)
Olive pomace	Total phenolic content	**ChCl: citric acid,** ChCl: lactic acid, ChCl: maltose, ChCl: glycerol	30 minutes, 60°, homogenate-assisted extraction 20% v/v water	35 mg/g	(Chanioti and Tzia 2018)

Table 2 contd. ...

...Table 2 contd.

Plant Material	Bioactive Compound	Type of DES	Extraction Conditions	Extraction Efficiency	Reference
Spent Coffee grounds	Total phenolic content	ChCl: urea, ChCl: acetamide, ChCl: glycerol, ChCl: sorbitol, ChCl: ethylene glycol, ChCl: 1,4-Butanediol, **ChCl: 1,6-hexanediol,** ChCl: malonic acid, ChCl: citric acid, ChCl: fructose, ChCl: xylose, ChCl: sucrose, ChCl: glucose	45 minutes, ultrasonic assisted extraction DES and 30% water	15 mg/g	(Yoo et al. 2018)
dittany, fennel, marjoram, mint and sage	Phenolic compounds	LA:ChCl, LA:sodium acetate, LA:ammonium acetate and **LA:glycine:water**	80°C for 90 minutes, in a temperature-controlled, sonication bath DES and 20% water	115.4 mg/g d.w for dittany, 34.72 mg/g d.w for fennel, 137.36 mg/g d.w for marjoram, 109.67 mg/g d.w for mint, 114.92 mg/g d.w for sage	(Bakirtzi et al. 2016)
CO leaves	Flavonoids (myricetin and amentoflavone)	ChCl: Ethyl glycol ChCl: Glycerol ChCl: 1,2-Butanediol ChCl: 1,3-Butanediol **ChCl: 1,4-Butanediol** ChCl: 2,3-Butanediol ChCl: 1,6-Hexanediol	70.0°C for 40.0 minutes and a solid/liquid ratio of 1/1 DES and 35 vol% water	0.031 and 0.518 mg/g of myricetin and amentoflavone	(Bi et al. 2013)

virgin olive oil	Phenolic compounds	ChCl: Glycerol ChCl: LA ChCl: Urea ChCl: Sucrose ChCl: 1,4-Butanediol **ChCl: Xylitol** **ChCl: 1,2-Propanediol** ChCl: Malonic acid ChCl: Urea:Glycerol Fructose:Glucose:Sucrose	40°C with agitation for 1 hour liquid/liquid ratio 1/1	higher proportions than with methanol/water, with an increased yield of 20–33% (Choline chloride:Xylitol) and 67.9–68.3% (Choline chloride:1,2-Propanediol)	(Garcia et al. 2016)
Equisetum palustre L.	flavonoids	**ChCl/betaine hydrochloride:ethylene glycol**	20% water content extraction pressure–0.07 MPa, extraction temperature 60°C, solvent/solid ratio 25:1 mL/g, and the extraction time, 20 min	content of flavonoids 0.17% to 3.37%, and the recovery yields 57.14% to 89.25%	(Qi et al. 2015)
Sanghuangporus baumii	Phenolic compounds	**ChCl/malic acid**	42 minutes, 58°C 1:34 solid/solvent ratio (mg/mL), 39% water content	Polyphenols yield 12.58 mg/g, Extraction yield 6.37 mg/g	(Zheng et al. 2021)
Hazelnut skin	Phenolic compounds	**ChCl/ LA (1:2)**	Water content 35%, 80°C, solid/solvent ratio 1:25g/mL	Phenolic recovery 39% more than organic solvent (12.92 gGAE/100 g skin)	(Zheng et al. 2021)
Sideritis scardica and *Plantago major* L.	Phenolic compounds	ChCl/Glycerol **ChCl/Glucose** ChCl/1,2-Propanediol Citric acid/1,2-Propanediol	Ultrasound bath for 1 hour, 1.34 mg/mL solid/ solvent ratio	Total phenolics: 6.30% for *Sideritis scardica* and 6.99% for *Plantago major* L.	(Grozdanova et al. 2020)

Table 2 contd. ...

...Table 2 contd.

Plant Material	Bioactive Compound	Type of DES	Extraction Conditions	Extraction Efficiency	Reference
Olive leaves	Phenolic compounds	Glucose/Fructose/Sucrose Glucose/Gructose Glucose/Sucrose Fructose/Sucrose ChCl/glucose **ChCl/fructose** ChCl/sucrose ChCl/LA ChCl/malonic acid ChCl/ethylene glycol ChCl/glycerol	Ultrasound bath at 65°C for 60 minutes, 8.61–90% DES, 1:50 mg/mL solid/solvent ratio	Total phenolics: 187.31 mg/g d.w and Total flavonoids: 12.75 mg apigenin equivalent/g d.w	(Ünlü 2021)
Polygonatum odoratum	Flavonoids	ChCl/malic acid **ChCl/LA** ChCl/citric acid Glycine betaine/malic acid Glycine betaine/LA Glycine betaine/citric acid	Ultrasound bath at 51°C, 27% DES water content, 22:1 mL/g solvents/solid ratio, for 21 minutes.	Total flavonoids: 11.47 mg/g	(Xia et al. 2021)
Phellodendri amurensis cortex	Bioactive alkaloids *berberine* and *palmatine*	ChCl/malic acid **ChCl/LA** ChCl/citric acid ChCl/laevulinic acid ChCl/succinic acid ChCl/ethylene glycol ChCl/glycerol ChCl/sucrose ChCl/xylitol	Ultrasound bath at 60°C, 30% DES water content, 20:1 mL/g solvents/solid ratio, for 30 minutes.	*Berberine* yield: 9.60 mg/g; *Palmatine* yield: 4.27 mg/g	(Li et al. 2020)

ChCl: Choline chloride, L.A: Lactic acid; optimal DES shown in bold compared to organic acids-based DESs. DESs can be used in conventional extraction methods (maceration) and also in new assisted methods, such as microwave-assisted, ultrasound-assisted, homogenate-assisted and high hydrostatic pressure (Chanioti and Tzia 2018). Recovery of bioactive compounds from DES extracts has also been studied with solvent back-extraction, such as a washing step in resins with water and ethanol (Procentese et al. 2015), column chromatography with resins (Zhuang et al. 2017) and countercurrent chromatography procedures (CCS) (Liu et al. 2016).

Khezeli et al. (2016) studied the extraction of three phenolic acids (ferulic, caffeic, and cinnamic) using ultrasound and eutectic solvents. Bakirtzi et al. (2016) studied the recovery of bioactive components from Greek medicinal herbs using eutectic solvents (lactic acid as a hydrogen bond acceptor) and ultrasonic extraction and concluded that lactic acid/glycine provided the best results for the total phenolic content relative to conventional solvents, i.e., water and 60% ethanol. Dai et al. (2016) replaced the conventional solvents with eutectic solvents for the recovery of anthocyanins from the plant *Catharanthus roseus*. Das et al. (2016) used eutectic solvents with choline chloride as a hydrogen bond acceptor for the extraction of k-carrageenan from the algae Kappaphycusalvarezii. Also, Peng et al. (2016) studied the extraction of five phenolic acids (chlorogenic acid, caffeic acid, 3,5-di-caffeoyltinic acid, 3,4-dicobuccinic acid, and 4,5-nicosilinoquinic acid) from *Lonicerae japonicae Flos* using eutectic solvents. Ozturk et al. (2018) used eutectic solvents with choline chloride as a hydrogen bond acceptor, and glycerol and ethanediol as hydrogen bond donors to extract phenolic components from orange by-products. García et al. (2016) studied the recovery of phenolic components from extra virgin olive oil using eutectic solvents (choline chloride as a hydrogen bond acceptor and various hydrogen bond donors, i.e., acids, carbohydrates, and urea) and concluded that the use of ChCl/xylitol and ChCl/1,2-propylene glycol showed the best results in terms of the total phenolic content and yielded in extracts with the highest concentration in oleicagel and oleasin than the ones obtained by the conventional solvent, which is 80% methanol.

In addition, Cui et al. (2015) studied the recovery of bioactive components from peas using eutectic solvents (1,6-hexanediol/ChCl) and microwave extraction (MAE) and achieved the optimization of the yield of target compounds in significantly reduced extraction time compared to the extraction using ultrasound or by conventional extraction and eutectic solvents. Alañón et al. (2018) combined microwave extraction using choline chloride as a hydrogen bond acceptor and various hydrogen bond donors (four polyols, three organic acids, one carbohydrate, and urea) to recover 48 phenolic compounds from olive leaves. Also, the combination of eutectic solvents and the use of microwaves were utilized as a green approach for the extraction of phenolic components from grape waste in which the eutectic solvents based on choline chloride and oxalic acid as a hydrogen bond donor were found to be the best choice for the recovery of the phenolic components (Cvjetko Bubalo et al. 2016).

Researchers have used DESs to extract tocopherols from crude palm oil (Abdul Hadi et al. 2015), anthocyanins from wine (Radošević et al. 2016), genistin, genistein, and apigenin from Pigeon pea roots (Cui et al. 2015), lignin from rice straw (Kumar et al. 2016, Hou et al. 2018), flavonoids from *Polygonatum odoratum* (Xia et al. 2021), and bioactive alkaloids such as berberine and palmatine (Li et al. 2020). Also, DESs have been applied for the extraction of polyphenols from lemon peels, olive leaves (Ünlü 2021), onion solid wastes, red grape pomace and wheat bran (Mouratoglou et al. 2016), grape skins (Radošević et al. 2016), *Cajanus cajan* leaves (Wei et al. 2015), *Morus alba* L. leaves (Zhou et al. 2018), olive pomace (Chanioti and Tzia 2018), coffee grounds (Yoo et al. 2018), *Sanghuangporus baumii* (Zheng et al. 2021), Hazelnut skin (Fanali et al. 2021), and *Sideritis scardica* and *Plantago major* L. (Grozdanova et al. 2020). Table 2 summarized the applications of DESs in the extraction of bioactive compounds from plant materials.

2.3 Bio-Based Solvents

Bio-based solvents are called solvents that are produced from renewable biomass sources, such as energy crops, forest products, aquatic biomass, and waste materials. They could be alcohols (ethanol), glycerols, terpenes, esters (ethyl lactate), furfurals (furfural, furfural alcohol, and levulinic acid), and furan (Li et al. 2016). Although they have low viscosity and are easy-handled for extraction processes, their use is limited to lab-scale or pilot plants due to the small number of biorefineries.

Bio-based ethanol is derived from sources, like sugar, starch, animal fats and vegetable oil, lignocellulosic materials, microalgae (Chemat et al. 2012). Bio-based glycerol is a by-product of biodiesel production. The bio-based methanol can be also produced, but it has toxicity issues. Ethyl acetate is a non-toxic, fully biodegradable, and industrially relevant ester (Chan and Su 2008). Bio-based ethyl acetate is mainly produced by esterification of acetic acid and ethanol in the liquid or vapor phase, acetylation of ethylene, and ethanol dehydrogenation (Santaella et al. 2015). Ethyl acetate is also produced from the conversion of sugar by yeasts, such as *Saccharomyces cerevisiae, Wickerhamomyces anomalus* and *Kluyveromyces marxianus* (Kruis et al. 2017). Ethyl lactate can be also used as a pharmaceutical and food additive according to the FDA (Bermejo et al. 2013).

Terpenes are very common components in plants. α-Pinene, a bicyclic monoterpene hydrocarbon, is one of the most abundant components in the essential oils of various plant species (Kim et al. 2018). It can be used in the flavor industries, pharmaceuticals, and fine chemistry. Also, D-Limonene is a colorless liquid cyclic terpene, and it is extracted from orange peels in the orange juice industry. It can be used for cosmetics and food (Chemat et al. 2012). Finally, p-Cymene is used for the synthesis of p-cresol and fine chemicals for perfumes, fungicides and pesticides and as a solvent for dyes and varnishes. It is made from the conversion of limonene into p-cymene but also is present in pine trees (Yao et al. 2019).

A recent green innovation is the use of cyclodextrins in water solutions that form inclusion complexes between bioactive compounds and their peculiar hydrophobic cavity. The ultrasound-assisted extraction of resveratrol and other phenolic compounds from the milled roots of *Polygonum cuspidatum* has been efficiently carried out in a water solution of β-cyclodextrin (1.5%). The selective inclusion properties of cyclodextrins toward phenolic stilbenes gave a much cleaner analytical extract profile if compared with that obtained with methanol. Thanks to phenolic compounds encapsulation, this extract showed excellent water dispersibility, higher stability, and antioxidant power (Mantegna et al. 2012).

Bioactive compounds' extraction from plant materials using bio-based solvents has limited applications. Bio-based solvents have been used to extract rosmarinic and caffeic acids from basil wastewater (Pagano et al. 2018), phenolic compounds, flavonoids and sinapine from seeds of rapeseed, mustard crambe and sunflower (Matthäus 2002), polyphenols, flavonoids, anthocyanins and ellagic acid from pomegranate peel (Masci et al. 2016), oil from rice bran (Liu and Mamidipally 2005), carotenoids and phenols from tomato waste (Silva et al. 2018), and volatile compounds from Cooperage woods in winemaking (Alañón et al. 2017). Ethyl lactate and ethyl acetate in water solutions are the most commonly used bio-based solvents applied at high temperatures (up to 170ºC) in order to achieve a high recovery of bioactive compounds. Virot et al. (2008) proposed the use of D-Limonene in olive oil extraction and concluded that it increased the oil yield by 8.3% more than that of hexane (also increased the extraction yield by 6% more than that of hexane in the extraction of rice bran oil was achieved). A-pinene and d-limonene were yielded in extracts with a higher concentration of carotenoids from carrot than that obtained by hexane (95.4, 94.8, and 78.1% respectively) (Varón et al. 2017).

3. Innovative Extraction Methods

The low yield of extracts together with the toxicity and the presence of the solvent residues in the extracts has led to the development of alternative extraction techniques, which can minimize or eliminate the use of organic solvents. These techniques are also known as green extraction methods, where the stability of the obtained bioactive compounds is maintained and the required energy for the extraction process is reduced (Tiwari 2015). The definition of green extraction process can be defined as follows: "Green extraction is based on the discovery and the design of extraction processes which reduce the energy consumption, allow the use of alternative solvents and renewable natural products, and ensure a safe and high quality final extract/product" (Chemat et al. 2019). The main objective of a green extraction process is to achieve a faster extraction rate, increased mass and heat transfer rate, more effective energy consumption, reduced equipment size, and reduced processing steps in a shorter time with minimal consumption of extraction solvents (Jacotet-Navarro et al. 2016, Mustafa and Turner 2011). A number of new alternatives to conventional techniques have been proposed to extract target compounds from various plant materials. These innovative methods include ultrasound-assisted extraction (UAE), microwave-assisted extraction (MAE), sub- and supercritical fluid extraction (SFE), pressurized liquid extraction (PLE), pulsed electric fields (PEF), high hydrostatic pressure-assisted extraction, and high-speed homogenization-assisted extraction (Ahmad-qasem et al. 2013, Barba et al. 2016, Roselló-Soto et al. 2015, Giacometti et al. 2018). However, there are some limitations of innovative techniques since they are related to high investment costs, control of the variables during the process, and lack of regulatory approval delaying their implementation on an industrial scale (Chemat et al. 2011).

3.1 Microwave-Assisted Extraction

The main mechanism in microwave-assisted extraction is the use of microwave irradiation. In plant materials, microwave irradiation directly interacts with different molecules causing rapid rotation of the polarized molecules (water) and leading to accelerated molecules movement (vibration), creating heat generation (Vo et al. 2018). In the application of microwaves, the most important characteristic of solvents and plant materials is the dielectric constant. Solvents with lower dielectric constant absorb less microwave energy (Sparr Eskilsson and Björklund 2000). Dielectric properties of plant materials depend on chemical composition, temperature, and frequency.

Microwave-assisted extraction is characterized by numerous advantages, which are less extraction time, faster heating, high energy efficiency, precise process control, selective heating, and significantly preserving the nutritional quality of the target compounds. It is important to optimize the microwave extraction for the recovery of bioactive compounds by combining the processing parameters, such as temperature, time, microwave power, type and solvent volume, solid/liquid ratio and the humidity of the material (Sparr Eskilsson and Björklund 2000).

The implementation of the microwave extraction process is distinguished according to the type of vessel used, which are (i) closed vessels, where pressure and temperature are controlled and (ii) open vessels, operating at atmospheric pressure. In closed vessels, the increased temperature and the pressure accelerate microwave-assisted extraction is due to the ability of the solvent to absorb microwave energy. Although the closed-vessel system offers an efficient extraction with less solvent consumption, it is susceptible to losses of volatile compounds with limited sample throughput (Chanioti et al. 2014). When using open vessels, the maximum temperature reached in the vessel is determined by the boiling point of the solvent at atmospheric pressure. Due to solvent evaporation, a reflux system is generally connected to the vessel. Compared to closed-vessel extractions, open vessels offer safer handling and allow the processing of larger amounts of samples.

Numerous studies have been applied for the extraction of oil from oilseeds and bioactive compounds, such as phenolic compounds, from different plant materials by using microwaves (Ekezie et al. 2017, Dai et al. 2010, Hao et al. 2002, Jiao et al. 2014, Chan et al. 2011, Chanioti et al. 2016, Pilar et al. 2021, Brantsen et al. 2021, Hiew et al. 2021). Besides essential oils and volatile compounds, bioactive compounds, such as total phenols, also show a possibility for extraction by microwaves. Table 3 performs the implementation of microwave extractions for the recovery of oil and bioactive compounds from plant materials.

Several researchers have observed that the increase of the solvent volume can initially improve the extraction process; however, in some cases, an excess solvent may adversely affect the phenomenon leading to excessive bulging and destruction of the plant material due to temperature increase (Xiao et al. 2009). Different extraction conditions are proposed by targeting different substances. The recovery of the key substances is influenced by the structure of the plant material, the solvent diffusion inside the plant material and the way that the heat transfer takes place.

Table 3. Applications of extraction assisted by microwaves.

Plant Material	Bioactive Compound	Solvent	Microwave Power (W)	Time (minutes)	Solid/Liquid Ratio (kg/L)	Reference
Tiger nut	Oil	Petroleum ether and acetone 2:1	420	55	1:7	(Hu et al. 2018)
Cottonseed	Cottonseed oil	Hexane	900	3;57	1:4	(Taghvaei et al. 2014)
Pongamia pinnata seeds	Oil	Hexane	600	14	-	(Kumar et al. 2018)
Moringaoleifera seeds	Oil	Petroleum ether	300	7	1:10	(Zhong et al. 2018)
Castor bean	Castor oil	5% Ethanol in hexane	300	20	1:20	(Ibrahim and Zaini 2018)
Olive pomace	Olive pomace oil	Hexane	287	16	1:10	(Yanik 2017)
Green tea leaves	Polyphenols	50% Ethanol	200	4	1:20	(Pan et al. 2003)
Olive pomace Olive leaves	Phenolic compounds	Buffer pH = 4.5	200	30	1:12.5	(Chanioti et al. 2016)
Grass leaves	Polyphenols	Acetone	200	15	1:10	(Grigonis et al. 2005)
Aronia	Polyphenols	25–75% Ethanol	300–600	5–15	-	(Simić et al. 2016)
Tomato by-products	Polyphenols	Methanol	100	45	1:50	(Li et al. 2012)
Olive leaves	Polyphenols	50% Ethanol	200	15	1:12	(Taamalli et al. 2012)
Green tea	Flavonols	Water	600	30	1:20	(Nkhili et al. 2009)
Black tea	Polyphenols	60% Ethanol	600	10	1: 12	(Wang et al. 2010)
Citrus mandarin pomace	Phenolic acid	66% Methanol	152	49 (s)	1:16	(Hayat et al. 2010)
Spices	Phenolic acid, bioactive compounds	50% Ethanol	200	18	1:20	(Gallo et al. 2010)
Syzygium cumini (L.)	Phenolic compounds, tannins	38% Ethanol	300	8	1:34	(Silva et al. 2021)

Table 3 contd. ...

...Table 3 contd.

Plant Material	Bioactive Compound	Solvent	Microwave Power (W)	Time (minutes)	Solid/Liquid Ratio (kg/L)	Reference
Ficus racemosa	Phenolic acid, bioactive compounds	Water at 3.5 pH	360	30 (s)	1:15	(Sharma et al. 2020)
Sorghum brans	Tannins	1% HCl in methanol	600	5	1:40	(Brantsen et al. 2021)
Garcinia mangostana L.	Phenolic compounds	Water	270	9	1:15	(Hiew et al. 2021)
Garlic skins	Tannins	70% ethanol	100	3	1:8	(Pardede et al. 2020)
Mango peel	Phenolic compounds	50% Methanol	800	98 (s)	1:47	(Pilar et al. 2021)

3.2 Ultrasound-Assisted Extraction

Ultrasound-assisted extraction produces a cavitation phenomenon, which leads to the production, growth, and collapse of bubbles (Azmir et al. 2013). UAE accelerates heat and mass transfer via the disruption of plant cell walls, leading to improved release of the target compounds from several plant materials (Roselló-Soto et al. 2015). The temperature, the frequency, the pressure, and the sonication time are the main parameters affecting the action of ultrasounds (Rajha et al. 2015, Vinatoru et al. 2017). It is characterized as an easy to use method, which is versatile, flexible, and requires low investment compared with other extraction techniques. Ultrasound has been used in various applications, such as the extraction of various biomaterials/substances including oils, polysaccharides, proteins, pigments, and bioactive compounds (Briones-Labarca et al. 2015, Tiwari 2015).

Ultrasound equipment consists of an inverter that converts electrical energy into acoustic energy, causing mechanical vibrations at ultrasonic frequencies. The resulting ultrasound is irradiated by the transmitter, which is also known as a reactor amplifying the waves. Ultrasound systems have been developed for both laboratory and industrial scales. On a laboratory scale, the most common equipment is the ultrasonic bath, which is low cost and is used for the dispersion of solids in a solvent and the ultrasonic treatment of liquid samples in containers by immersion into the bath. However, it has certain disadvantages, such as reduced power over time, lack of uniformity of the ultrasound energy distribution, and reduced reproducibility and repeatability of the experiments. For smaller volumes, closed type extractors are equipped with an ultrasound transducer and are considered more potent as the ultrasound intensity is emitted from a small surface where the probe is submerged directly into the flask avoiding attenuation (Pingret et al. 2013, Azmir et al. 2013, Takeuchi et al. 2009, Mason et al. 2005).

In both devices the main purpose is the satisfactory disruption of the cell walls and adequate mass transfer, leading to the enhancement of ultrasonic extraction yield (Briars and Paniwnyk 2013). Compared to other conventional extraction methods, the solute can be diffused through the cell walls into the solvent in a shorter time (Chemat et al. 2012).

Table 4 performs some applications of UAE for the recovery of extracts of plant materials. Similarly, UAE enhances the recovery of bioactive compounds from plant sources. Chung et al. (2010) combined aqueous acetone and ethanol solvents with ultrasound and they found that the use of ultrasound improved the extraction of phenolic compounds from soybean compared to the conventional Soxhlet extraction. Recently, the UAE has also been proposed as an alternative green extraction method for a variety of bioactive compounds from grape (Ghafoor et al. 2009, Bajerová et al. 2014, Samaram et al. 2015), apple pomace (Minjares-Fuentes et al. 2014), mandarin pomace (Dahmoune et al. 2013), pomegranate seeds (Tabaraki et al. 2012), olive leaves (Şahin and Shamli 2013a), *Eucalyptus globulus* leaves (Palma et al. 2021), *Moroccan Lavandula stoechas* L. (Ez et al. 2021), *Empetrum nigrum* (Gao et al. 2021), *Moringa Oleifera* leaves (Daghaghele et al. 2021), and black quinoa (Melini and Melini 2021). UAE is considered a flexible technique that can be scaled up benefiting the industry.

Table 4. Applications of extraction assisted by ultrasounds.

Plant Materials	Bioactive Compounds	Solvent	Ultrasound Intensity	Time (minutes)	Temperature (°C)	Reference
Soyabean, Sunflower	Oil	Hexane	-	10	-	(Luque-García and Luque de Castro 2004)
Almond	Oil	Hexane	40 kHz	40–60	40–60	(Zhang et al. 2009)
Papaya seeds	Oil	Hexane	40 kHz	5–30	25–50	(Samaram et al. 2015)
Pistachio	Oil	Hexane	30 kHz	10	30–50	(Hashemi et al. 2015)
Flaxseed	Oil	Hexane	20 kHz	30	30	(Zhang et al. 2008)
Soyabean	Oil	Hexane	-	30	-	(Li et al. 2004)
Pomegranate seeds	Oil	Hexane	20–60% of 20 kHz	2–40	20–80	(Goula 2013)
Isatis indigotica Fort	Oil	Hexane	40 kHz	30–50	30–50	(Li et al. 2012)
Pomegranate seeds	Oil	Hexane, ethyl acetate, diethyl ether acetone, isopropanol	140–180 W	20–40	35–45	(Tian et al. 2013)
Hazelnut	Oil	Hexane	120 W	48	-	(Wen et al. 2018)
Rapeseed	Oil	Hexane	20 kHz	15	40	(Sicaire et al. 2016)
Olive	Oil	Hexane	40 kHz	10	-	(Juliano et al. 2017)
Olive pomace	Oil	Hexane	30 kHz	1	40–60	(Chanioti and Tzia 2017)
Grape seeds	Phenolic compounds	Water	24 kHz	0,5–20	20, 35, 50	(Samaram et al. 2015)
Grape seeds	Phenolic compounds	50% Ethanol	25	-	20–50	(Bajerová et al. 2014)
Olive leaves	Phenolic compounds	0-100% Ethanol		20–60		(Şahin and Şamli 2013)
Grape seeds	Total phenols	53% Ethanol	40 kHz	29	50–60	(Ghafoor et al. 2009)
Citrus peels	Total phenols	63,93% Ethanol	77,79%	15,05	25	(Dahmoune et al. 2013)
Olive leaves	Total phenols	50% Ethanol	50 Hz	60	25	(Şahin and Şamli 2013)

Pomegranate peels	Total phenols	70% Ethanol	20–100 kHz	30	30–60	(Tabaraki et al. 2012)
Rosemary	Rosmarinic acid carnosic acid	Ethanol, hexane, acetone, water	19.5 kHz	10	-	(Bellumori et al. 2016)
Eucalyptus globulus leaves	Terpenes, Phenolic compounds, flavonoids	15–30% Ethanol	40–120 W	5–10	30–60	(Palma et al. 2021)
Moroccan Lavandula stoechas L.	Phenolic compounds	40% Ethanol	35 kHz	33	25	(Ez et al. 2021)
Empetrum nigrum	Phenolic compounds	62% Ethanol	250 W	21	42	(Gao et al. 2021)
Moringa Oleifera Leaves	Phenolic compounds	100% Ethanol	200 W	15	25	(Daghaghele et al. 2021)
Black Quinoa	Phenolic compounds	80% Methanol	37 kHz	10	30	(Melini and Melini 2021)

3.3 Homogenization-Assisted Extraction (HAE)

The extraction technique using a high-speed homogenizer is an alternative method for the recovery of bioactive compounds from plant materials. The application of high mechanical shear results in the mixing of the plant material with the solvent and the cell walls disruption without pressure (Guo et al. 2017, Liu et al. 2013b, Zhu et al. 2014). The effectiveness of the method has been documented in the extraction of camptothecin and hydroxycamptothecin from *Camptotheca acuminata* leaves (Shi et al. 2009) and isoflavones from soy flour (Zhu et al. 2011). In addition, Pereira et al. (2017) applied HAE to extract phenolic compounds from the banana peel, Zhu et al. (2014) for the extraction of a yellow pigment from gardenia *Gardenia jasminoides Ellis* and Liu et al. (2013b) for the recovery of the active substance shikini from *Arnebia euchroma*. Tong et al. (2018) used a high-speed homogenizer to extract the crocin from saffron and optimized the process by specifying the following optimal conditions, which include ethanol concentration of 70%, the temperature of 57°C, and time of 40 s. Ke and Chen (2016) applied a high-speed homogenizer to recover polysaccharides from a *Lentinusedodes* mushroom and comparing the results with those obtained by a conventional method, it was found that the optimum extraction conditions were at pH value of 10, liquid: solid ratio of 30:1 mL/g and extraction time of 66s. The application of these extraction conditions resulted in the optimal yield of polysaccharides which was found to be approximately 29.82% higher than that of conventional extraction. Recently, Eyiz et al. (2019) used HAE to extract bioactive compounds, such as polyphenols from red grape pomace and demonstrated that the optimum extraction conditions were glycerol concentration of 50% (w/v) and liquid:solid ratio of 22.4:1 mL/g. Zuin et al. (2020) applied HAE to recover mangiferin and hyperoside from mango waste and optimized the process by specifying ethanol concentration of 68% and 70%, extraction time of 4.5 and 5.0 min and liquid: solid ratio of 29% and 28% v/w for mangiferin and hyperoside extraction, respectively.

3.4 Enzyme-Assisted Extraction (EAE)

Enzyme-assisted extraction (EAE) is a promising method of using enzymes for the recovery of oils from oilseeds and other bioactive compounds, such as phenolic compounds from plant materials. Water is mainly used as the main solvent of the EAE and has many advantages over conventional solvents. Enzymes are also used for the preparation of plant material prior to extraction by conventional and/ or innovative methods. Various enzymes, such as cellulases, pectinases, proteases, polygalacturonases and hemicellulases, are often used to disrupt the plant cell wall and thus improve the extraction of bioactive compounds from the plant materials. These enzymes hydrolyze the cell wall of the plants, thereby increasing plant cell permeability and leading to a higher extraction yield of the target substances. Enzymes may be derived from bacteria, fungi, or vegetable/fruit extracts. In order to use enzymes effectively in extraction processes, it is important to know their mode of action, the optimal operation conditions, and the type of enzyme or combination of enzymes that is suitable for the selected plant material (Puri et al. 2012, Marić et al. 2018).

Enzyme-assisted extraction (EAE) depends on several factors, such as incubation time, enzyme type and concentration, pH value, extraction temperature and particle size of plant material (Poojary et al. 2017, Roselló-Soto et al. 2016).

With regard to the pH value, the mixing of the cellulase, pectinase and hemicellulase enzymes (1:1:1) at pH 4.5–5.0 studied by Long and Abdelkader (2011) and Tabtabaei and Diosady (2013) gave a higher yield of linseed and mustard seed, respectively, which is relative to the yield of each individual enzyme at different pH values. Apart from pH, the temperature is another parameter that affects EAE. The optimum incubation temperature varies for the different oilseeds and different enzymes. Thus, in addition to the yield of the oil, quality characteristics must also be taken into account, as well as the fact that high temperatures can also induce enzyme deactivation. On the other hand, low temperatures result in slow enzyme activity and in low oil extraction rate. The optimal temperature range is between 45–55°C for enzymatic hydrolysis (Rui et al. 2009).

During the extraction, the improvement of the liquid to solid ratio is beneficial for separating the oil, while reducing the amount of liquid waste. A high liquid-to-solid ratio leads to an increase in extraction rate and yield. However, the high liquid-to-solid ratio may lead to the increased liquid waste and the dysfunction of the emulsion degradation which increases the cost of the treatment. Therefore, the oil yield, the amount of liquid waste, and the de-emulsification should be taken into account when determining the ratio of liquid to solid (Liu et al. 2016).

The incubation time also varies for different oilseeds and enzymes. Oil yield, its quality, the recovery rate, and the cost should be taken into account when determining the optimal incubation time. The long incubation time increases the oil de-emulsification dysfunction and the short incubation time reduces the efficiency of the enzymatic hydrolysis of the oilseeds, thus leading to low oil yield. Extending the incubation time may enhance the degradation of cell wall components (Abdulkarim et al. 2006). Passos et al. (2009) reported that a mixture of cellulase, protease, xylanase, and pectinase enzymes used for 120 hours resulted in an oil yield of 3.8%, which was higher than that obtained when the enzyme mixture was used for 24 hours. However, the use of such long incubation times is not applied to the industry as it causes degradation of oil quality and requires high energy consumption (Abdulkarim et al. 2006, Jiang et al. 2010).

The EAE is widely used for laboratory and industrial-scale extractions. Because of its benefits, including being environmentally friendly and efficient, the EAE is proposed as an emerging technology. In foods, enzymes are used industrially for the extraction, clarification and concentration of fruit juices (Nakkeeran et al. 2011), pectin extraction (Ptichkina et al. 2008), oil extraction (Mishra et al. 2005), and the recovery of aromatic compounds, pigments, and dyes from plant sources (Sowbhagya and Chitra 2010). Compared to conventional extraction methods, EAE can be used for oil extraction, but also for the recovery of bioactive compounds, such as phenolic components from plant materials minimizing the use of solvents and the required energy (Puri et al. 2012, Heemann et al. 2019, Nishad et al. 2019, Vardakas

Table 5. Applications of extraction assisted by enzymes.

Product	Plant material	Enzyme	Reference
Oil	Mustard seeds	Viscozyme L, Pectinex Ultra SP-L, Celluclast 1.5 L	(Tabtabaei and Diosady 2013)
	Rapeseed	Protex 7L, Multifect Pectinase FE, Multifect CX 13L, Natuzyme	(Latif et al. 2008)
	Moringa oleifera seeds	Neutrase 0.8L (neutral protease), Termamyl 120L, type L (a-amylase), Pectinex Ultra SP-L (pectinase), Celluclast 1.5L FG (cellulase)	(Abdulkarim et al. 2006)
	Soyabean	Protex 6L	(de Moura et al. 2016)
	Sunflower seeds	Protex 7L, Alcalase 2.4L, Viscozyme L	(Latif and Anwar 2009)
	Olive	Pectinase and β-glucanase	(García et al. 2001)
	Olive	Pectinex Ultra SP-L, Pectinase 1.6021	(Najafian et al. 2009)
	Palmfruit	Celluclast 1.5 L, Pectinase, Multieffect FE	(Teixeira et al. 2013)
	Grape pomace	Cellulase, protease, xylanase, pectinase	(Puri et al. 2012)
	Peanut	Proteases	(Sharma et al. 2002)
	Olive	Pectinex Ultra SP-L and Pectinase 1.6021	(Najafian et al. 2009)
	Isatis indigotica seeds	Cellulase, protease, pectinase	(Gai et al. 2013)
	Bayberry (*Myrica rubra*) kernels	Cllulase/protease	(Zhang et al. 2012)
	Cardamom	Celluclast 1.5 L, Pectinex Ultra SP.L, ViscozymeL and Protease	(Baby and Ranganathan 2016)
Phenolic compounds	Pomegranate by-products	4% (vol.) pectinase και 4% (vol.) cellulase	(Alexandre et al. 2018)
	Olive pomace Olive leaves	Novozym 33095 (pectinase and polygalactorunase)	(Chanioti et al. 2016b)
	Peanut	Cellulases	(Zhang et al. 2011)
	Citrus peels	Celluzyme MX	(Li et al. 2006)
	Cajanuscajan (L.) Millsp. leaves	Cellulase, β-glucosidase και pectinase	(Fu et al. 2008)
	Saffron (*Crocus Sativus* L.)	Pectinex Ultra color, Celluclast BG, Xylanase AN	(Vardakas et al. 2021)
	Green yerba mate	Viscozyme® L (blend of arabanase, beta-glucanase, cellulase, hemicellulase and xylanase)	(Heemann et al. 2019)
	Citrus sinensis (cv. Malta) peel	Viscozyme® L (blend of arabanase, beta-glucanase, cellulase, hemicellulase and xylanase)	(Nishad et al. 2019)

et al. 2021). Enzymes used for oil extraction are usually cellulase, α-amylase and pectinase. The incorporation of the enzyme into the olive oil production process produces an oil with a high content of antioxidant compounds. A list of extracts resulting from enzyme assisted extraction in recent years is presented in Table 5.

3.5 *High Hydrostatic Pressure-Assisted Extraction (HHPAE)*

High hydrostatic pressure treatment is a method of food processing, which finds many applications in the food industry, such as the deactivation of microorganisms activity, protein denaturation, and extending the shelf life of food (López-Fandiño 2006, San Martín et al. 2002).

The application of high-pressure treatment for the extraction of bioactive compounds from plant materials is a new emerging technique known as High Hydrostatic Pressure Assisted Extraction (HHPAE). The extraction process is performed at high pressure (usually from 100 to 600 MPa) and low temperature (usually up to 60°C), using a small volume of solvents, while providing a recovery yield similar to other techniques (Shouqin et al. 2004). It offers high extraction yield, low energy consumption and does not degrade the bioactive compounds. The main parameters that significantly affect the extraction process are the type of solvent, pressure, temperature, time and the number of cycles (Corrales et al. 2008).

Pressure is considered to be one of the most important factors of high-pressure extraction and is directly related to the solubility of bioactive compounds. In high-pressure extraction, the mass transfer rate of the bioactive compounds from the raw material to the solvent is increased, confirming the theory that high-pressure extraction favors mass transfer resulting in enhanced mass transfer rates due to the variation of the coefficient of diffusion (Knorr 1993, Sánchez-Moreno et al. 2009, Barbosa-Canovas et al. 1998). The variation of the diffusion coefficient is mainly attributed to the change in pressure caused to the plant cell membranes, thereby increasing their permeability and thus facilitating the penetration of the solvent into the cells of the plant materials.

The application of high pressure inactivates degradative enzymes, offering a higher extraction yield compared to other techniques (Ahmed and Ramaswamy 2006). High pressure can also reduce the pH of the solvent during extraction, enhancing the efficiency of extraction of the bioactive compounds since most of these compounds are more stable at a pH of less than 4 (Corrales et al. 2008).

HHPAE equipment consists of a high-pressure chamber and its sealing system, a pressure production system, a temperature control system, and a material management system (Chen et al. 2009). Pressure in a container can be applied in different ways. In direct compression, fluid food is compressed directly from the smaller side of a piston. The larger side of the piston is connected to a low-pressure pump, thus allowing the pressure to multiply. Direct compression allows the application of very fast compression but cannot be applied to large-scale containers due to potential and inadequate leak-proof of the piston. During indirect compression, a pressure exerted fluid (water or glycerol) is pumped into a hermetically sealed chamber by a pressure booster. Most isostatic compression systems use indirect compression. Finally, compression can be achieved by heating the pressure fluid, which expands. This method is appropriate in cases where a combination of pressure and high temperature is required and needs precise control of temperature.

The above systems of high pressure are applied to discontinuous systems where food is processed in batches. Unlike continuous systems, the design of the equipment is simpler. Until now, only discontinuous or semi-continuous systems are used in the industry (Barbosa-Canovas et al. 1998).

The technique of high-pressure was successfully used in the extraction of anthocyanin (Corrales et al. 2008), catechins and polyphenols (Xi 2013), flavonoids, and phenolic compounds (Prasad et al. 2010, Šeremet et al. 2021) as well carotenoids. Recent applications have demonstrated the advantage of the application of HHP for the extraction of other bioactive components, such as lycopene (Xi 2006), caffeine (Xi 2009), lignans (Liu et al. 2009), polysaccharides (Sun and Wei 2020). Table 6 represents some applications of extraction assisted by high hydrostatic pressure.

Table 6. Applications of extraction assisted by high hydrostatic pressure.

Bioactive compound	Solvent	Pressure (MPa)	Time (minutes)	Reference
Phenolic compounds				
Phenolic compounds from *nasturtium*	100% Ethanol	600	3,1	(Pinela et al. 2017)
Phenolic compounds from pomegranate by-products	Water	300	15	(Alexandre et al. 2018)
Phenolic compounds from rice	Water	100	1440	(Young et al. 2017)
Phenolic compounds from sour cherry	20% Ethanol (1:6.6)	200	25	(Adil et al. 2008)
Phenolic compounds from peanut shells	70% Ethanol (1:40)	400	1	(Yu et al. 2008)
Phenolic compounds from green tea leaves	50% Ethanol (1:20)	500	1	(Xi 2013)
Phenolic compounds from longan fruit	50% Ethanol (1:50)	500	2.5	(Prasad et al. 2009)
Phenolic compounds from longan fruit	Ethanol: HCl (85:15)	500	2.5	(Prasad et al. 2010)
Phenolic compounds from spent coffee grounds	80% Methanol	500	15	(Okur et al. 2021)
Phenolic compounds from different types of tea	100% Ethanol	200	5	(Šeremet et al. 2021)
Others				
Lycopene from tomato seeds	75% Ethanol (1:6)	500	1	(Xi 2006)
Caffeine from green tea	50% Ethanol (1:20)	500	1	(Xi 2009)
Lignans from *Schisandra chinensis* baill	90% Ethanol (1:90)	400	5	(Liu et al. 2009)
Polysaccharides from *Huangshan* Stone Ear	Water	303	14	(Sun and Wei 2020)

3.6 Supercritical Fluid Extraction (SFE)

Superficial Fluid Extraction (SFE) is an environmentally friendly extraction method providing high selectivity, short extraction time, prevention of environmental pollution and the use of non-toxic organic solvents (Wang et al. 2008). SFE is based on selected fluid properties, such as density, diffusion capacity, dielectric constant and viscosity, and their ability to create a supercritical fluid (Sihvonen et al. 1999) with a suitable change in pressure and temperature (Azmir et al. 2013). The characteristics of supercritical fluids are mentioned above.

In the process of supercritical extraction, the raw material (plant materials, food processing by-products, algae, micro-algae, etc.) is placed in the extraction vessel where the temperature and pressure are adjusted to the desired conditions. The vessel is subjected to pressure from the fluid through a pump. Subsequently, the fluid and the dissolved components are transferred to the separators, where they are collected from the bottom of them, while the fluid is regenerated and recycled or released into the environment (Sihvonen et al. 1999).

Efficient extraction of bioactive components from plant materials is based on various parameters of SFE, such as temperature, pressure, particle size, raw material moisture, extraction time, CO_2 flow rate and solid/liquid ratio (Soquetta et al. 2018).

The advantages of using SFE for the extraction of bioactive compounds can be understood by considering the following:

1) The supercritical fluid leads to greater penetration into the solid matrix and enhances the mass transfer. So, the extraction time is significantly reduced by the application of SFE compared to conventional methods.

2) The separation of the solute from the solvent is bypassed by decompression of the supercritical fluid, thus providing time-saving.

3) SFE operates at room temperature, making the method ideal for isolating thermosensitive compounds.

4) SFE uses a small amount of organic solvent and is considered to be an environmentally friendly method.

5) Recycling and reuse of supercritical fluid are possible, thus minimizing the generation of solvent waste.

6) With appropriate design, SFE can be used to extract bioactive compounds both on a laboratory scale and on an industrial scale (Azmir et al. 2013).

However, the use of SFE has two major disadvantages that limited its wide use, including the high processing costs and the complex industrial equipment. Albarelli et al. (2018) reported an increase of the total investment by 71% and high energy demand and operational costs for a sugarcane-microalgae biorefinery using CO_2-SFE with respect to a traditional biorefinery. Moreover, De Marco et al. (2018) presented that the environmental impact of caffeine extraction from coffee beans using SFE was increased due to the high temperatures and pressure applied.

The use of CO_2 is an alternative method for extracting oil and antioxidant compounds from plant materials (Knez et al. 2014, da Silva et al. 2016, Cabeza et al. 2017, Djas and Henczka 2018, Meneses et al. 2015), fruits (Viganó et al. 2016), vegetables (Bagheri et al. 2016) and microorganisms (Yen et al. 2015). SFE has been used to extract phenolic compounds from Vietnamese *Callisia fragrans leaves* (Thao et al. 2020), red propolis (Barreto et al. 2020), passion fruit seeds (Oliveira et al. 2017) and grape seeds (Pérez et al. 2015), phytochemical compounds from soy bean expeller (Alvarez et al. 2019), essential oil from orange peel (Xhaxhiu and Wenclawiak 2015), phenols from olive oil mill waste (Lafka et al. 2011), phytosterol from roselle seeds (Nyam et al. 2010), solanesol from tobacco waste (Wang and Gu 2018), oil from the plant *Cannabis sativa* L. (Grijó et al. 2019), rapeseed oil (Martin et al. 2015), Butter of cupuacu (Cavalcanti et al. 2016), carotenoids and chlorophyll from *Scenedes musobliquus* (Guedes et al. 2013), Pistaciakhinjuk fruit oil (Sodeifian et al. 2016), lycopene from industrial tomato by-products (mixture of peels and seeds) (Nobre et al. 2012), olive oil (Stavroulias and Panayiotou 2005) and saponins from Agave salmiana bagasse (Santos-Zea et al. 2019). The use of co-solvents, such as ethanol, for improving recoveries of polar and medium polar compounds has been studied by Radzali et al. (2020), Xhaxhiu and Wenclawiak (2015) and Campone et al. (2018), who concluded in high recovery yield of the target substances.

3.7 Pulsed Electric Field Assisted Pulse Extraction (PEF)

The pulsed electric field (PEF) was identified as a useful treatment to improve drying, extraction, and diffusion processes over the last decade (Soquetta et al. 2018, Azmir et al. 2013, Vorobiev and Lebovka 2009). The principle of the method is based on the electroporation of the cell membrane to increase the extraction. During the application of a cell to the electric field, an electrical potential passes through the membrane of the cell. According to the bipolar nature of the membrane molecules, the electrical potential separates the molecules, depending on their charge, into the cell membrane. Upon a critical value of about 1V of transmembrane potential, obstruction is induced between the charged molecules forming pores in the weak areas of the membrane and is performed a dramatic increase of permeability (Soquetta et al. 2018). The PEF equipment consists of a chamber carrying two electrodes in which the plant materials are placed. Depending on the design of the PEF chamber, the process may be continuous or discontinuous (in batches). The efficiency of PEF processing depends on various parameters, such as electric field strength, input energy, pulse number, processing temperature, and properties of the material to be treated (Puertolas et al. 2016).

PEF can increase mass transfer during extraction due to the destruction of the cell membrane of plant materials, enhancing process yield and reducing the extraction time. PEF was applied to improve the release of target compounds from plant tissue by helping to increase cell membrane permeability (Toepfl et al. 2014). PEF processing in a moderate electric field (500 and 1,000 V/cm for 10^{-4}–10^{-2} s) was found to cause damage to the cell membrane of the plant tissue with a slight increase in temperature. For this reason, PEF can cause minimal degradation of the thermosensitive compounds (Ade-Omowaye et al. 2001). Also, PEF can be applied to plant materials as a pretreatment procedure prior to conventional extraction.

PEF has been used for the recovery of procyanidins from *Vitis amurensis* seeds (Dong et al. 2020), polyphenols from *Vitis vinifera, Sideritis scardica* and *Crocus sativus* (Lakka et al. 2021), blueberry pomace (Lončarić et al. 2020), potato peels (Lončarić et al. 2020), carotenoids and phenolic compounds from carrots (López-Gámez et al. 2021), phytosterols from maize and isoflavones from soy (Guderjan et al. 2005), the extraction of polyphenols and anthocyanins from Merlot grapes (Delsart et al. 2014), the release of anthocyanins from grape juice (Leong et al. 2016), the extraction of polyphenols from *Borago officinalis* leaves (Segovia et al. 2015) and the apple juice extraction (Jaeger et al. 2012).

3.8 *Pressurized Liquid Extraction (PLE) Extraction*

PLE—also referred to as accelerated or hypothermic solvent extraction—uses organic liquid solvents at temperatures of 50–200°C and pressures of 99–148 atm in order to achieve a rapid extraction rate of bioactive compounds (Mustafa and Turner 2011). When the solvent is pure water, the process is called superheated water extraction, hypothermic water extraction or extraction using hot water under pressure (Pronyk and Mazza 2009).

The high extraction temperature favors (i) the increase of the ability of the solvent to dissolve the recovered compounds, (ii) the increase of the diffusion coefficients, (iii) the breaking of the cohesion forces between the compounds and the raw material, (iv) reduces the viscosity of the solvent and (v) reduces surface tension (Ramos et al. 2002). As the temperature rises, the dielectric constant of the solvent decreases, resulting in a decrease in its polarity. The temperature regulates the polarity of the solvent so as to approximate the polarity of the recovered compounds (Dunford et al. 2010). For example, the value of the dielectric constant of water is about 80 at room temperature, while at 250°C it is reduced to 30. Under these conditions, the value of the dielectric constant is similar to that of organic solvents, such as ethanol or methanol. The same is observed with the solubility which is decreased approaching the value of the reduced polarity components (Adil et al. 2008). Therefore, this method can be used to extract non-polar components and replace organic solvents. The high pressure applied by the liquid solvent on the solid material facilitates its faster penetration into the cells and further improves the extraction of the target compounds (Soquetta et al. 2018).

The advantages of PLE with respect to conventional extraction are processing speed and allowing the use of smaller amounts of solvent and higher yield. The main disadvantage of the method is its inappropriateness for thermally sensitive components since high temperature may degrade their structure and functionality (Azmir et al. 2013).

PLE has been used to receive extracts of bioactive compounds from berry by-products (Machado et al. 2015), to recover polysaccharides from blackcurrant (Xu et al. 2016), to extract bioactive compounds from peppers (Bajer et al. 2015), to extract a large number of functional components, including carotenoids from algae *Himanthalia elongata* and microalgae *Synechocystis* sp. (Plaza and Turner 2015), to release phenolic compounds from olive leaves (*Olea europaea*, cv. Oblica) (Putnik et al. 2017), skin and seeds of *Vitis vinifera* L. cv. *Negra Criolla* pomace (Allcca-alca et al. 2021), *Thymus serpyllum* herbal dust (Mrkonjić et al. 2021) and Goldenberry (Osmar et al. 2018) and to recover oil from the safflower (Conte et al. 2016).

4. Combination Techniques for Improved Extraction Efficiency

The combination of the extraction techniques mentioned above is the modern effective approach to recovering bioactive compounds that can significantly reduce extraction time by consuming smaller amounts of solvents and lead to higher extraction yield than conventional extraction.

Many studies have examined the combination of various techniques for the recovery of target substances, such as the combination using ultrasound, microwave, supercritical CO_2, pulsed electric fields, and ultra-high pressure. Corrales et al. (2008) studied the extraction of anthocyanins from grape by-products using a combination of ultrasound, pulsed electric field, and high hydrostatic pressure. Garcia-Mendoza et al. (2015) applied two sequential steps for the extraction of mangos (*Mangifera indica* L.) bioactive compounds. Initially, they used supercritical CO_2 and then applied high pressure (30 MPa) with solvent ethanol. Dias et al. (2016) evaluated the effect of the extraction of bioactive compounds from red pepper (*Capsicum baccatum* L.) with a combination of ultrasound and supercritical CO_2. Also, Alexandre et al. (2018) combined the use of enzymes and supercritical CO_2 to obtain phenolic compounds from pomegranates. In addition, the combined effect of the use of enzymes and microwaves known as microwave enzymatic-assisted extraction and finds many applications for the recovery of oil from oilseeds, such as yellow pumpkin seeds (Jiao et al. 2014) from seeds of the *Isatis indigotica* plant (Gai et al. 2013), *Xanthoceras sorbifolia* Bunge (Li et al. 2013) and lavender (Rashed et al. 2017) and bioactive ingredients from plant sources, such as polyphenols from olive pomace and olive leaves (Chanioti et al. 2016), polysaccharides from *Schisandra chinensis* (Cheng et al. 2015) and *Pyropia yezoensis* (Lee et al. 2016). Also, Yang et al. (2010) optimized the microwave-assisted enzymatic extraction of corilagin and geraniin from *Geranium sibiricum Linne* with good results.

In other research, an ultrasound-assisted method combined with enzymes was applied to obtain extracts from the stems of *Trapa quadrispinosa* Roxb. residues with a high yield of phenolic content, a strong antioxidant, and antitumor activities (Li et al. 2017). Ionic liquid-based enzyme-assisted extraction combined with cellulase treatment and *in situ* hydrolysis process for obtaining genipin from *Eucommia ulmoides* olive barks (Chen et al. 2018). Another research clearly showed that a high voltage electrical discharges (HVED) pre-treatment of orange peels was efficient to enhance the accessibility of cellulosic biomass to enzymes (El Kantar et al. 2018).

Yang and Wei (2015) developed an efficient method to extract bioactive compounds from *Rabdosia rubescens* by combining heat reflux extraction using ethanol as solvent and ultrasound-assisted extraction. Using pulsed electric field and high voltage techniques to recover bioactive compounds from mango, the results demonstrated the feasibility of a combination of the pulsed electric field and high voltage techniques to recover antioxidants and proteins from mango skin (Parniakov et al. 2016).

A combination of SFE and pressing (Gas Assisted Mechanical Expression, GAME) has been recently investigated to increase oil extraction yield from various seeds (Voges et al. 2008). Also, nanofiltration coupled to CO_2-SFE for the purification of beta-carotene from carrot oil and fractionation of fish oil triglycerides (Sarrade et al. 1998).

The combination of natural eutectic solvent with a new assisted extraction method is a promising approach. The combination of NADES/water/microwave was used to extract soluble sugars from banana puree (Gomez et al. 2018). Also, the use of natural deep eutectic solvents (NADES) with microwave-assisted extraction (MAE) was investigated to extract bioactive compounds from *Pseudevernia furfuracea* (Parmeliaceae) (Ezra et al. 2016). Chanioti and Tzia (2018) applied successfully to extract phenolic compounds from olive pomace using natural deep eutectic solvents based on choline chloride combined with homogenate-, microwave-, ultrasound- or high hydrostatic pressure-assisted extractions. The combination of NADES with ultrasound-assisted extraction (UAE) has also been investigated (Grozdanova et al. 2020, Ünlü 2021, Li et al. 2020).

References

Abdul Hadi, N., Ng, M., Choo, Y., Hashim, M. and Jayakumar, N. 2015. Performance of choline-based deep eutectic solvents in the extraction of tocols from crude palm oil. JAOCS, Journal of the American Oil Chemists' Society 92(11–12): 1709–1716. https://doi.org/10.1007/s11746-015-2720-6.

Abdulkarim, S., Lai, O., Muhammad, S. and Long, H. 2006. Use of enzymes to enhance oil recovery during aqueous extraction of *Moringa oleifera* seed oil. Journal of Food Lipids 13: 113–130.

Ade-Omowaye, B., Angersbach, A., Taiwo, K. and Knorr, D. 2001. Use of pulsed electric field pre-treatment to improve dehydration characteristics of plant based foods. Trends in Food Science and Technology 12(8): 285–295. https://doi.org/10.1016/S0924-2244(01)00095-4.

Adil, I., Yener, E. and Bayindirli, A. 2008. Extraction of total phenolics of sour cherry pomace by high pressure solvent and subcritical fluid and determination of the antioxidant activities of the extracts. Separation Science and Technology 43(5): 1091–1110.

Ahmad-qasem, M., Hussam, Barrajon-catalan, E., Vicente, M., Cárcel, J. and Garcia-perez, J. 2013. Influence of air temperature on drying kinetics and antioxidant potential of olive pomace. Journal of Food Engineering 119(3): 516–524. https://doi.org/10.1016/j.jfoodeng.2013.06.027.

Ahmed, J. and Ramaswamy, H. 2006. High pressure processing of fruit and vegetables. Stewart Postharvest Review 1: 1–10.

Alañón, M.E., Alarcón, M., Marchante, L., Díaz-Maroto, M.C. and Pérez-Coello, M.S. 2017. Extraction of natural flavorings with antioxidant capacity from cooperage by-products by green extraction procedure with subcritical fluids. Industrial Crops and Products 103(April): 222–232. https://doi.org/10.1016/j.indcrop.2017.03.050.

Alañón, M., Ivanović, M., Gómez-Caravaca, A., Arráez-Román, D. and Segura-Carretero, A. 2018. Choline chloride derivative-based deep eutectic liquids as novel green alternative solvents for extraction of phenolic compounds from olive leaf. Arabian Journal of Chemistry. https://doi.org/10.1016/J.ARABJC.2018.01.003.

Albarelli, J., Santos, D., Ensinas, A., Maréchal, F., Cocero, M. and Meireles, M.A. 2018. Comparison of extraction techniques for product diversification in a supercritical water gasification-based sugarcane-wet microalgae biorefinery: Thermoeconomic and environmental analysis. Journal of Cleaner Production 201: 697–705. https://doi.org/10.1016/J.JCLEPRO.2018.08.137.

Alexandre, E., Silva, S., Santos, S., Silvestre, A., Duarte, M. and Saraiva, J. 2018. Emerging technologies to extract high added value compounds from fruit residues: Sub/supercritical, ultrasound-, and enzyme-assisted extractions. Food Research International 34: 581–612. https://doi.org/10.1016/j.foodres.2018.08.044.

Allcca-alca, E.E., Le, N.C., Luque-vilca, O.M., Mart, M., Ricardo, P. and Leander, N. 2021. Hot pressurized liquid extraction of polyphenols from the skin and seeds of *Vitis vinifera* L. cv. Negra Criolla pomace a peruvian native pisco industry waste. Agronomy 11: 866.

Alvarez, M., Cabred, S., Ramirez, C. and Fanovich, M. 2019. Valorization of an agroindustrial soybean residue by supercritical fluid extraction of phytochemical compounds. The Journal of Supercritical Fluids 143: 90–96. https://doi.org/10.1016/J.SUPFLU.2018.07.012.

Baby, K. and Ranganathan, T. 2016. Effect of enzyme pre-treatment on extraction yield and quality of cardamom (Elettaria cardamomum maton.) volatile oil. Industrial Crops and Products 89: 200–206. https://doi.org/10.1016/J.INDCROP.2016.05.017.

Bagheri, H., Yamini, Y., Safari, M., Asiabi, H., Karimi, M. and Heydari, A. 2016. Simultaneous determination of pyrethroids residues in fruit and vegetable samples via supercritical fluid extraction coupled with magnetic solid phase extraction followed by HPLC-UV. The Journal of Supercritical Fluids 107: 571–580. https://doi.org/10.1016/J.SUPFLU.2015.07.017.

Bajer, T., Bajerová, P., Kremr, D., Eisner, A. and Ventura, K. 2015. Central composite design of pressurised hot water extraction process for extracting capsaicinoids from chili peppers. Journal of Food Composition and Analysis 40: 32–38. https://doi.org/10.1016/J.JFCA.2014.12.008.

Bajerová, P., Adam, M., Bajer, T. and Ventura, K. 2014. Comparison of various techniques for the extraction and determination of antioxidants in plants. Journal of Separation Science 37: 835–844.

Bakirtzi, C., Triantafyllidou, K. and Makris, D. 2016. Novel lactic acid-based natural deep eutectic solvents: Efficiency in the ultrasound-assisted extraction of antioxidant polyphenols from common native Greek medicinal plants. Journal of Applied Research on Medicinal and Aromatic Plants 3: 120–127.

Barba, F.J., Zhu, Z., Koubaa, M., de Souza Sant'Ana, A. and Orlien, V. 2016. Green alternative methods for the extraction of antioxidant bioactive compounds from winery wastes and by-products: A review. Trends in Food Science and Technology 49: 96–109.

Barbosa-Canovas, G., Pothakamury, U., Palou, E. and Swanson, B. 1998. Nonthermal Preservation of Foods. Marcel Dekker.

Barreto, G.D.A., Pereira, J. and Moraes, L. 2020. Supercritical extraction of red propolis: operational conditions and chemical characterization. Molecules 25: 4816.

Bellumori, M., Innocenti, M., Binello, A., Boffa, L., Mulinacci, N. and Cravotto, G. 2016. Selective recovery of rosmarinic and carnosic acids from rosemary leaves under ultrasound-and microwave-assisted extraction procedures. Comptes Rendus Chimie 19(6): 699–706. https://doi.org/10.1016/j.crci.2015.12.013.

Bermejo, D., Luna, P., Manic, M., Najdanovic-Visak, V., Reglero, G. and Fornari, T. 2013. Extraction of caffeine from natural matter using a bio-renewable agrochemical solvent. Food and Bioproducts Processing 91(4): 303–309. https://doi.org/10.1016/J.FBP.2012.11.007.

Bi, W, Tian, M. and Ho, K. 2013. Evaluation of alcohol-based deep eutectic solvent in extraction and determination of flavonoids with response surface methodology optimization. Journal of Chromatography A 1285: 22–30. https://doi.org/10.1016/j.chroma.2013.02.041.

Brantsen, J.F., Herrman, D.A., Ravisankar, S. and Awika, J.M. 2021. Effect of tannins on microwave-assisted extractability and color properties of sorghum 3-deoxyanthocyanins. Food Research International 148(July): 110612. https://doi.org/10.1016/j.foodres.2021.110612.

Briars, R. and Paniwnyk, L. 2013. Effect of ultrasound on the extraction of artemisinin from Artemisia annua. Industrial Crops and Products 42: 595–600. https://doi.org/10.1016/J.INDCROP.2012.06.043.

Briones-Labarca, V., Plaza-Morales, M., Giovagnoli-Vicuna, C. and Fabiola, F. 2015. High hydrostatic pressure and ultrasound extractions of antioxidant compounds, sulforaphane and fatty acids from Chilean papaya (Vasconcellea pubescens) seeds: Effects of extraction conditions and methods. LWT - Food Science and Technology Journal 60: 525–534. https://doi.org/10.1016/j.lwt.2014.07.057.

Bubalo, M., Toma, M., Kovac, K., Bubalo, M. and Natka, C. 2016. Green extraction of grape skin phenolics by using deep eutectic solvents Green extraction of grape skin phenolics by using deep eutectic solvents. Food Chemistry 200: 156–166. https://doi.org/10.1016/j.foodchem.2016.01.040.

Byrne, F., Jin, S., Paggiola, G., Petchey, T., Clark, J., Farmer, T., Hunt, A., M, R. and Sherwood, J. 2016. Tools and techniques for solvent selection: green solvent selection guides. Sustainable Chemical Processes 4(1). https://doi.org/10.1186/s40508-016-0051-z.

Cabeza, L., de Gracia, A., Fernández, A.I. and Farid, M. 2017. Supercritical CO_2 as heat transfer fluid: A review. Applied Thermal Engineering 125: 799–810. https://doi.org/10.1016/J.APPLTHERMALENG.2017.07.049.

Campone, L., Celano, R., Lisa Piccinelli, A., Pagano, I., Carabetta, S., Sanzo, R., Russo, M., Ibañez, E., Cifuentes, A. and Rastrelli, L. 2018. Response surface methodology to optimize supercritical carbon dioxide/co-solvent extraction of brown onion skin by-product as source of nutraceutical compounds. Food Chemistry 269: 495–502. https://doi.org/10.1016/J.FOODCHEM.2018.07.042.

Cavalcanti, R., Albuquerque, C. and Meireles, M. 2016. Supercritical CO_2 extraction of cupuassu butter from defatted seed residue: Experimental data, mathematical modeling and cost of manufacturing. Food and Bioproducts Processing 97: 48–62. https://doi.org/10.1016/J.FBP.2015.10.004.

Chan, C., Yusoff, R., Ngoh, G. and Kung, F. 2011. Microwave-assisted extractions of active ingredients from plants. Journal of Chromatography A 1218(37): 6213–6225. https://doi.org/10.1016/J.CHROMA.2011.07.040.

Chan, W. and Su, M. 2008. Biofiltration of ethyl acetate and amyl acetate using a composite bead biofilter. Bioresource Technology 99(17): 8016–8021. https://doi.org/10.1016/J.BIORTECH.2008.03.045.

Chanioti, S., Liadakis, G. and Tzia, C. 2014. Solid-liquid extraction. pp. 253–286. *In*: Varzakas, T. and Tzia, C. (eds.). Food Engineering Handbook. CRC Press.

Chanioti, S., Siamandoura, P. and Tzia, C. 2016a. Evaluation of extracts prepared from olive oil by-products using microwave-assisted enzymatic extraction: effect of encapsulation on the stability of final products. Waste and Biomass Valorization 7(4). https://doi.org/10.1007/s12649-016-9533-1.

Chanioti, S., Siamandoura, P. and Tzia, C. 2016b. Evaluation of extracts prepared from olive oil by-products using microwave-assisted enzymatic extraction: effect of encapsulation on the stability of final products. Waste and Biomass Valorization 7(4). https://doi.org/10.1007/s12649-016-9533-1.

Chanioti, S. and Tzia, C. 2017. Optimization of ultrasound-assisted extraction of oil from olive pomace using response surface technology: Oil recovery, unsaponifiable matter, total phenol content and antioxidant activity. LWT - Food Science and Technology 79: 178–189. https://doi.org/10.1016/j.lwt.2017.01.029.

Chanioti, S. and Tzia, C. 2018. Extraction of phenolic compounds from olive pomace by using natural deep eutectic solvents and innovative extraction techniques. Innovative Food Science and Emerging Technologies 48: 228–239. https://doi.org/10.1016/J.IFSET.2018.07.001.

Cheaib, D., El Darra, N., Rajha, H., El-Ghazzawi, I., Maroun, R. and Louka, N. 2018. Effect of the extraction process on the biological activity of lyophilized apricot extracts recovered from apricot pomace. Antioxidants 7(1): 11. https://doi.org/10.3390/antiox7010011.

Chemat, F., Vian, M. and Cravotto, G. 2012. Green extraction of natural products: concept and principles. International Journal of Molecular Sciences 13: 8615–8627.

Chen, G., Sui, X., Liu, T., Wang, H., Zhang, J., Sun, J. and Xu, T. 2018. Application of cellulase treatment in ionic liquid based enzyme-assisted extraction in combine with *in-situ* hydrolysis process for obtaining genipin from Eucommia ulmoides Olive barks. Journal of Chromatography A 1569: 26–35. https://doi.org/10.1016/J.CHROMA.2018.07.063.

Chen, R., Meng, F., Zhang, S. and Liu, Z. 2009. Effects of ultrahigh pressure extraction conditions on yields and antioxidant activity of ginsenoside from ginseng. Separation and Purification Technology 66(2): 340–346. https://doi.org/10.1016/J.SEPPUR.2008.12.026.

Cheng, Z., Song, H., Yang, Y., Liu, Y., Liu, Z., Hu, H. and Zhang, Y. 2015. Optimization of microwave-assisted enzymatic extraction of polysaccharides from the fruit of *Schisandra chinensis* Baill. International Journal of Biological Macromolecules 76: 161–168. https://doi.org/10.1016/J.IJBIOMAC.2015.01.048.

Cláudio, A., Cognigni, A., de Faria, E., Silvestre, A., Zirbs, R., Freire, M. and Bica, K. 2018. Valorization of olive tree leaves: Extraction of oleanolic acid using aqueous solutions of surface-active ionic liquids. Separation and Purification Technology, 204, 30–37. https://doi.org/10.1016/J.SEPPUR.2018.04.042

Conte, R., Gullich, L., Bilibio, D., Zanella, O., Bender, J., Carniel, N. and Priamo, W. 2016. Pressurized liquid extraction and chemical characterization of safflower oil : A comparison between methods. Food Chemistry 213: 425–430. https://doi.org/10.1016/j.foodchem.2016.06.111.

Cooney, M. and Benjamin, K. 2016. Ionic liquids in lipid extraction and recovery. Ionic Liquids in Lipid Processing and Analysis 279–316. https://doi.org/10.1016/B978-1-63067-047-4.00009-X.

Corrales, M., Toepfl, S., Butz, P., Knorr, D. and Tauscher, B. 2008. Extraction of anthocyanins from grape by-products assisted by ultrasonics, high hydrostatic pressure or pulsed electric fields: A comparison. Innovative Food Science and Emerging Technologies 9(1): 85–91. https://doi.org/10.1016/J.IFSET.2007.06.002.

Cui, Q., Peng, X., Yao, X., Wei, Z., Luo, M., Wang, W., Zhao, C., Fu, Y. and Zu, Y. 2015. Deep eutectic solvent-based microwave-assisted extraction of genistin, genistein and apigenin from pigeon pea roots. Separation and Purification Technology 150: 63–72. https://doi.org/10.1016/J. SEPPUR.2015.06.026.

Cvjetko Bubalo, M., Ćurko, N., Tomašević, M., Kovačević Ganić, K. and Radojčić Redovniković, I. 2016. Green extraction of grape skin phenolics by using deep eutectic solvents. Food Chemistry 200: 159–166. https://doi.org/10.1016/J.FOODCHEM.2016.01.040.

Cvjetko Bubalo, Marina, Vidović, S., Radojčić Redovniković, I. and Jokić, S. 2018. New perspective in extraction of plant biologically active compounds by green solvents. Food and Bioproducts Processing 109: 52–73. https://doi.org/10.1016/j.fbp.2018.03.001.

da Silva, R., Rocha-Santos, T. and Duarte, A. 2016. Supercritical fluid extraction of bioactive compounds. TrAC Trends in Analytical Chemistry 76: 40–51. https://doi.org/10.1016/J.TRAC.2015.11.013.

Daghaghele, S., Kiasat, A.R., Mohammad, S. and Ardebili, S. 2021. Intensification of extraction of antioxidant compounds from *Moringa oleifera* leaves using ultrasound-assisted approach: BBD-RSM design intensification of extraction of antioxidant compounds from *Moringa oleifera* leaves using ultrasound-assisted approach. International Journal of Fruit Science 21(1): 693–705. https://doi.org/10.1080/15538362.2021.1926396.

Dahmoune, F., Boulekbache, L., Moussi, K., Aoun, O. and Spigno, G. 2013. Valorization of Citrus limon residues for the recovery of antioxidants : Evaluation and optimization of microwave and ultrasound application to solvent extraction. Industrial Crops and Products 50: 77–87. https://doi.org/10.1016/j. indcrop.2013.07.013.

Dai, J., Orsat, V., Raghavan, G.S.V. and Yaylayan, Y. 2010. Investigation of various factors for the extraction of peppermint (Mentha piperita L.) leaves. Journal of Food Engineering 96(4): 540–543.

Dai, Y., Spronsen, J. and Witkamp, G. 2013. Natural deep eutectic solvents as new potential media for green technology. Analytica Chimica Acta 766: 61–68. https://doi.org/10.1016/j.aca.2012.12.019.

Dai, Y., Rozema, E., Verpoorte, R. and Choi, Y. 2016. Application of natural deep eutectic solvents to the extraction of anthocyanins from *Catharanthus roseus* with high extractability and stability replacing conventional organic solvents. Journal of Chromatography A 1434: 50–56. https://doi.org/10.1016/j. chroma.2016.01.037.

Das, A., Sharma, M., Mondal, D. and Prasad, K. 2016. Deep eutectic solvents as efficient solvent system for the extraction of κ-carrageenan from Kappaphycus alvarezii. Carbohydrate Polymers 136: 930–935. https://doi.org/10.1016/J.CARBPOL.2015.09.114.

Das, S., Mondal, A. and Balasubramanian, S. 2017. Recent advances in modeling green solvents. Current Opinion in Green and Sustainable Chemistry 5: 37–43. https://doi.org/10.1016/J. COGSC.2017.03.006.

De Marco, I., Riemma, S. and Iannone, R. 2018. Life cycle assessment of supercritical CO_2 extraction of caffeine from coffee beans. The Journal of Supercritical Fluids 133: 393–400. https://doi. org/10.1016/J.SUPFLU.2017.11.005.

de Moura, J., Maurer, D., Yao, L., Wang, T., Jung, S. and Johnson, L. 2016. Characteristics of oil and skim in enzyme-assisted aqueous extraction of soybeans. Journal of the American Oil Chemists' Society 90: 1079–1088.

Delsart, C., Cholet, C., Ghidossi, R., Grimi, N., Gontier, E., Gény, L., Vorobiev, E. and Mietton-Peuchot, M. 2014. Effects of pulsed electric fields on cabernet sauvignon grape berries and on the characteristics of wines. Food and Bioprocess Technology 7(2): 424–436.

DeSimone, J. 2002. Practical approaches to green solvents\r10.1126/science.1069622. Science 297(5582): 799–803. http://www.sciencemag.org/cgi/content/abstract/297/5582/799.

Dias, A., Arroio Sergio, C., Santos, P., Barbero, G., Rezende, C. and Martínez, J. 2016. Effect of ultrasound on the supercritical CO_2 extraction of bioactive compounds from dedo de moça pepper (Capsicum baccatum L. var. pendulum). Ultrasonics Sonochemistry 31: 284–294. https://doi.org/10.1016/J. ULTSONCH.2016.01.013.

Djas, M. and Henczka, M. 2018. Reactive extraction of carboxylic acids using organic solvents and supercritical fluids: A review. Separation and Purification Technology 201: 106–119. https://doi. org/10.1016/J.SEPPUR.2018.02.010.

Dong, Z.Y., Wang, H.H., Li, M.Y., Liu, W. and Zhang, T.H. 2020. Optimization of high-intensity pulsed electric field-assisted extraction of procyanidins from *Vitis amurensis* seeds using response surface methodology. E3S Web of Conferences 189: 02029.

Dunford, N., Irmak, S. and Jonnala, R. 2010. Pressurised solvent extraction of policosanol from wheat straw, germ and bran. Food Chemistry 119(3): 1246–1249. https://doi.org/10.1016/J. FOODCHEM.2009.07.039.

Ekezie, F., Sun, D. and Cheng, J. 2017. Acceleration of microwave-assisted extraction processes of food components by integrating technologies and applying emerging solvents: A review of latest developments. Trends in Food Science and Technology 67: 160–172. https://doi.org/10.1016/j. tifs.2017.06.006.

El Kantar, S., Boussetta, N., Rajha, H.N., Maroun, R.G., Louka, N. and Vorobiev, E. 2018. High voltage electrical discharges combined with enzymatic hydrolysis for extraction of polyphenols and fermentable sugars from orange peels. Food Research International 107(November 2017): 755–762. https://doi.org/10.1016/j.foodres.2018.01.070.

Eyiz, V., Tontul, I. and Turker, S. 2019. Optimization of green extraction of phytochemicals from red grape pomace by homogenizer assisted extraction. Journal of Food Measurement and Characterization 0123456789. https://doi.org/10.1007/s11694-019-00265-7.

Ez, Y., Fadil, M., Bousta, D., El, A., Lalami, O., Lachkar, M. and Farah, A. 2021. Ultrasound-assisted extraction of phenolic compounds from moroccan Lavandula stoechas L.: optimization using response surface methodology. Journal of Chemistry 2021: 1–11.

Ezra, L., Paquin, L., Sauvager, A., Tomasi, S. and Mulholland, S. 2016. Natural deep eutectic solvents as a new extraction media to extract compounds from Pseudevernia furfuracea (Parmeliaceae). Planta Medica 81: S1–S381.

Fanali, C., Gallo, V., Della Posta, S., Dugo, L., Mazzeo, L., Cocchi, M., Piemonte, V. and De Gara, L. 2021. Choline chloride – lactic acid-based NADES as an extraction medium in a response surface methodology-optimized method for the extraction of phenolic compounds from hazelnut skin. Molecules 26: 2652.

Fu, Y., Liu, W., Zu, Y., Tong, M., Li, S., Yan, M., Efferth, T. and Luo, H. 2008. Enzyme assisted extraction of luteolin and apigenin from pigeonpea [Cajanuscajan (L.) Millsp.] leaves. Food Chemistry 111(2): 508–512. https://doi.org/10.1016/J.FOODCHEM.2008.04.003.

Gai, Q., Jiao, J., Mu, P., Wang, W., Luo, M., Li, C., Zu, Y., Wei, F. and Fu, Y. 2013. Microwave-assisted aqueous enzymatic extraction of oil from Isatis indigotica seeds and its evaluation of physicochemical properties, fatty acid compositions and antioxidant activities. Industrial Crops and Products 45: 303–311. https://doi.org/10.1016/J.INDCROP.2012.12.050.

Gallo, M., Ferracane, R., Graziani, G., Ritieni, A. and Fogliano, V. 2010. Microwave assisted extraction of phenolic compounds from four different spices. Molecules 15: 6365–6374.

Gao, Y., Wang, S., Dang, S., Han, S., Yun, C., Wang, W. and Wang, H. 2021. Optimized ultrasound-assisted extraction of total polyphenols from Empetrum nigrum and its bioactivities. Journal of Chromatography B 1173(January): 122699. https://doi.org/10.1016/j.jchromb.2021.122699.

Garcia-Mendoza, M., Paula, J., Paviani, L., Cabral, F. and Martinez-Correa, H. 2015. Extracts from mango peel by-product obtained by supercritical CO_2 and pressurized solvent processes. LWT - Food Science and Technology 62(1): 131–137. https://doi.org/10.1016/J.LWT.2015.01.026.

Garcia-Salas, P., Morales-Soto, A., Segura-Carretero, A. and Fernández-Gutiérrez, A. 2010. Phenolic-compound-extraction systems for fruit and vegetable samples. Molecules 15(12): 8813–8826. https://doi.org/10.3390/molecules15128813.

García, A., Brenes, M., José Moyano, M., Alba, J., García, P. and Garrido, A. 2001. Improvement of phenolic compound content in virgin olive oil by using enzymes during malaxation. Journal of Food Engineering 48(3): 189–194. https://doi.org/10.1016/S0260-8774(00)00157-6.

García, A., Rodríguez-juan, E., Rodríguez-gutiérrez, G., Rios, J. and Fernández-bolaños, J. 2016. Extraction of phenolic compounds from virgin olive oil by deep eutectic solvents (DESs). Food Chemistry 197: 554–561. https://doi.org/10.1016/j.foodchem.2015.10.131.

Ghafoor, K., Choi, Y., Jeon, J. and Jo, I. 2009. Optimization of ultrasound-assisted extraction of phenolic compounds, antioxidants, and anthocyanins from grape (*Vitis vinifera*) seeds. Journal of Agricultural and Food Chemistry 57(11): 4988–4994. https://doi.org/10.1021/jf9001439.

Giacometti, J., Bursać Kovačević, D., Putnik, P., Gabrić, D., Bilušić, T., Krešić, G., Stulić, V., Barba, F., Chemat, F., Barbosa-Cánovas, G. and Režek Jambrak, A. 2018. Extraction of bioactive compounds and essential oils from mediterranean herbs by conventional and green innovative techniques: A review. Food Research International 113(March): 245–262. https://doi.org/10.1016/j.foodres.2018.06.036.

Gomez, A., Biswas, A., Tadini, C., Vermillion, K., Buttrum, M. and Cheng, H. 2018. Effects of microwave and water incorporation on natural deep eutectic solvents (NADES) and their extraction properties. Advances in Food Science and Engineering 2. https://doi.org/10.22606/afse.2018.24004.

Goula, A. 2013. Ultrasound-assisted extraction of pomegranate seed oil - Kinetic modeling. Journal of Food Engineering 117(4): 492–498. https://doi.org/10.1016/j.jfoodeng.2012.10.009.

Grigonis, D., Venskutonis, P., Sivik, B., Sandahl, M. and Eskilsson, C. 2005. Comparison of different extraction techniques for isolation of antioxidants from sweet grass (Hierochloë odorata). The Journal of Supercritical Fluids 33(3): 223–233. https://doi.org/10.1016/J.SUPFLU.2004.08.006.

Grijó, D., Piva, G., Osorio, I. and Cardozo-Filho, L. 2019. Hemp (*Cannabis sativa* L.) seed oil extraction with pressurized n-propane and supercritical carbon dioxide. The Journal of Supercritical Fluids 143: 268–274. https://doi.org/10.1016/J.SUPFLU.2018.09.004.

Grozdanova, T., Trusheva, B., Alipieva, K., Popova, M., Dimitrova, L., Najdenski, H., Zaharieva, M.M., Ilieva, Y., Vasileva, B., Miloshev, G. and Georgieva, M. 2020. Extracts of medicinal plants with natural deep eutectic solvents: enhanced antimicrobial activity and low genotoxicity. BMC Chemistry 14(73): 1–9. https://doi.org/10.1186/s13065-020-00726-x.

Guderjan, M., Töpfl, S., Angersbach, A. and Knorr, D. 2005. Impact of pulsed electric field treatment on the recovery and quality of plant oils. Journal of Food Engineering 67(3): 281–287. https://doi.org/10.1016/J.JFOODENG.2004.04.029.

Guedes, A., Gião, M., Matias, A., Nunes, A., Pintado, M., Duarte, C. and Malcata, X. 2013. Supercritical fluid extraction of carotenoids and chlorophylls a, b and c, from a wild strain of *Scenedesmus obliquus* for use in food processing. Journal of Food Engineering 116(2): 478–482. https://doi.org/10.1016/J.JFOODENG.2012.12.015.

Guo, X., Zhao, W., Liao, X., Hu, X., Wu, J. and Wang, X. 2017. Extraction of pectin from the peels of pomelo by high-speed shearing homogenization and its characteristics. LWT - Food Science and Technology 79: 640–646. https://doi.org/10.1016/J.LWT.2016.12.001.

Gutiérrez-Arnillas, E., Álvarez, M., Deive, F., Rodríguez, A. and Sanromán, M. 2016. New horizons in the enzymatic production of biodiesel using neoteric solvents. Renewable Energy 98: 92–100. https://doi.org/10.1016/J.RENENE.2016.02.058.

Hao, J., Han, W., Huang, S., Xue, B. and Deng, X. 2002. Microwave-assisted extraction of artemisinin from *Artemisia annua* L. S. Separation and Purification Technology 28: 191–196.

Hashemi, S., Michiels, J., Asadi Yousefabad, S. and Hosseini, M. 2015. Kolkhoung (Pistacia khinjuk) kernel oil quality is affected by different parameters in pulsed ultrasound-assisted solvent extraction. Industrial Crops and Products 70: 28–33. https://doi.org/10.1016/J.INDCROP.2015.03.023.

Hayat, K., Zhang, X., Farooq, U., Abbas, S., Xia, S., Jia, C., Zhong, F. and Zhang, J. 2010. Effect of microwave treatment on phenolic content and antioxidant activity of citrus mandarin pomace. Food Chemistry 123(2): 423–429. https://doi.org/10.1016/J.FOODCHEM.2010.04.060.

Heemann, A.C.W., Heemann, R., Kalegari, P., Spier, M.R. and Santin, E. 2019. Enzyme-assisted extraction of polyphenols from green yerba mate. Brazilian Journal of Food Technology 22: e2017222.

Henderson, R., Jiménez-González, C., Constable, D., Alston, S., Inglis, G., Fisher, G., Sherwood, J., Binks, S. and Curzons, A. 2011. Expanding GSK's solvent selection guide - Embedding sustainability into solvent selection starting at medicinal chemistry. Green Chemistry 13(4): 854–862. https://doi.org/10.1039/c0gc00918k.

Herrero, M., Cifuentes, A. and Ibañez, E. 2006. Sub- and supercritical fluid extraction of functional ingredients from different natural sources: Plants, food-by-products, algae and microalgae: A review. Food Chemistry 98(1): 136–148. https://doi.org/10.1016/J.FOODCHEM.2005.05.058.

Hiew, C., Lee, L., Junus, S., Tan, Y., Chai, T. and Ee, K. 2021. Optimization of microwave-assisted extraction and the effect of microencapsulation on mangosteen (*Garcinia mangostana* L.) rind extract. Food Science and Technology 1–10.

Hou, Q., Li, W., Ju, M., Liu, L., Chen, Y., Yang, Q. and Wang, J. 2015. Separation of polysaccharides from rice husk and wheat bran using solvent system consisting of BMIMOAc and DMI. Carbohydrate Polymers 133: 517–523. https://doi.org/10.1016/J.CARBPOL.2015.07.059.

Hou, X., Lin, K., Li, A., Yang, L. and Fu, M. 2018. Effect of constituents molar ratios of deep eutectic solvents on rice straw fractionation efficiency and the micro-mechanism investigation. Industrial Crops and Products 120: 322–329. https://doi.org/10.1016/J.INDCROP.2018.04.076.

Hu, B., Zhou, K., Liu, Y., Liu, A., Zhang, Q., Han, G. and Liu, S. 2018. Optimization of microwave-assisted extraction of oil from tiger nut (*Cyperus esculentus* L.) and its quality evaluation. Industrial Crops and Products 115(February): 290–297. https://doi.org/10.1016/j.indcrop.2018.02.034.

Hu, X., Zhang, B., Dong, S., Zhao, Y. and Gao, Y. 2014. Hydrolisis of soybean by-products to prepare reducing sugar in ionic liquids. Asian Journal of Chemistry 26(24): 8475–8478.

Huang, Y, Feng, F., Jiang, J., Qiao, Y., Wu, T., Voglmeir, J. and Chen, Z. 2016. Green and efficient extraction of rutin from tartary buckwheat hull by using natural deep eutectic solvents. Food Chemistry 221: 1400–1407. https://doi.org/10.1016/j.foodchem.2016.11.013.

Huang, Yao, Feng, F., Jiang, J., Qiao, Y., Wu, T., Voglmeir, J. and Chen, Z.G. 2017. Green and efficient extraction of rutin from tartary buckwheat hull by using natural deep eutectic solvents. Food Chemistry. https://doi.org/10.1016/j.foodchem.2016.11.013.

Ibrahim, N. and Zaini, M. 2018. Microwave-assisted solvent extraction of castor oil from castor seeds. Chinese Journal of Chemical Engineering 26(12): 2516–2522. https://doi.org/10.1016/J.CJCHE.2018.07.009.

Jacotet-Navarro, M., Rombaut, N., Deslis, S., Fabiano-Tixier, A.-S., Pierre, F.-X., Bily, A. and Chemat, F. 2016. Towards a "dry" bio-refinery with-out solvents or added water using microwaves and ultrasound for total valorization of fruit and vegetable by-products. Green Chemistry 18(10): 3106–3115.

Jaeger, H., Schulz, M., Lu, P. and Knorr, D. 2012. Adjustment of milling, mash electroporation and pressing for the development of a PEF assisted juice production in industrial scale. Innovative Food Science and Emerging Technologies 14: 46–60. https://doi.org/10.1016/J.IFSET.2011.11.008.

Jerman, T., Trebse, P. and Vodopivec, M. 2010. Ultrasound-assisted solid liquid extraction (USLE) of olive fruit (*Olea europaea*) phenolic compounds. Food Chemistry 123: 175–182. https://doi.org/10.1016/j.foodchem.2010.04.006.

Jiang, L., Hua, D., Wang, Z. and Xu, S. 2010. Aqueous enzymatic extraction of peanut oil and protein hydrolysates. Food and Bioproducts Processing 88(2–3): 233–238. https://doi.org/10.1016/J.FBP.2009.08.002.

Jiao, J., Li, Z., Gai, Q., Li, X., Wei, F., Fu, Y. and Ma, W. 2014. Microwave-assisted aqueous enzymatic extraction of oil from pumpkin seeds and evaluation of its physicochemical properties, fatty acid compositions and antioxidant activities. Food Chemistry 147: 17–24. https://doi.org/10.1016/J.FOODCHEM.2013.09.079.

Juliano, P., Bainczyk, F., Swiergon, P., Supriyatna, M., Guillaume, C., Ravetti, L., Canamasas, P., Cravotto, G. and Xu, X. 2017. Extraction of olive oil assisted by high-frequency ultrasound standing waves. Ultrasonics Sonochemistry 38: 104–114. https://doi.org/10.1016/J.ULTSONCH.2017.02.038.

Ke, L. and Chen, H. 2016. Homogenate extraction of crude polysaccharides from *Lentinus edodes* and evaluation of the antioxidant activity. Food Science and Technology 36(3): 533–539. https://doi.org/10.1590/1678-457x.00916.

Khan, A., Man, Z., Bustam, M., Nasrullah, A., Ullah, Z., Sarwono, A., Shah, F. and Muhammad, N. 2018. Efficient conversion of lignocellulosic biomass to levulinic acid using acidic ionic liquids. Carbohydrate Polymers 181(October 2017): 208–214. https://doi.org/10.1016/j.carbpol.2017.10.064.

Khezeli, T., Daneshfar, A. and Sahraei, R. 2016. A green ultrasonic-assisted liquid–liquid microextraction based on deep eutectic solvent for the HPLC-UV determination of ferulic, caffeic and cinnamic acid from olive, almond, sesame and cinnamon oil. Talanta 150: 577–585.

Kim, M., Sowndhararajan, K., Park, S. and Kim, S. 2018. Effect of inhalation of isomers, (+)-α-pinene and (+)-β-pinene on human electroencephalographic activity according to gender difference. European Journal of Integrative Medicine 17(October 2017): 33–39. https://doi.org/10.1016/j.eujim.2017.11.005.

Kiyonga, A.N., Hong, G., Kim, H.S., Suh, Y.G. and Jung, K. 2021. Facile and rapid isolation of oxypeucedanin hydrate and byakangelicin from Angelica dahurica by using [Bmim] Tf2N ionic liquid. Molecules 26(4): 830.

Knez, Ž., Markočič, E., Leitgeb, M., Primožič, M., Knez Hrnčič, M. and Škerget, M. 2014. Industrial applications of supercritical fluids: A review. Energy 77: 235–243. https://doi.org/10.1016/J. ENERGY.2014.07.044.

Knorr, D. 1993. Effects of high-hydrostatic-pressure processes on food safety and quality. Food Technology 47(6): 156–161.

Kruis, A., Levisson, M., Mars, A., van der Ploeg, M., Garcés Daza, F., Ellena, V., Kengen, S.W.M., van der Oost, J. and Weusthuis, R.A. 2017. Ethyl acetate production by the elusive alcohol acetyltransferase from yeast. Metabolic Engineering 41: 92–101. https://doi.org/10.1016/j.ymben.2017.03.004.

Kumar, A.K., Parikh, B.S. and Pravakar, M. 2016. Natural deep eutectic solvent mediated pretreatment of rice straw: bioanalytical characterization of lignin extract and enzymatic hydrolysis of pretreated biomass residue. Environmental Science and Pollution Research 23(10): 9265–9275. https://doi. org/10.1007/s11356-015-4780-4.

Kumar, C., Benal, M.M., Prasad, B.D., Krupashankara, M.S. and Kulkarni, R.S. 2018. Microwave assisted extraction of oil from pongamia pinnata seeds. Materials Today: Proceedings 5(1): 2960–2964. https://doi.org/10.1016/j.matpr.2018.01.094.

Lafka, T., Lazou, A., Sinanoglou, V. and Lazos, E. 2011. Phenolic and antioxidant potential of olive oil mill wastes. Food Chemistry 125(1): 92–98. https://doi.org/10.1016/J.FOODCHEM.2010.08.041.

Lakka, A., Bozinou, E., Makris, D.P. and Lalas, S.I. 2021. Evaluation of pulsed electric field polyphenol extraction from *Vitis vinifera, Sideritis scardica* and *Crocus sativus*. ChemEngineering 5: 25.

Larriba, M., Omar, S., Navarro, P., García, J., Rodrígueza, F. and Gonzalez-Miquel, M. 2016. Recovery of tyrosol from aqueous streams using hydrophobic ionic liquids: a first step towards developing sustainable processes for olive mill wastewater (OMW) management. RSC Advances 6(23): 18751–18762.

Latif, S., Diosady, L. and Anwar, F. 2008. Enzyme-assisted aqueous extraction of oil and protein from canola (*Brassica napus* L.) seeds. European Journal of Lipid Science and Technology 110: 887–892.

Latif, S. and Anwar, F. 2009. Effect of aqueous enzymatic processes on sunflower oil quality. JAOCS, Journal of the American Oil Chemists' Society 86: 393–400.

Lee, J., Kim, H., Ko, J., Jang, J., Kim, G., Lee, J., Nah, J. and Jeon, Y. 2016. Rapid preparation of functional polysaccharides from Pyropia yezoensis by microwave-assistant rapid enzyme digest system. Carbohydrate Polymers 153: 512–517. https://doi.org/10.1016/J.CARBPOL.2016.07.122.

Leong, S., Burritt, D. and Oey, I. 2016. Evaluation of the anthocyanin release and health-promoting properties of Pinot Noir grape juices after pulsed electric fields. Food Chemistry 196: 833–841. https://doi.org/10.1016/J.FOODCHEM.2015.10.025.

Li, B., Smith, B. and Hossain, M. 2006. Extraction of phenolics from citrus peels: II. Enzyme-assisted extraction method. Separation and Purification Technology 48(2): 189–196. https://doi. org/10.1016/J.SEPPUR.2005.07.019.

Li, C., Wang, Q. and Zhao, Z. 2008. Acid in ionic liquid: An efficient system for hydrolysis of lignocellulose. Green Chemistry 10(2): 177–182. https://doi.org/10.1039/b711512a.

Li, F., Mao, Y.-D., Wang, Y.-F., Raza, A., Qiu, L.-P. and Xu, X.-Q. 2017. Optimization of ultrasonic-assisted enzymatic extraction conditions for improving total phenolic content, antioxidant and antitumor activities *in vitro* from Trapa quadrispinosa Roxb. residues. Molecules 22: 396. https:// doi.org/10.3390/molecules22030396.

Li, H., Pordesimo, L. and Weiss, J. 2004. High intensity ultrasound-assisted extraction of oil from soybeans. Food Research International 37(7): 731–738. https://doi.org/10.1016/j.foodres.2004.02.016.

Li, H., Deng, Z., Wu, T., Liu, R., Loewen, S. and Tsao, R. 2012. Microwave-assisted extraction of phenolics with maximal antioxidant activities in tomatoes. Food Chemistry 130(4): 928–936. https://doi.org/10.1016/J.FOODCHEM.2011.08.019.

Li, J., Zu, Y., Luo, M., Gu, C., Zhao, C., Efferth, T. and Fu, Y. 2013. Aqueous enzymatic process assisted by microwave extraction of oil from yellow horn (*Xanthoceras sorbifolia* Bunge.) seed kernels and its quality evaluation. Food Chemistry 138(4): 2152–2158. https://doi.org/10.1016/j. foodchem.2012.12.011.

Li, T., Qu, X., Zhang, Q. and Wang, Z. 2012. Ultrasound-assisted extraction and profile characteristics of seed oil from Isatis indigotica Fort. Industrial Crops and Products 35(1): 98–104. https://doi.org/10.1016/j.indcrop.2011.06.013.

Li, Y., Pan, Z., Wang, B., Yu, W., Song, S., Feng, H. and Zhang, J. 2020. Ultrasound-assisted extraction of bioactive alkaloids from Phellodendri amurensis cortex using deep eutectic solvent aqueous solutions. New Journal of Chemistry 44(22): 9172–9178. https://doi.org/10.1039/D0NJ00877J.

Li, Z., Smith, K. and Stevens, G. 2016. The use of environmentally sustainable bio-derived solvents in solvent extraction applications—A review. Chinese Journal of Chemical Engineering 24(2): 215–220. https://doi.org/10.1016/J.CJCHE.2015.07.021.

Liu, C., Zhang, S. and Wu, H. 2009. Non-thermal extraction of effective ingredients from Schisandra chinensis baill and the antioxidant activity of its extract. Natural Product Research 23: 1390–1401.

Liu, J., Gasmalla, M., Li, P. and Yang, R. 2016. Enzyme-assisted extraction processing from oilseeds: Principle, processing and application. Innovative Food Science and Emerging Technologies. https://doi.org/10.1016/j.ifset.2016.05.002.

Liu, S. and Mamidipally, P. 2005. Quality comparison of rice bran oil extracted with d-limonene and hexane. Cereal Chemistry 82(2): 209–215. https://doi.org/10.1094/CC-82-0209.

Liu, T., Ma, C., Yang, L., Wang, W., Sui, X., Zhao, C. and Zu, Y. 2013a. Optimization of shikonin homogenate extraction from. Molecules 18: 466–481. https://doi.org/10.3390/molecules18010466.

Liu, T., Ma, C., Yang, L., Wang, W., Sui, X., Zhao, C. and Zu, Y. 2013b. Optimization of Shikonin homogenate extraction from Arnebia euchroma using response surface methodology. Molecules 18(1): 466–481.

Liu, Y., Garzon, J., Friesen, J.B., Zhang, Y., McAlpine, J.B., Lankin, D.C., Chen, S.N. and Pauli, G.F. 2016. Countercurrent assisted quantitative recovery of metabolites from plant-associated natural deep eutectic solvents. Fitoterapia 112: 30–37. https://doi.org/10.1016/j.fitote.2016.04.019.

Liza, M.S., Abdul Rahman, R., Mandana, B., Jinap, S., Rahmat, A., Zaidul, I.S.M. and Hamid, A. 2010. Supercritical carbon dioxide extraction of bioactive flavonoid from Strobilanthes crispus (Pecah Kaca). Food and Bioproducts Processing 88(2–3): 319–326. https://doi.org/10.1016/J.FBP.2009.02.001.

Lončarić, A., Celeiro, M., Jozinovic, A., Jelinić, J., Kovač, T., Jokić, S. and Lores, M. 2020. Green extraction methods for extraction of polyphenolic compounds from blueberry pomace. Foods 9(11): 1521.

Long, R. and Abdelkader, E. 2011. Mixed-polarity azeotropic solvents for efficient extraction of lipids from nannochloropsis. Microalgae 7(2): 70–73.

López-Fandiño, R. 2006. Functional improvement of milk whey proteins induced by high hydrostatic pressure. Critical Reviews in Food Science and Nutrition 46: 351–363.

López-Gámez, G., Elez-Martínez, P., Quiles-Chuliá, A., Martín-Belloso, O., Hernando-Hernando, I. and Soliva-Fortuny, R. 2021. Effect of pulsed electric fields on carotenoid and phenolic bioaccessibility and their relationship with carrot structure. Food and Function 12(6): 2772–2783.

Luque-García, J. and Luque de Castro, M. 2004. Ultrasound-assisted Soxhlet extraction: an expeditive approach for solid sample treatment. Application to the extraction of total fat from oleaginous seeds. Journal of Chromatography A 1034: 237–242.

Machado, A., Pasquel-Reátegui, J., Barbero, G. and Martínez, J. 2015. Pressurized liquid extraction of bioactive compounds from blackberry (*Rubus fruticosus* L.) residues: a comparison with conventional methods. Food Research International 77: 675–683. https://doi.org/10.1016/J.FOODRES.2014.12.042.

Mantegna, S., Binello, A., Boffa, L., Giorgis, M., Cena, C. and Cravotto, G. 2012. A one-pot ultrasound-assisted water extraction/cyclodextrin encapsulation of resveratrol from Polygonum cuspidatum. Food Chemistry 130(3): 746–750. https://doi.org/10.1016/j.foodchem.2011.07.038.

Marić, M., Grassino, A., Zhu, Z., Barba, F., Brnčić, M. and Rimac Brnčić, S. 2018. An overview of the traditional and innovative approaches for pectin extraction from plant food wastes and by-products: Ultrasound-, microwaves-, and enzyme-assisted extraction. Trends in Food Science and Technology 76: 28–37. https://doi.org/10.1016/J.TIFS.2018.03.022.

Martin, L., Skinner, C. and Marriott, R. 2015. Supercritical extraction of oil seed rape: Energetic evaluation of process scale. The Journal of Supercritical Fluids 105: 55–59. https://doi.org/10.1016/J.SUPFLU.2015.04.017.

Martins, P., Braga, A. and de Rosso, V. 2017. Can ionic liquid solvents be applied in the food industry? Trends in Food Science and Technology 66: 117–124. https://doi.org/10.1016/J.TIFS.2017.06.002.

Masci, A., Coccia, A., Lendaro, E., Mosca, L., Paolicelli, P. and Cesa, S. 2016. Evaluation of different extraction methods from pomegranate whole fruit or peels and the antioxidant and antiproliferative activity of the polyphenolic fraction. Food Chemistry 202: 59–69. https://doi.org/10.1016/J.FOODCHEM.2016.01.106.

Mason, T., Riera, E., Vercet, A. and Lopez-Buesa, P. 2005. Application of ultrasound. pp. 323–343. *In*: Sun, D. (ed.). Emerging Technologies for Food Process. Elsevier Academic Press.

Matthäus, B. 2002. Antioxidant activity of extracts obtained from residues of different oilseeds. Journal of Agricultural and Food Chemistry 50(12): 3444–3452. https://doi.org/10.1021/jf011440s.

Melini, V. and Melini, F. 2021. Modelling and optimization of ultrasound-assisted extraction of phenolic compounds from black quinoa by response surface methodology. Molecule 26: 3616.

Meneses, M., Caputo, G., Scognamiglio, M., Reverchon, E. and Adami, R. 2015. Antioxidant phenolic compounds recovery from Mangifera indica L. by-products by supercritical antisolvent extraction. Journal of Food Engineering 163: 45–53. https://doi.org/10.1016/J.JFOODENG.2015.04.025.

Minjares-Fuentes, R., Femenia, A., Garau, M.C., Meza-Velázquez, J.A., Simal, S. and Rosselló, C. 2014. Ultrasound-assisted extraction of pectins from grape pomace using citric acid: A response surface methodology approach. Carbohydrate Polymers 106(1): 179–189. https://doi.org/10.1016/j.carbpol.2014.02.013.

Mishra, D., Shukla, A., Dixit, A. and Singh, K. 2005. Aqueous enzymatic extraction of oil from mandarin peels. Journal of Oleo Science 54: 355–359.

Mouratoglou, E., Malliou, V. and Makris, D. 2016. Novel glycerol-based natural eutectic mixtures and their efficiency in the ultrasound-assisted extraction of antioxidant polyphenols from agri-food waste biomass. Waste and Biomass Valorization 7(6): 1377–1387. https://doi.org/10.1007/s12649-016-9539-8.

Mrkonjić, Ž., Rakić, D., Kaplan, M., Teslić, N., Zeković, Z. and Pavlić, B. 2021. Pressurized-liquid extraction as an efficient method for valorization of Thymus serpyllum herbal dust towards sustainable production of antioxidants. Molecules 26(9): 2548.

Mustafa, A. and Turner, C. 2011. Pressurized liquid extraction as a green approach in food and herbal plants extraction: A review. Analytica Chimica Acta 703(1): 8–18. https://doi.org/10.1016/j.aca.2011.07.018.

Najafian, L., Ghodsvali, A., Haddad Khodaparast, M. and Diosady, L. 2009. Aqueous extraction of virgin olive oil using industrial enzymes. Food Research International 42(1): 171–175. https://doi.org/10.1016/J.FOODRES.2008.10.002.

Nakkeeran, E., Umesh-Kumar, S. and Subramanian, R. 2011. Aspergillus carbonarius polygalacturonases purified by integrated membrane process and affinity precipitation for apple juice production. Bioresource Technology 102(3): 3293–3297. https://doi.org/10.1016/J.BIORTECH.2010.10.048.

Nishad, J., Saha, S. and Kaur, C. 2019. Enzyme- and ultrasound-assisted extractions of polyphenols from Citrus sinensis (cv. Malta) peel: A comparative study. Journal of Food Processing and Preservation 43(8): e14046. https://doi.org/10.1111/jfpp.14046.

Nkhili, E., Tomao, V., El Hajji, H., El Boustani, E., Chemat, F. and Dangles, O. 2009. Microwave-assisted water extraction of green tea polyphenols. Phytochemical Analysis 20: 408–415.

Nobre, B., Gouveia, L., Matos, P., Cristino, A., Palavra, A. and Mendes, R. 2012. Supercritical extraction of lycopene from tomato industrial wastes with ethane. Molecules 17(7): 8397–8407.

Nyam, K.L., Tan, C.P., Lai, O.M., Long, K. and Che Man, Y.B. 2010. Optimization of supercritical fluid extraction of phytosterol from roselle seeds with a central composite design model. Food and Bioproducts Processing 88(2–3): 239–246. https://doi.org/10.1016/J.FBP.2009.11.002.

Oliveira, D.A., Mezzomo, N., Gomes, C. and Ferreira, S.R.S. 2017. Encapsulation of passion fruit seed oil by means of supercritical antisolvent process. The Journal of Supercritical Fluids 129: 96–105. https://doi.org/10.1016/J.SUPFLU.2017.02.011.

Osmar, G., Bilibio, D., Zanella, O., Luis, A., Paulo, J., Carniel, N., Pereira, P. and Priamo, W.L. 2018. Pressurized liquid extraction of polyphenols from Goldenberry: Influence on antioxidant activity and chemical composition. Food and Bioproducts Processing 112: 63–68. https://doi.org/10.1016/j.fbp.2018.09.001.

Ozturk, B., Parkinson, C. and Gonzalez-miquel, M. 2018. Extraction of polyphenolic antioxidants from orange peel waste using deep eutectic solvents. Separation and Purification Technology 206: 1–13. https://doi.org/10.1016/j.seppur.2018.05.052.

Pagano, I., Sánchez-Camargo, A., Mendiola, J., Campone, L., Cifuentes, A., Rastrelli, L. and Ibañez, E. 2018. Selective extraction of high-value phenolic compounds from distillation wastewater of basil (*Ocimum basilicum* L.) by pressurized liquid extraction. Electrophoresis 39(15): 1884–1891. https://doi.org/10.1002/elps.201700442.

Palma, A., Díaz, M.J., Ruiz-montoya, M. and Morales, E. 2021. Ultrasonics sonochemistry ultrasound extraction optimization for bioactive molecules from Eucalyptus globulus leaves through antioxidant activity. Ultrasonics Sonochemistry 76: 105654. https://doi.org/10.1016/j.ultsonch.2021.105654.

Pan, X., Niu, G. and Liu, H. 2003. Microwave-assisted extraction of tea polyphenols and tea caffeine from green tea leaves. Chemical Engineering and Processing: Process Intensification 42(2): 129–133. https://doi.org/10.1016/S0255-2701(02)00037-5.

Paradiso, V., Clemente, A., Summo, C., Pasqualone, A. and Caponio, F. 2016. Towards green analysis of virgin olive oil phenolic compounds: Extraction by a natural deep eutectic solvent and direct spectrophotometric detection. Food Chemistry 212: 43–47. https://doi.org/10.1016/j.foodchem.2016.05.082.

Pardede, C., Tambun, R., Fitri, M.D. and Husna, R. 2020. Extraction of tannin from garlic skins by using microwave with ethanol as solvent. IOP Conference Series: Materials Science and Engineering 801(1): 012054. https://doi.org/10.1088/1757-899X/801/1/012054.

Parniakov, O., Barba, F., Grimi, N., Lebovka, N. and Vorobiev, E. 2016. Extraction assisted by pulsed electric energy as a potential tool for green and sustainable recovery of nutritionally valuable compounds from mango peels. Food Chemistry 192: 842–848. https://doi.org/10.1016/J.FOODCHEM.2015.07.096.

Passos, C., Yilmaz, S., Silva, C. and Coimbra, M. 2009. Enhancement of grape seed oil extraction using a cell wall degrading enzyme cocktail. Food Chemistry 115(1): 48–53. https://doi.org/10.1016/J.FOODCHEM.2008.11.064.

Passos, H., Freire, M. and Coutinho, J. 2014. Ionic liquid solutions as extractive solvents for value-added compounds from biomass. Green Chemistry 16(12): 4786–4815. https://doi.org/10.1039/c4gc00236a.

Peng, X., Duan, M., Yao, X., Zhang, Y., Zhao, C., Zu, Y. and Fu, Y. 2016. Green extraction of five target phenolic acids from Lonicerae japonicae Flos with deep eutectic solvent. Separation and Purification Technology 157: 249–257. https://doi.org/10.1016/J.SEPPUR.2015.10.065.

Pereira, G.A., Molina, G., Arruda, H.S. and Pastore, G.M. 2017. Optimizing the homogenizer-assisted extraction (HAE) of total phenolic compounds from banana peel. Journal of Food Process Engineering 40(3): e12438.

Pérez, C., Ruiz Del Castillo, M., Gil, C., Blanch, G. and Flores, G. 2015. Supercritical fluid extraction of grape seeds: Extract chemical composition, antioxidant activity and inhibition of nitrite production in LPS-stimulated Raw 264.7 cells. Food and Function 6(8): 2607–2613. https://doi.org/10.1039/c5fo00325c.

Pilar, S., Buelvas-puello, L.M., Martinez-correa, H.A., Cifuentes, A., Ferreira, S.R.S. and Guti, L. 2021. Microwave-assisted extraction of phenolic compounds with antioxidant and anti-proliferative activities from supercritical CO_2 pre-extracted mango peel as valorization strategy. 137(October 2020). https://doi.org/10.1016/j.lwt.2020.110414.

Pinela, J., Prieto, M.A., Barros, L., Carvalho, A., Beatriz, M.P.P., Saraiva, J. and Ferreira, I. 2017. Cold extraction of phenolic compounds from watercress by high hydrostatic pressure: Process modelling and optimization. Separation and Purification Technology 192: 501–512. https://doi.org/10.1016/j.seppur.2017.10.007.

Pingret, D., Fabiano-Tixier, A. and Chemat, F. 2013. Ultrasound-assisted extraction. pp. 89–105. *In*: Rostagno, M., Prado, J. and Kraus, G. (eds.). Natural Product Extraction: Principles and Applications. Royal Society of Chemistry.

Plaza, M. and Turner, C. 2015. Pressurized hot water extraction of bioactives. TrAC Trends in Analytical Chemistry 71: 39–54. https://doi.org/10.1016/J.TRAC.2015.02.022.

Poojary, M., Orlien, V., Passamonti, P. and Olsen, K. 2017. Enzyme-assisted extraction enhancing the umami taste amino acids recovery from several cultivated mushrooms. Food Chemistry 234: 236–244. https://doi.org/10.1016/J.FOODCHEM.2017.04.157.

Prasad, K.N., Yang, E., Yi, C., Zhao, M. and Jiang, Y. 2009. Effects of high pressure extraction on the extraction yield, total phenolic content and antioxidant activity of longan fruit pericarp. Innovative Food Science and Emerging Technologies 10: 155–159. https://doi.org/10.1016/j.ifset.2008.11.007.

Prasad, K., Yang, B., Zhao, M., Sun, J., Wei, X. and Jiang, Y. 2010. Effects of high pressure or ultrasonic treatment on extraction yield and antioxidant activity of pericarp tissues of longan fruit. Journal of Food Biochemistry 34: 838–855.

Procentese, A., Johnson, E., Orr, V., Garruto Campanile, A., Wood, J., Marzocchella, A. and Rehmann, L. 2015. Deep eutectic solvent pretreatment and subsequent saccharification of corncob. Bioresource Technology 192: 31–36. https://doi.org/10.1016/j.biortech.2015.05.053.

Pronyk, C. and Mazza, G. 2009. Design and scale-up of pressurized fluid extractors for food and bioproducts. Journal of Food Engineering 95(2): 215–226. https://doi.org/10.1016/J.JFOODENG.2009.06.002.

Ptichkina, N., Markina, O. and Rumyantseva, G. 2008. Pectin extraction from pumpkin with the aid of microbial enzymes. Food Hydrocolloids 22(1): 192–195. https://doi.org/10.1016/J.FOODHYD.2007.04.002.

Puertolas, E., Saldana, G. and Raso, J. 2016. Pulsed electric field treatment for fruit and vegetable processing. *In*: Miklavcic, D. (ed.). Handbook of Electroporation. Springer International Publishing.

Puri, M., Sharma, D. and Barrow, C. 2012. Enzyme-assisted extraction of bioactives from plants. Trends in Biotechnology 30(1): 37–44. https://doi.org/10.1016/J.TIBTECH.2011.06.014.

Putnik, P., Barba, F., Španić, I., Zorić, Z., Dragović-Uzelac, V. and Bursać Kovačević, D. 2017. Green extraction approach for the recovery of polyphenols from Croatian olive leaves (*Olea europea*). Food and Bioproducts Processing 106: 19–28. https://doi.org/10.1016/J.FBP.2017.08.004.

Putnik, P., Lorenzo, J., Barba, F., Roohinejad, S., Režek Jambrak, A., Granato, D., Montesano, D. and Bursać Kovačević, D. 2018. Novel food processing and extraction technologies of high-added value compounds from plant materials. Foods 7(7): 106. https://doi.org/10.3390/foods7070106.

Qi, X.L., Peng, X., Huang, Y.Y., Li, L., Wei, Z.F., Zu, Y.G. and Fu, Y.J. 2015. Green and efficient extraction of bioactive flavonoids from Equisetum palustre L. by deep eutectic solvents-based negative pressure cavitation method combined with macroporous resin enrichment. Industrial Crops and Products 70: 142–148. https://doi.org/10.1016/j.indcrop.2015.03.026.

Radošević, K., Ćurko, N., Gaurina Srček, V., Cvjetko Bubalo, M., Tomašević, M., Kovačević Ganić, K. and Radojčić Redovniković, I. 2016. Natural deep eutectic solvents as beneficial extractants for enhancement of plant extracts bioactivity. LWT 73: 45–51. https://doi.org/10.1016/J.LWT.2016.05.037.

Radzali, S.A., Markom, M. and Saleh, N.M. 2020. Co-solvent selection for supercritical fluid extraction (SFE) of phenolic compounds from Labisia pumila. Molecules 25(24): 5859.

Rajha, H.N., Boussetta, N., Louka, N., Maroun, R.G. and Vorobiev, E. 2015. Effect of alternative physical pretreatments (pulsed electric field, high voltage electrical discharges and ultrasound) on the dead-end ultrafiltration of vine-shoot extracts. Separation and Purification Technology 146: 243–251.

Ramos, L., Kristenson, E. and Brinkman, Uat. 2002. Current use of pressurised liquid extraction and subcritical water extraction in environmental analysis. Journal of Chromatography A 975(1): 3–29. https://doi.org/10.1016/S0021-9673(02)01336-5.

Rashed, M., Tong, Q., Nagi, A., Li, J., Khan, N., Chen, L., Rotail, A. and Bakry, A. 2017. Isolation of essential oil from Lavandula angustifolia by using ultrasonic-microwave assisted method preceded by enzymolysis treatment, and assessment of its biological activities. Industrial Crops and Products 100: 236–245. https://doi.org/10.1016/J.INDCROP.2017.02.033.

Roselló-Soto, E., Koubaa, M., Moubarik, A., Lopes, R.P., Saraiva, J.A., Boussetta, N. and Barba, F.J. 2015. Emerging opportunities for the effective valorization of wastes and by-products generated during olive oil production process: Non-conventional methods for the recovery of high-added value compounds. Trends in Food Science and Technology 45(2): 296–310.

Roselló-Soto, E., Parniakov, O., Deng, Q., Patras, A., Koubaa, M., Grimi, N., Boussetta, N., Brijesh, K., Tiwari, E., Vorobiev, E., Lebovka, N. and Barba, F. 2016. Application of non-conventional extraction methods: toward a sustainable and green production of valuable compounds from mushrooms. Food Engineering Reviews 8: 214–234.

Ruesgas-Ramon, M., Figueroa-Espinoza, M. and Durand, E. 2017. Application of deep eutectic solvents (DES) for phenolic compounds extraction: overview, challenges, and opportunities. Journal of Agricultural and Food Chemistry 65: 3591–3601. https://doi.org/10.1021/acs.jafc.7b01054.

Rui, H., Zhang, L., Li, Z. and Pan, Y. 2009. Extraction and characteristics of seed kernel oil from white pitaya. Journal of Food Engineering 93(4): 482–486. https://doi.org/10.1016/j.jfoodeng.2009.02.016.

Saha, K., Dasgupta, J., Chakraborty, S., Antunes, F., Sikder, J., Curcio, S., dos Santos, J., Arafat, H. and da Silva, S. 2017. Optimization of lignin recovery from sugarcane bagasse using ionic liquid aided pretreatment. Cellulose 24(8): 3191–3207. https://doi.org/10.1007/s10570-017-1330-x.

Şahin, S. and Şamli, R. 2013. Optimization of olive leaf extract obtained by ultrasound-assisted extraction with response surface methodology. Ultrasonics Sonochemistry 20(1): 595–602. https://doi.org/10.1016/j.ultsonch.2012.07.029.

Samaram, S., Mirhosseini, H., Tan, C., Ghazali, H., Bordbar, S. and Serjouie, A. 2015. Optimisation of ultrasound-assisted extraction of oil from papaya seed by response surface methodology: Oil recovery, radical scavenging antioxidant activity, and oxidation stability. Food Chemistry 172: 7–17. https://doi.org/10.1016/j.foodchem.2014.08.068.

San Martín, M., Barbosa-Cánovas, G. and Swanson, B. 2002. Food processing by high hydrostatic pressure. Critical Reviews in Food Science and Nutrition 42: 627–645.

Sánchez-Moreno, C., de Ancos, B., Plaza, L., Elez-Martínez, P. and Cano, M. 2009. Nutritional approaches and health-related properties of plant foods processed by high pressure and pulsed electric fields. Critical Reviews in Food Science and Nutrition 49(6): 552–576.

Santaella, M., Orjuela, A. and Narváez, P. 2015. Comparison of different reactive distillation schemes for ethyl acetate production using sustainability indicators. Chemical Engineering and Processing: Process Intensification 96: 1–13. https://doi.org/10.1016/J.CEP.2015.07.027.

Santos-Zea, L., Gutiérrez-Uribe, J. and Benedito, J. 2019. Effect of ultrasound intensification on the supercritical fluid extraction of phytochemicals from Agave salmiana bagasse. The Journal of Supercritical Fluids 144: 98–107. https://doi.org/10.1016/J.SUPFLU.2018.10.013.

Sarrade, S.J., Rios, G.M. and Carlès, M. 1998. Supercritical CO_2 extraction coupled with nanofiltration separation: Applications to natural products. Separation and Purification Technology 14(1–3): 19–25. https://doi.org/10.1016/S1383-5866(98)00056-2.

Satlewal, A., Agrawal, R., Bhagia, S., Sangoro, J. and Ragauskas, A. 2018. Natural deep eutectic solvents for lignocellulosic biomass pretreatment: Recent developments, challenges and novel opportunities. Biotechnology Advances 36(8): 2032–2050. https://doi.org/10.1016/J.BIOTECHADV.2018.08.009.

Segovia, F., Luengo, E., Corral-Pérez, J., Raso, J. and Almajano, M. 2015. Improvements in the aqueous extraction of polyphenols from borage (*Borago officinalis* L.) leaves by pulsed electric fields: Pulsed electric fields (PEF) applications. Industrial Crops and Products 65: 390–396. https://doi.org/10.1016/J.INDCROP.2014.11.010.

Šeremet, D., Karlović, S., Vojvodić Cebin, A., Mandura, A., Ježek, D. and Komes, D. 2021. Extraction of bioactive compounds from different types of tea by high hydrostatic pressure. Journal of Food Processing and Preservation e15751.

Shankar, M., Chhotaray, P., Agrawal, A., Gardas, R., Tamilarasan, K. and Rajesh, M. 2017. Protic ionic liquid-assisted cell disruption and lipid extraction from fresh water Chlorella and Chlorococcum microalgae. Algal Research 25: 228–236. https://doi.org/10.1016/J.ALGAL.2017.05.009.

Sharma, A., Khare, S. and Gupta, M. 2002. Enzyme-assisted aqueous extraction of peanut oil. Journal of the American Oil Chemists' Society 79: 215–218.

Sharma, B.R., Kumar, V., Kumar, S. and Panesar, P.S. 2020. Microwave assisted extraction of phytochemicals from *Ficus racemosa*. Current Research in Green and Sustainable Chemistry 3: 100020. https://doi.org/10.1016/j.crgsc.2020.100020.

Shi, W.G., Zu, Y.G., Zhao, C. and Yang, L. 2009. Homogenate extraction technology of camptothecine and hydroxycamptothecin from Camptotheca acuminata leaves. Journal of Forest Research 20(2): 168–170.

Shouqin, Z., Junjie, Z. and Changzhen, W. 2004. Novel high pressure extraction technology. International Journal of Pharmaceutics 278(2): 471–474. https://doi.org/10.1016/J.IJPHARM.2004.02.029.

Sicaire, A., Vian, M., Fine, F., Carré, P., Tostain, S. and Chemat, F. 2016. Ultrasound induced green solvent extraction of oil from oleaginous seeds. Ultrasonics Sonochemistry 31: 319–329. https://doi.org/10.1016/J.ULTSONCH.2016.01.011.

Sihvonen, M., Järvenpää, E., Hietaniemi, V. and Huopalahti, R. 1999. Advances in supercritical carbon dioxide technologies. Trends in Food Science and Technology 10(6–7): 217–222. https://doi. org/10.1016/S0924-2244(99)00049-7.

Silva, C.C.A.R., Gomes, C.L., Danda, L.J.A., M, A.N.A.E., Carvalho, A.N.A.M.R.D.E., Ximenes, E.C.P.A., Da, R.M.F., Angelos, M.A., Rolim, L.A. and Neto, P.J.R. 2021. Optimized microwave-assisted extraction of polyphenols and tannins from Syzygium cumini (L.) Skeels leaves through an experimental design coupled to a desirability approach. Annals of the Brazilian Academy of Sciences 93(2): 1–13. https://doi.org/10.1590/0001-3765202120190632.

Silva, Y., Ferreira, T., Celli, G. and Brooks, M. 2018. Optimization of lycopene extraction from tomato processing waste using an eco-friendly ethyl lactate–ethyl acetate solvent: a green valorization approach. Waste and Biomass Valorization 0(0): 1–11. https://doi.org/10.1007/s12649-018-0317-7.

Simić, V., Rajković, K., Stojičević, S., Veličković, D., Nikolić, N., Lazić, M. and Karabegović, I. 2016. Optimization of microwave-assisted extraction of total polyphenolic compounds from chokeberries by response surface methodology and artificial neural network. Separation and Purification Technology 160: 89–97. https://doi.org/10.1016/J.SEPPUR.2016.01.019.

Sodeifian, G., Ghorbandoost, S., Sajadian, S. and Saadati Ardestani, N. 2016. Extraction of oil from Pistacia khinjuk using supercritical carbon dioxide: Experimental and modeling. The Journal of Supercritical Fluids 110: 265–274. https://doi.org/10.1016/J.SUPFLU.2015.12.004.

Soquetta, M., Terra, L. and Bastos, C. 2018. Green technologies for the extraction of bioactive compounds in fruits and vegetables. CYTA - Journal of Food 16(1): 400–412. https://doi.org/10.1080/1947633 7.2017.1411978.

Sowbhagya, H. and Chitra, V. 2010. Enzyme-assisted extraction of flavorings and colorants from plant materials. Critical Reviews in Food Science and Nutrition 50: 146–161.

Sparr Eskilsson, C. and Björklund, E. 2000. Analytical-scale microwave-assisted extraction. Journal of Chromatography A 902(1): 227–250. https://doi.org/10.1016/S0021-9673(00)00921-3.

Stavroulias, S. and Panayiotou, C. 2005. Determination of optimum conditions for the extraction of squalene. Chemical and Biochemical Engineering Quarterly 19(4): 373–381.

Sun, K.K. and Wei, J.A. 2020. Study on ultra-high pressure assisted extraction of polysaccharides from Umbilicaria in yellow mountain. Materials Science Forum 980: 187–196. https://doi.org/10.4028/ www.scientific.net/MSF.980.187.

Taamalli, A., Arráez-Román, D., Barrajón-Catalán, E., Ruiz-Torres, V., Pérez-Sánchez, A., Herrero, M., Ibañez, E., Micol, V., Zarrouk, M., Segura-Carretero, A. and Fernández-Gutiérrez, A. 2012. Use of advanced techniques for the extraction of phenolic compounds from Tunisian olive leaves: Phenolic composition and cytotoxicity against human breast cancer cells. Food and Chemical Toxicology 50(6): 1817–1825. https://doi.org/10.1016/J.FCT.2012.02.090.

Tabaraki, R., Heidarizadi, E. and Benvidi, A. 2012. Optimization of ultrasonic-assisted extraction of pomegranate (*Punica granatum* L.) peel antioxidants by response surface methodology. Separation and Purification Technology 98: 16–23. https://doi.org/10.1016/J.SEPPUR.2012.06.038.

Tabtabaei, S. and Diosady, L. 2013. Aqueous and enzymatic extraction processes for the production of food-grade proteins and industrial oil from dehulled yellow mustard flour. Food Research International 52(2): 547–556. https://doi.org/10.1016/J.FOODRES.2013.03.005.

Taghvaei, M., Mahdi, S., Assadpoor, E., Nowrouzieh, S. and Alishah, O. 2014. Optimization of microwave-assisted extraction of cottonseed oil and evaluation of its oxidative stability and physicochemical properties. Food Chemistry 160: 90–97. https://doi.org/10.1016/j.foodchem.2014.03.064.

Takeuchi, T., Pereira, C., Braga, M., Marostica, M., Leal, P. and Meireles, A. 2009. Low-pressure solvent extraction (solid–liquid extraction, microwave assisted, and ultrasound assisted) from condimentary plants. pp. 140–171. *In*: Meireles, A. (ed.). Extracting Bioactive Compounds for Food Products: Theory and Applications. CRC Press.

Teixeira, C., Macedo, G., Macedo, J., da Silva, L. and Rodrigues, A. 2013. Simultaneous extraction of oil and antioxidant compounds from oil palm fruit (*Elaeis guineensis*) by an aqueous enzymatic process. Bioresource Technology 129: 575–581. https://doi.org/10.1016/J.BIORTECH.2012.11.057.

Thao, L., Phan, M., Thien, K., Nguyen, P., Vuong, H.T., Tran, D.D., Xuan, T., Nguyen, P., Hoang, M.N., Mai, T.P. and Nguyen, H.H. 2020. Supercritical fluid extraction of polyphenols from vietnamese Callisia fragrans leaves and antioxidant activity of the extract. Journal of Chemistry 2020: 1–7.

Tian, Y., Xu, Z., Zheng, B. and Martin, L. 2013. Optimization of ultrasonic-assisted extraction of pomegranate (*Punica granatum* L.) seed oil. Ultrasonics Sonochemistry 20: 202–208.

Tiwari, B. 2015. Ultrasound: A clean, green extraction technology. Trends in Analytical Chemistry 71: 100–109. https://doi.org/10.1016/j.trac.2015.04.013.

Toepfl, S., Siemer, C. and Heinz, V. 2014. Effect of high-intensity electric field pulses on solid foods. Emerging Technologies for Food Processing 147–154. https://doi.org/10.1016/B978-0-12-411479-1.00008-5.

Tong, Y., Yan, Y., Jiang, S., Wang, P., Jiang, Y., Bathaie, S., Tong, Y., Guo, D. and Lu. 2018. Homogenate extraction of crocins from saffron optimized by response surface methodology. Journal of Chemistry 2018: 1–6. https://doi.org/10.1155/2018/9649062.

Ünlü, A.E. 2021. Green and non-conventional extraction of bioactive compounds from olive leaves : screening of novel natural deep eutectic solvents and investigation of process parameters. Waste and Biomass Valorization. https://doi.org/10.1007/s12649-021-01411-3.

Uzel, R. 2018. Microwave-assisted green extraction technology for sustainable food processing. pp. 159–177. *In*: You, K. (ed.). Emerging Microwave Technologies in Industrial, Agricultural, Medical and Food Processing Resources. InTech. https://doi.org/http://dx.doi.org/10.5772/intechopen.76140 177.

Vardakas, A.T., Shikov, V.T., Dinkova, R.H. and Mihalev, K.M. 2021. Optimisation of the enzyme-assisted extraction of polyphenols from saffron (*Crocus sativus* L.) tepals. Acta Scientiarum Polonorum. Technologia Alimentaria 20(3): 359–367.

Ventura, S., Silva, F., Quental, M., Mondal, D., Freire, M. and Coutinho, J. 2017. Ionic-liquid-mediated extraction and separation processes for bioactive compounds: past, present, and future trends. Chemical Reviews 117: 6984–7052.

Vian, M., Breil, C., Vernes, L., Chaabani, E. and Chemat, F. 2017. Green solvents for sample preparation in analytical chemistry. Current Opinion in Green and Sustainable Chemistry 5: 44–48. https://doi.org/10.1016/J.COGSC.2017.03.010.

Viganó, J., Coutinho, J., Souza, D., Baroni, N., Godoy, H., Macedo, J. and Martínez, J. 2016. Exploring the selectivity of supercritical CO_2 to obtain nonpolar fractions of passion fruit bagasse extracts. The Journal of Supercritical Fluids 110: 1–10. https://doi.org/10.1016/J.SUPFLU.2015.12.001.

Vinatoru, M., Mason, T.J. and Calinescu, I. 2017. Ultrasonically assisted extraction (UAE) and microwave assisted extraction (MAE) of functional compounds from plant materials. TrAC - Trends in Analytical Chemistry 97: 159–178. https://doi.org/10.1016/j.trac.2017.09.002.

Virot, M., Tomao, V., Ginies, C. and Chemat, F. 2008. Total lipid extraction of food using d-limonene as an alternative to n-hexane. Chromatographia 68(3–4): 311–313. https://doi.org/10.1365/s10337-008-0696-1.

Vo, S., Brn, M. and Rimac-brn, S. 2018. New trends in food technology compounds from plant materials. pp. 1–36. *In*: Grumezescu, A. and Holban, A. (eds.). Role of Materials Science in Food Bioengineering. Academic Press. https://doi.org/10.1016/B978-0-12-811448-3/00001-2.

Voges, S., Eggers, R. and Pietsch, A. 2008. Gas assisted oilseed pressing. Separation and Purification Technology 63(1): 1–14. https://doi.org/10.1016/J.SEPPUR.2008.03.039.

Vorobiev, E. and Lebovka, N. 2009. Pulsed-electric-fields-induced effects in plant tissues: fundamental aspects and perspectives of applications. pp. 39–81. *In*: Vorobiev, E. and Lebovka, N. (eds.). Electrotechnologies for Extraction from Food Plants and Biomaterial. Springer International Publishing.

Wang, L., Weller, C., Schlegel, V., Carr, T. and Cuppett, S. 2008. Supercritical CO_2 extraction of lipids from grain sorghum dried distillers grains with solubles. Bioresource Technology 99(5): 1373–1382. https://doi.org/10.1016/J.BIORTECH.2007.01.055.

Wang, L., Qin, P. and Hu, Y. 2010. Study on the microwave-assisted extraction of polyphenols from tea. Frontiers of Chemical Engineering in China 4: 307–313.

Wang, Y. and Gu, W. 2018. Study on supercritical fluid extraction of solanesol from industrial tobacco waste. The Journal of Supercritical Fluids 138: 228–237. https://doi.org/10.1016/J.SUPFLU.2018.05.001.

Wei, Z, Qi, X., Li, T., Luo, M., Wang, W., Zu, Y. and Fu, Y. 2015. Application of natural deep eutectic solvents for extraction and determination of phenolics in Cajanus cajan leaves by ultra performance liquid chromatography. Separation and Purification Technology 149: 237–244. https://doi.org/10.1016/j.seppur.2015.05.015.

Wei, Zuofu, Qi, X., Li, T., Luo, M., Wang, W., Zu, Y. and Fu, Y. 2015. Application of natural deep eutectic solvents for extraction and determination of phenolics in Cajanus cajan leaves by ultra performance liquid chromatography. Separation and Purification Technology 149: 237–244. https://doi.org/10.1016/j.seppur.2015.05.015.

Wen, C., Zhang, J., Zhang, H., Dzah, C., Zandile, M., Duan, Y., Ma, H. and Luo, X. 2018. Advances in ultrasound assisted extraction of bioactive compounds from cash crops—A review. Ultrasonics - Sonochemistry 48: 538–549. https://doi.org/10.1016/j.ultsonch.2018.07.018.

Xhaxhiu, K. and Wenclawiak, B. 2015. Comparison of supercritical CO_2 and ultrasonic extraction of orange peel essential oil from Albanian moro cultivars. Journal of Essential Oil-Bearing Plants 18(2): 289–299. https://doi.org/10.1080/0972060X.2015.1010603.

Xi, J. 2006. Application of high hydrostatic pressure processing of food to extracting lycopene from tomato paste waste. High Pressure Researc 26: 33–41.

Xi, J. 2009. Caffeine extraction from green tea leaves assisted by high pressure processing. Journal of Food Engineering 94(1): 105–109. https://doi.org/10.1016/J.JFOODENG.2009.03.003.

Xi, J. 2013. High-pressure processing as emergent technology for the extraction of bioactive ingredients from plant materials. Critical Reviews in Food Science and Nutrition 53: 837–852.

Xia, G., Li, X. and Jiang, Y. 2021. Deep eutectic solvents as green media for flavonoids extraction from the rhizomes of Polygonatum odoratum. Alexandria Engineering Journal 60(2): 1991–2000. https://doi.org/10.1016/j.aej.2020.12.008.

Xiao, X., Wang, J., Wang, G., Wang, J. and Li, G. 2009. Evaluation of vacuum microwave-assisted extraction technique for the extraction of antioxidants from plant samples. Journal of Chromatography A 1216(51): 8867–8873. https://doi.org/10.1016/J.CHROMA.2009.10.087.

Xu, Y., Cai, F., Yu, Z., Zhang, L., Li, X., Yang, Y. and Liu, G. 2016. Optimisation of pressurised water extraction of polysaccharides from blackcurrant and its antioxidant activity. Food Chemistry 194: 650–658. https://doi.org/10.1016/J.FOODCHEM.2015.08.061.

Yang, R., Li, W., Zhu, C. and Zhang, Q. 2010. Effects of ultra-high hydrostatic pressure on foaming and physical-chemistry properties of egg white. Journal of Biomedical Science and Engineering 02(08): 617–620. https://doi.org/10.4236/jbise.2009.28089.

Yang, Y., Li, J., Zu, Y., Fu, Y., Luo, M., Wu, N. and Liu, X.-L. 2010. Optimisation of microwave-assisted enzymatic extraction of corilagin and geraniin from Geranium sibiricum Linne and evaluation of antioxidant activity. Food Chemistry 122: 373–380. https://doi.org/10.1016/j.foodchem.2010.02.061.

Yang, Y. and Wei, M. 2015. Kinetic and characterization studies for three bioactive compounds extracted from Rabdosia rubescens using ultrasound. Food and Bioproducts Processing 94: 101–113. https://doi.org/10.1016/J.FBP.2015.02.001.

Yanik, D. 2017. Alternative to traditional olive pomace oil extraction systems: Microwave-assisted solvent extraction of oil from wet olive pomace. LWT - Food Science and Technology 77: 45–51. https://doi.org/10.1016/j.lwt.2016.11.020.

Yao, G., Wang, L., Chen, X., Liao, D., Wei, X., Liang, J. and Tong, Z. 2019. Measurement and correlation of vapor–liquid equilibrium data for binary and ternary systems composed of (−)-β-caryophyllene, p-cymene and 3-carene at 101.33 kPa. The Journal of Chemical Thermodynamics 128: 215–224. https://doi.org/10.1016/J.JCT.2018.08.015.

Yao, X., Zhang, D., Duan, M., Cui, Q., Xu, W., Luo, M. and Li, C. 2015. Preparation and determination of phenolic compounds from Pyrola incarnata Fisch. with a green polyols based-deep eutectic solvent. Separation and Purification Technology 149: 116–123. https://doi.org/10.1016/j.seppur.2015.03.037.

Yara Varón, E., Li, Y., Balcells, M., Canela, R., Fabiano-Tixier, A.-S. and Chemat, F. 2017. Vegetable oils as alternative solvents for green oleo-extraction, purification and formulation of food and natural products. Molecules 22: 1474. https://doi.org/10.3390/molecules22091474.

Yen, H., Yang, S., Chen, C. and Chang, J. 2015. Supercritical fluid extraction of valuable compounds from microalgal biomass. Bioresource Technology 184: 291–296. https://doi.org/10.1016/J.BIORTECH.2014.10.030.

Yoo, D.E., Jeong, K.M., Han, S.Y., Kim, E.M., Jin, Y. and Lee, J. 2018. Deep eutectic solvent-based valorization of spent coffee grounds. Food Chemistry 255(February): 357–364. https://doi.org/10.1016/j.foodchem.2018.02.096.

Young, M., Hoon, S., Yeong, G., Li, M., Ri, Y., Lee, J. and Sang, H. 2017. Changes of phenolic-acids and vitamin E profiles on germinated rough rice (*Oryza sativa* L.) treated by high hydrostatic pressure. Food Chemistry 217: 106–111. https://doi.org/10.1016/j.foodchem.2016.08.069.

Yu, Y., Shouqin, Z. and Gao, F. 2008. Comparation of the polyphenol yield and antioxidant activity of peanut shell extract made by different solvent extraction methods. Journal of Biotechnology 136(1): 508–509.

Zahari, S., Amin, A., Halim, N., Rosli, F., Halim, W., Samsukamal, N., Sasithran, B., Ariffin, N., Azman, H., Hassan, N. and Othman, Z. 2018. Deconstruction of Malaysian agro-wastes with inexpensive and bifunctional triethylammonium hydrogen sulfate ionic liquid. AIP Conference Proceedings 1972(June). https://doi.org/10.1063/1.5041245.

Zhang, Q., Zhang, Z., Yue, X., Fan, X., Li, T. and Chen, S. 2009. Response surface optimization of ultrasound-assisted oil extraction from autoclaved almond powder. Food Chemistry 116(2): 513–518. https://doi.org/10.1016/j.foodchem.2009.02.071.

Zhang, Q., De Oliveira Vigier, K., Royer, S. and Jérôme, F. 2012. Deep eutectic solvents: syntheses, properties and applications. Chemical Society Reviews 41(21): 7108–7146. https://doi.org/10.1039/c2cs35178a.

Zhang, S., Lu, Q., Yang, H., Li, Y. and Wang, S. 2011. Aqueous enzymatic extraction of oil and protein hydrolysates from roasted peanut seeds. Journal of the American Oil Chemists' Society 88: 727–732.

Zhang, Y., Li, S., Yin, C., Jiang, D., Yan, F. and Xu, T. 2012. Response surface optimisation of aqueous enzymatic oil extraction from bayberry (Myrica rubra) kernels. Food Chemistry 135(1): 304–308. https://doi.org/10.1016/J.FOODCHEM.2012.04.111.

Zhang, Z., Wang, L., Li, D., Jiao, S., Chen, X. and Mao, Z. 2008. Ultrasound-assisted extraction of oil from flaxseed. Separation and Purification Technology 62(1): 192–198. https://doi.org/10.1016/J.SEPPUR.2008.01.014.

Zheng, N., Ming, Y., Chu, J., Yang, S., Wu, G. and Li, W. 2021. Optimization of extraction process and the antioxidant activity of phenolics from Sanghuangporus baumii. Molecules 26: 3850.

Zhong, J., Wang, Y., Yang, R., Liu, X., Yang, Q. and Qin, X. 2018. The application of ultrasound and microwave to increase oil extraction from *Moringa oleifera* seeds. Industrial Crops and Products 120(April): 1–10. https://doi.org/10.1016/j.indcrop.2018.04.028.

Zhou, P., Wang, X., Liu, P., Huang, J., Wang, C., Pan, M. and Kuang, Z. 2018. Enhanced phenolic compounds extraction from *Morus alba* L. leaves by deep eutectic solvents combined with ultrasonic-assisted extraction. Industrial Crops and Products 120(May): 147–154. https://doi.org/10.1016/j.indcrop.2018.04.071.

Zhu, X., Mang, Y., Shen, F. and Xie, J. 2014. Homogenate extraction of gardenia yellow pigment from Gardenia Jasminoides Ellis fruit using response surface methodology. LWT - Food Science and Technology 51: 1575–1581. https://doi.org/10.1007/s13197-012-0683-2.

Zhu, X.Y., Lin, H.M., Xie, J., Chen, S.S. and Wang, P. 2011. Homogenate extraction of isoflavones from soybean meal by orthogonal design. Journal of Scientific and Industrial Research 70: 455–460.

Zhuang, B., Dou, L.L., Li, P. and Liu, E.H. 2017. Deep eutectic solvents as green media for extraction of flavonoid glycosides and aglycones from Platycladi Cacumen. Journal of Pharmaceutical and Biomedical Analysis 134. https://doi.org/10.1016/j.jpba.2016.11.049.

Žlabur, J., Voća, S., Dobričević, N., Pliestić, S., Galić, A., Boričević, A. and Borić, N. 2016. Ultrasound-assisted extraction of bioactive compounds from lemon balm and peppermint leaves. International Agrophysics 30(1): 95–104. https://doi.org/10.1515/intag-2015-0077.

Zuin, V.G., Segatto, M.L. and Zanotti, K. 2020. Towards a green and sustainable fruit waste valorisation model in Brazil: optimisation of homogenizer-assisted extraction of bioactive compounds from mango waste using a response surface methodology. Pure and Applied Chemistry 92(4): 617–629. https://doi.org/10.1515/pac-2019-1001.

CHAPTER 9

Food Chain and Green Chemistry

Pritha Chakraborty,[1,*] *Marlia Mohd Hanafiah*[2,3] and
Sivarama Krishna Lakkaboyana[4]

1. Introduction

Food is one of the key components of human survival. Human civilization was shaped and grown based on the availability of food and a functioning supply chain. In the last several decades, as the global population increases, food has become cheaper and readily available, though there is growing inequality and a rise in hunger due to the lack of an effective global food system. The growing demand for food causes competition for land, water, energy and other resources. According to Garcia and You (2016), in 2050 global population will increase by 40% and the total population is expected to be 9 billion. A nexus by United Nations predicted that the population of Europe will decline, while Africa and South East Asia will see a boom in the population (Godfray et al. 2010). Even though food production is expected to rise by 60% (OCED 2013), a report by FAO (2010) showed that approximately 1.3 billion tonnes of food is wasted every year and at the same time 1.2 billion people do not have access to clean water (Matharu et al. 2017).

The highly symbiotic phenomenon of natural resource use can be described by a five node system, which is the water, energy, food, land and mineral resources nexus or simply it is also called 'resources nexus'. The resource nexus of food chain

[1] Assistant Professor, Department of Clinical Assistance and Diagnosis, School of Allied Healthcare and Sciences, Jain (Deemed To Be) University, Bengaluru-560066, Karnataka, India.
[2] Associate Professor and Head, Center for Earth Sciences and Environment, Faculty of Science and Technology, Universiti Kebangsaan Malaysia, 43600 Bangi UKM, Selangor, Malaysia.
[3] Centre for Tropical Climate Change System, Institute of Climate Change, Universiti Kebangsaan Malaysia, 43600 Bangi UKM, Selangor, Malaysia.
[4] Department of Chemistry, Vel Tech Rangarajan Dr. Sagunthala R&D Institute of Science and Technology, Avadi, Chennai 600062, India.
Email: mhmarlia@ukm.edu.my
* Corresponding author: prithachakraborty7@gmail.com

supply shows that it consumes 30% and 70% of all produced energy and distributed water, respectively (Golden and Handfield). Approximately one-third of all food produced for human and animal consumption is wasted per year due to coordination problems in the supply chain, which is equivalent to a carbon footprint of 3.3 billion tonnes of CO_2 (Matharu et al. 2017). Aggressive use of land and fossil fuel, biodiversity reduction, eutrophication of water bodies, water shortages, eco-toxicity and bioaccumulation of pesticides are some of the serious environmental threats of overproduction of crops and a dysfunctional food chain. The need of the hour is to develop waste and energy-efficient technologies and unique processes which maintain the sustainability of food production (Bond et al. 2013).

The term 'Green Chemistry' is defined as the design of chemical products and processes that reduce or eliminate the use and generation of hazardous substances (Anastas and Warner 1998). In green chemistry, the chemicals are produced and used in a safer and more efficient way, mainly through trans-materialization and dematerialization of energy and resources. In trans-materialization, the production process uses non-hazardous and renewable materials, whereas in dematerialization the consumption of energy and material is minimized. These strategies offer a technological and intellectual framework that not only guarantees the physical, chemical and toxicological properties of chemicals and the production process but also takes care of their safe handling and disposal. Green chemistry promotes new technologies and opportunities for innovations to achieve social, economic, political and technological benefits in full potential (Mulvihill et al. 2011).

This chapter briefly discusses all the aspects, challenges and environmental impacts of the food chain. The green technologies used in the food industry are discussed in detail. The practice of organic farming, the application of different enzymes and bio-refineries are identified as the need of the hour as they consume lower energy, guarantee safety, ensure the nutritional quality of food, consumer health and deal with the waste generation problem more sustainably.

2. Food Chain

The beginning of agriculture was the onset of a new era in human history. The social structure was completely transformed afterwards as the prevalence of food was the most important factor for survival. The earliest form of the food chain was very simple and limited to only seasonal farming and consumption of a particular food. But this simple structure was suitable for rural areas and small towns (Gendron and Audet 2012). After a significant change in socio-economic structure was brought in by industrialization, retailers were introduced between producers and customers. The demand for quality food with long shelf life and 'ready to eat' or pre-cooked meals was increased as the standard of living improved for urban workers in large cities. Food processing techniques were developed to fill this gap in the food supply chain. As the food supply chain become larger and more complicated, large industries were drawn into the global market. Since then the agricultural industry has seen tremendous growth, and in 2009 it was worth 920 billion US dollars according to a report by Ferguson 2012. This agri-food supply chain is also referred to as the 'Farm to Fork' system. This system produces consumer-oriented products and consists of different steps, such as production, processing, distribution, and retail. The main challenges of this system include keeping up the supply and demand of seasonal food products, managing risks to health, nutrition and safety and several serious environmental impacts of the food production process, such as aggressive use of resources, land, water, and greenhouse gas emission. Long shelf life, good taste, texture, safety, non-GMO and different nutritional benefits are some of the factors that determine the selection and choices of food (Mena and Stevens 2010). The pressure of higher production pushes the farms to use sophisticated machinery, fertilizers, pesticides and genetically modified varieties. This generates a monoculture of a single variety and emerges as a threat to biodiversity (Boye and Arcand 2013).

2.1 Key Drivers of the Food Chain

The main drivers of the food chain are the demand and supply of food. The deciding factors of these two key drivers are discussed in this section. Population size, average age, gender, education and location (rural or urban) greatly influence the food chain and relatively affect the food demand. Consumption of food also differs among high income and low-income families. A study by Lutz et al. (2010) showed that when the undernourished population can access more food calories, their choice of diet changes from grain, tuber, root, pulses to meat and food rich in vegetable oil and sugar. This nutritional transition drives a change in the food supply chain as well as increases health risks, like obesity and other chronic diseases. Urbanization, economic growth, regulatory liberation and globalization encouraged the development of the fast food and supermarket sector (Popkin 1998). This transition in diet brings concerning changes in agricultural trends, exploitation of resources, consumption and supply of food as well as consumer health and safety. With increasing demand, the pressure on supply and production is increased (Kennedy and Shetty 2004, Satterthwaite et al. 2010). Conventional breeding in animals, the introduction of modern genomic approaches and other robust growth models are used to optimize livestock production. The burgeoning growth in livestock production also produces a considerable amount of greenhouse gases and results in climate change. The aquatic food industry is a highly complex and vulnerable food system (Gornall et al. 2010). A report by Garcia and Rosenberg (2010) showed that overfishing is a serious risk where marine fisheries have been increased but the production has decreased continuously. The oil obtained from fish and fish-based meals are in high demand, which increases the production of farm fish maintaining nutritional qualities. Another important element of the food supply chain is wild food, which includes animals and plants that is not a part of traditional agricultural ecosystems and is not greatly nutritionally beneficial (Bharucha and Pretty 2010). Though the use of these wild foods is limited to local communities, it does not pressurize the land, energy use and biodiversity.

2.2 Environmental Impact of Food Chain

Humans are living off natural resources for centuries. As a result, the natural resources are shrinking while the accumulation of waste is exceeding at a speeding rate. Wastes are generated at different steps of the food chain. Quantity, type and source of food waste differ among developed and developing nations. Developing countries produce waste from poor handling of post-harvest, while in developed countries maximum waste comes from the post-consumption step (FAO 2013). Food waste generated during primary, secondary processing steps and post-consumption is a great source of bio-based chemicals and primary resources for bioenergy. The generation of food waste unveils the unsustainable side of the food chain (Vilariño et al. 2017). The environmental impact and waste produced at different stages of the food chain can be defined with a tool named life cycle assessment or LCA. To evaluate the environmental impact of a product, LCA disintegrates the whole sequence into different sub-process, such as the desired products, by-products and unwanted outputs to tract and tally the resources and final waste generated. This phenomenon is called 'life cycle inventory' or 'LCI' (Woodhouse et al. 2018).

Plant origin foods are either gathered from natural habitats or collected from plantations that need minerals for growth (Brentrup et al. 2012). Studies have shown that the LCA tool quantifies and identifies eutrophication, acidification and global warming as the significant effect of fertilizer (phosphorus and nitrogen) overuse (FAO 2003, 2009). Formation of dust, reduction of biodiversity, extensive use of land, and toxicity are also some of the concerns accounted for by the LCA study. The LCA models can change tremendously for plant origin foods as agriculture takes place in an open space with uncontrolled parameters, like temperature and rainfall (Bouwman et al. 2002, Brentrup et al. 2000, Vayssie`res and Rufino 2012).

High demand for animal protein has led to high-density animal farming, genetic improvements, use of antibiotics and hormones (FAO 2009). According to a report by FAO (2010), consumption of meat has increased by 7% in developed nations and 100% in sub-Saharan Africa. Consumption of sea food has increased by 6.5-fold according to a study by WHO/FAO (2003). LCA of animal production counts the feed, land and water use. The land and water used for the production of feeds are also considered in this system as the environmental burden is huge (Verge et al. 2012, Conant and Paustian 2002, Delgado-Gutierrez and Bruhn 2008). Unorganized disposal of manure leads to water and air pollution. Besides that, enteric fermentation, manure and fertilizer degradation also contribute to the emission of three main greenhouse gases, i.e., carbon-di-oxide, nitrous oxide and methane (Verge et al. 2012, Smith et al. 2002, Steinfeld et al. 2006). Methane and nitrous oxide are 20–300 times more potential for global warming than carbon dioxide. The water consumption is estimated to be 15,500, 6,100, 4,800 and 3,900 L per kilo of beef, sheep, pork and poultry, respectively (Hoekstra and Chapagain 2006).

The most non-sustainable stages of food chains are food processing, packaging, transportation and extensive use of pesticides (Hunt et al. 1996). Raw materials used in food industries are resourced from agricultural origin, which used land, energy and water for their production. The food processing chain generates wastewater, packaging waste, organic residues, volatile organic compounds, particulate matter and greenhouse gases and reduces biodiversity, nitrogen, phosphorus and sulfur availability and toxicity (Boye and Arcand 2013).

Transportation of food products around the globe is responsible for greenhouse gas emissions. It is a highly complex network involving ships, trains, trucks, planes and warehouses. Maintaining the long shelf life of foods requires special packaging and is particularly challenging due to seasonality, freshness, spoilage and sanitary issues (Arcand et al. 2012, Boye and Arcand 2013).

3. Green Chemistry

The term 'green chemistry' was first coined by Anastas and Warner in 1998 and offers a new paradigm in agriculture. Green chemistry starts at a molecular level and introduces environment-friendly benign processes for agricultural and industrial production. The main framework follows 12 basic principles to provide design for the environment to ensure consumer health and safety of foods and crops produced following green methods (Anastas and Warner 1998). In the agri-food system, each step generates renewable biomass which can be used to generate bio-based food products, bioenergy and valuable compounds through suitable valorisation methods. Green chemistry chooses methods that ensure lower inputs, zero waste, sustainable social values and minimizes environmental impacts over chemical methods. The basic principles of green chemistry are discussed below (Bhandari and Kasana 2018, Mulvihill et al. 2011).

1. Prevention of waste generation
2. Atom economy
3. Synthesis of durable and less hazardous end products
4. Maximum and safer use of all materials to transform into products
5. Replacing toxic organic solvents with safer ionic liquids offers higher thermal stability and non-volatility and can dissolve both polar, non-polar organic and inorganic compounds.
6. The implication of safe and energy-efficient design
7. Selection of renewable and nontoxic feedstock
8. Reduction in harmful product derivatives by using blocking groups and temporary modification of the synthetic process
9. Broad use of bio-catalysis using enzymes, like transaminase, reductases, hydrolases and oxidases, etc., for economic benefits in the agri-food system
10. Maintaining the natural degradability of all products and by-products
11. Real-time monitoring, control and analysis of soil, water and air pollution
12. Replacing toxic pesticides with bio-pesticides and minimizing accidents and explosions, assuring safer chemistry.

The common goal of these basic principles is to develop maximize efficiency and minimize health and environmental hazards. The various industries modify and prioritize these metrics according to their need. This principle also allows big industries to evaluate their production and process in the context of economic, environmental and social factors, also known as the 'triple bottom line' (Elkington 1998, González 2009, Hjeresen et al. 2002). Expenses can be reduced by waste removal, protective equipment, regulatory compliance, decreased liability and manufacturing security.

As green chemistry eliminates toxic chemicals, it is also referred to as 'benign for design' (Wilson and Schwarzman 2009).

4. Green Technologies Used in Food Production

4.1 Organic Farming

The negative environmental impacts of the growing population and industrialization led to the development of green technologies. Organic farming has emerged as an alternative and environmentally friendly agriculture. Unlike conventional agricultural methods, organic farming does not use harmful pesticides and fertilizers, which disturb the soil nutrition cycle. The benefits of organic farming over conventional farming are discussed in Table 1. It follows cradle to cradle system and ensures soil fertility (Schader et al. 2012). Intensive agricultural cultivation reduces biodiversity significantly by the high nutrient influx, cutting frequencies and use of pesticides. Niggli et al. (2010) Knob and Boutin et al. (2008) reported that organic farming secured 50% greater biodiversity and higher species richness compared to conventional farming techniques. Organic farming is also beneficial for birds, insects, pests, soil organisms and weed flora. As organic farming encourages the cultivation of rare plant and animal species and diverse crop rotations, genetic diversity is positively influenced in turn (Norton et al. 2009). Organic farming is also more energy-efficient than conventional farming. Stolze et al. (2000) reported that product and area related energy input of permanent crops and arable crops is lesser in organic farming than in the conventional farming system. Haas et al. (2001) and Thomassen et al. (2008) reported approximately 45% lower energy use per hectare and per ton of milk in organic production due to concentrated use of feed. The total energy use of mixed farming is around 60 GJ/Hectare and for dairy, suckler cow, and speciality crop farms are around 20–30 GJ/Hectare. While energy consumption of the organic counterpart is reported to be only around 10–20 GJ/Hectare, which is almost 50% lesser than conventional farming and production (Schader et al. 2009a). Other than lower energy consumption, controlled use of soluble phosphorus and potassium fertilizer in organic farming leads to a reduction in nutrient resources depletion (Nemecek et al. 2005). Approximately 12–14% of all greenhouse gases are produced by conventional systems. Among these greenhouse gases, CH_4 and NO_2 emission are predominant while CO_2 emission caused by burning fossil fuel is negligible (Smith et al. 2007). Organic farming contributes lesser emission of CO_2 and NH_3 per hectare and per ton of product than conventional farming and less CH_4 because of lower stocking densities. Eutrophication of water bodies due to nitrogen and phosphorus accumulation mainly from the wide use of fertilizer is another negative environmental impact of conventional farming (Gattinger et al. 2010). Several studies reported a 40–65% reduction in nitrogen leaching and a 20% reduction in eutrophication in organic farming than in conventional farming (Osterburg and Runge 2007, Schader et al. 2009b). Pientel et al. (1995) and Lal et al. (2005) reported that between 1955 to 1995 intense agriculture caused the loss of one-third of fertile soil. Crop rotation and the use of organic fertilizers increase the organic matter content of the soil (Mader et al. 2002, Pientel et al. 2005). The higher

Table 1. Comparison of environmental impacts of organic farming and conventional farming (Sourced from Schader et al. 2012).

Environmental Impacts Organic Farming		Efficiency	
		Conventional Farming	
Landscape and Biodiversity	Genetic diversity	Equal efficiency	Equal efficiency
	Habitat diversity	Equal efficiency	Equal efficiency
	Floral diversity	High efficiency	--------------------
	Landscape	Equal efficiency	Equal efficiency
	Faunal diversity	High efficiency	--------------------
Depletion of Resources	Nutrient resources	High efficiency	--------------------
	Water resources	Equal efficiency	Equal efficiency
	Energy resources	High efficiency	--------------------
Air Pollution	Ch_4	Equal efficiency	Equal efficiency
	CO_2	High efficiency	--------------------
	N_2O	Equal efficiency	Equal efficiency
Soil and Water pollution	Pesticide leaching	Much higher efficiency	--------------------
	Nitrate leaching	High efficiency	--------------------
	Phosphorus emission	High efficiency	--------------------
Air Pollution	Pesticides	Much higher efficiency	--------------------
	Nh_3	High efficiency	--------------------
Fertility of Soil	Biological activity	Much higher efficiency	--------------------
	Soil erosion	High efficiency	--------------------
	Organic matter	High efficiency	--------------------
	Soil structure	Equal efficiency	Equal efficiency

organic content of soil leads to higher biological activity and increases the ability to capture nutrients (Koepke 2003). These activities improve soil structure and reduce soil erosion (Shepherd et al. 2002).

5. Green Technologies in Food Processing

5.1 Enzymes

Enzymes are widely used as biocatalysts in the food industry as they are safe and eco-friendly. Vasic-Racki (2006) reported that the industrial enzymes market in 2017 was worth 6.1 billion U.S. dollars and by 2022 it was estimated it will increase to 8.5 billion U.S. dollars. Bio-catalysis using enzymes is employed in various sectors of the food industry, such as baking, dairy, brewing, fats and oil, meat and functional food. Other than this, some food possesses natural enzymes which have both desirable (ripening, flavour and texture generation) and undesirable (discolouration, acidic taste) effects (Poulsen and Buchholz 2003, Schafer et al. 2006, Fernandes 2010). Enzyme kinetic models are used to explore the behaviour of enzymes and their interaction with the substrate to produce the best quality food products (Leisola et al. 2002, Norus 2006). More than 55 novel enzymes are used in the food industry at present (Berka and Cherry 2006), and they offer exclusive properties like thermos-stability, tolerance to high salinity and pressure and pH adaptation. The important enzymes and trends in enzymatic bio-processes are discussed in this section and also presented in Table 2 (Kirk et al. 2002).

α-amylase: Endo-1, 4-α-D-glucan glucanohydrolase EC 3.2.1.1 or α-amylase breaks down complex polysaccharides by cleaving α-1, 4 linkages between two adjacent sugar moieties into oligomers and α-dextrin. Sahnoun et al. (2015) reported that α-amylase has wide application in the brewery, baking, pulp, paper and pharmaceutical industry. Besides *Bacillus* sp. and *Aspergillus* sp., other bacteria and fungi are also used for industrial production (by submerged fermentation and solid-state fermentation) of α-amylase with different characteristics, like psycho-tolerance, thermostability, alkali stability and halotolerance (Prakash et al. 2009, Sen et al. 2014, Roohi and Kuddus 2014, Sundarram and Murthy 2014).

Amyloglucosidase: Amyloglucosidase cleaves α-1, 4 linkages to produce two glucose moieties and also reacts with glycogen, amylopectin and amylose by cleaving β-D glucose (James and Lee 1997, Espinosa-Ramírez et al. 2014). Most commonly amyloglucosidase is commercially produced from *Aspergillus oryzae* and *A. niger* (Espinosa-Ramírez et al. 2014), and it is mainly used in pastry, food and brewery industries (Diler et al. 2015) due to their maximum activity at slightly acidic pH (4.5–5.0) and 40–60°C (Kumar and Satyanarayana 2009).

Xylanase: Xylan is an important plant polysaccharide made of xylose residue and xylanase breaks down β-1, 4-glycosidic bond (Kumar et al. 2017). Though filamentous fungi reportedly secrete a great quantity of xylanase, different bacteria, fungi, plants, insects and crustaceans can also produce this enzyme (Knob et al. 2013). The enzyme complex of xylanase is composed of endoxylanases, β-xylosidases, ferulic acid esterase, p-coumaric acid esterase, acetylxylan esterase and α-glucuronidase as xylan is also a complex polysaccharide. It is widely used in bioethanol, biomedical, food and feed industries (Goswami and Pathak 2013, Ramalingam and Harris 2010, Das et al. 2012). It helps in gluten agglomeration, fruit juices clarification and texture improvement in baked products (Singhania et al. 2014).

Inulinase: Inulinase acts on insulin which is an important part of fructose and fructooligosaccharides used as a sweetener. Inulinase breaks down the substrate completely into fructose in comparison to other enzymes, which yields fructose (45%), glucose (50%) and little amount of other oligosaccharides. Both categories of inulinase, exo-inulinase and endo-inulinase are used to produce bioethanol, lactic acid, butanediol and citric acid (Vijayaraghavan et al. 2009). According to Chi et al. (2009), *Streptococcus salivarius*, *Actinomyces viscosus*, *Kluyveromyces fragilis*, *Chrysosporium pannorum*, *Penicillium* sp., and *Aspergillus niger* are used for the commercial production of this enzyme.

Lactase: The enzyme that hydrolyses lactose into glucose and galactose is known as β-D-galactohydrolase or lactase (Nakkharat and Haltrich 2006, Duan et al.

Table 2. A detailed list of enzymes and their role in the food and feed industry [Sourced from Fernandes 2010].

Different Classes of Enzymes	Members	Different Applications and Role
Oxidoreductase	Glucose oxidase	Enhances dough strength
	Laccase	Clarifies and enhances flavours in beer and juices
	Lipoxygenase	Enhances dough strength and whitens the dough
Transferase	Cyclodextrin	Carry out cyclodextrin production
	Glycosyltransferase	
	Fructosyltransferase	Directs fructose oligomers synthesis
	Transglutaminase	Used in dough and meat processing
Hydrolases	Amylases	Increasing shelf life by holding moisture, helps in saccharification, enhances bread volume and softness and fermentation.
	Galactosidases	Used to produce animal feed, boost digestibility.
	Glucanase	Used to produce animal feed, boost digestibility.
	Glucoamylase	Saccharification
	Lipase	Used in cheese flavour development, aromatic molecules synthesis helps in dough emulsification.
	Lactase	Hydrolysis of lactose
	Invertase	Used to invert sugar production by directing sucrose hydrolysis.
	Proteases	Tenderizes meats enhance flavour in cheese and milk and increases digestibility.
	Pectinase	Clarification of juice.
	Peptidase	Hydrolyses protein and helps in the ripening of cheese.
	Phospholipase	Dough conditioning
	Phytases	Enhances digestibility by releasing phosphate from phytate.
	Xylanases	Used in dough conditioning and enhances digestibility
Lyases	Acetolactate decarboxylase	Beer mutation
Isomerases	Xylose isomerase	Isomerizes glucose to fructose.

2014). Lactose is a very important enzyme for the dairy and food industry and is best produced by *Kluyveromyces lactis. Candida pseudotropicalis* and *Trichoderma viride* are also known to produce lactase which has a wide pH range (3.0–7.5) (Seyis and Aksoz 2004). Fusarium moniliforme are reported to produce extracellular lactase from wheat bran, molasses and whey (Macris 1981).

β-glucanase: Endo-1,3-1,4-β-glucanases or β-glucanases act on 1,3-1,4-β-Glucans and breaks down into 3-O-β-D-cellotriosyl-D-glucose and cellobiosyl-Dglucose (Celestino et al. 2006). This enzyme is very critical for wine production as 1,3-1,4-β-glucans are present in the cell walls of a number of cereals, such as barley, rye, sorghum and oats. To achieve the best result, β-glucanases are used in combination with xylanase and cellulose which help in viscosity reduction in fluids and consume less water and energy (Tang et al. 2004).

Invertase and Pectinase: Hydrolysis of sucrose into glucose and fructose is catalysed by β-fructofuranosidase or invertase. *Saccharomyces cerevisiae* is the source of industrial invertase which acts at 4.5 pH and 55°C. It produces invert sugar more efficiently than acid hydrolysis (Kulshrestha et al. 2013). Pectin is a vital part of the plant cell wall and is hydrolyzed by the pectinase enzyme. Pectin esterase and propectinases are two categories of pectins with a different modes of action. Pectin esterase directs the de-esterification of methyl group and propectinases hydrolyses propectin; it is used in wine production for turbidity removal, clarification, filtration and colour formation (Ravindran and Jaiswal 2018).

Proteases: Proteases are one of the most important classes of enzymes used in bioprocessing due to their temperature (20–80°C) and pH (3–13) stability. Cathepsin, renin and pepsin are an example of aspartic proteases that directs the degradation of protein compounds (Sawant and Nagendran 2014). Their high specificity for dipeptide with hydrophobic nature and beta methylene group is due to the presence of two aspartate residues in the active site, and their activity is associated with a bound water molecule under acidic pH (Gupta et al. 2002). They are mostly used in both novel and traditional processes, like beverage clarification, wine preservation and cheese production. Cysteine protease is also known as bromelain and is another important protease that can be isolated from different parts of pineapple plants and has wide use in industrial and pharmaceutical industries (Pillai et al. 2011). They are generally used in extraction, concentration and drying processes, though the pharmaceutical industry requires a more pure form (Radha et al. 2011).

Transglutaminase: Transglutaminase is a member of the transferase enzymes class which helps in isopeptide bond formation between glutamine and lysine residues (Kieliszek and Misiewicz 2014). Motoki and Seguro (1998) reported that *Streptoverticillium mobaraense* is used for commercial production of this enzyme which is calcium-independent and has low molecular weight. *Escherichia coli*, *Corynebacterium glutamicum* and *Streptomyces lividans* are also reported to produce transglutaminase enzymes (Noda et al. 2013). Andersen et al. (2003) and Cortez et al. (2004) reported that this is an important enzyme in the bakery, dairy, meat, leather and cosmetic industry.

Cellulase: Cellulases are an important enzyme for bioethanol production, baking and brewery industry, textile and paper industry as it breaks down cellulose into glucose by cleaving glycosidic bonds (Singhania et al. 2014). Cellulases are complex enzymes comprised of endo-cellulases, exo-cellulases and cellobiohydrolases (Doi and Kosugi 2004, Ferreira et al. 2014). Several bacterial species (*Clostridium thermocellum, Bacillus circulans, Proteus vulgaris, Klebsiella pneumoniae, E. coli* and *Cellulomonas* sp.) (Juturu and Wu 2014) and fungal species (*Schizophyllum commune, Melanocarpus* sp., *Aspergillus* sp., *Penicillium* sp., and *Fusarium* sp.) are reported to produce cellulases for industrial application (Várnai et al. 2014).

Laccase: Laccase is a multi-copper oxidase enzyme that oxidizes phenolic compounds under aerobic conditions. Besides 20 different bacterial species, different classes of fungi, especially wood-rotting fungi (white-rot fungi, brown rot fungi and soft rot fungi) are reported to produce this enzyme (El-Batal et al. 2015). Even though bacterial laccases can be found in periplasm fungal laccases are extracellular (Mate and Alcalde 2015). As it degrades phenolic, non-phenolic lignin and pollutants, they are extensively used in the degradation of xenobiotic compounds, decolourization of dyes, baking, bioleaching, bio-pulping and effluent treatment (Couto and Herrera 2006, Shekher et al. 2011).

Lipases: Lipases mainly hydrolyzes lipids present in the cell wall into fatty acids and glycerols. This enzyme directs an array of reactions, like esterification, transesterification, aminolysis, hydrolysis, acidolysis and alcoholysis. These unique characteristics make this enzyme one of the chief enzymes in the dairy, oil, pharmaceutical, biopolymer and bakery industry (Aravindan et al. 2007). *Thermomyces lanuginosus* and *Rhizomucor miehei* produce lipases that are unstable and only used in the immobilized form (Fernandez-Lafuente 2010, Mohammadi et al. 2014). Other bacterial species (*Bacillus*, *Serratia*, *Pseudomonas* and *Staphylococcus*) also produces more stable form of lipases (Prasad and Manjunath 2011).

6. Green Technologies Used in Food Processing

Before commercialization, food products requires to be processed or preserved through different methods, such as frying, drying, cooking, and filtering. But conventional, thermal, chemical and microbiological process techniques are sensitive to food ingredients which leads to loss of nutrients, high energy consumption, prolonged time requirement and lower production efficiency. This phenomenon leads to the development of sustainable processing techniques supercritical extraction and processing, ultrasound-assisted processing, microwave processing, pulsed electric field and controlled pressure drop process. In this section, green techniques for preservation, transformation and extraction processes are discussed and presented in Table 3.

6.1 Instant Controlled Pressure Drop Technology

This process is first developed by Allaf and Vidal (1989), and the principle is based on the thermodynamics of auto-vaporization and instantaneity combined with hydro-thermo-mechanical evolution. The swelling phenomenon, auto-vaporization of water and instant cooling create mechanical stress on the cell wall resulting in cell wall disruption and metabolite secretion. This processing technique offers a change in texture, higher porosity and increased surface area of the sample. This treatment is widely used in cosmetic, food and pharmaceutical industries for extraction, preservation and transportation (Allaf and Allaf 2014, Mounir and Allaf 2008) as an alternative process for freeze-drying and hot air drying (Alonzo-Macías et al. 2013). Studies showed that the controlled pressure drop technique enhanced jatropha, rapeseed and olive oil extraction by 300% without any major modification of fatty acids compared to the conventional extraction procedure (Van 2010). Apart from decontamination, this technique is used to produce ready to meals, dairy products

Table 3. A detailed comparison of different green technologies used in the food and feed industry (Resourced from Chemat et al. 2017).

	Instant Controlled Pressure Drop Technology		Pulsed Electric Field		Supercritical Fluids		Microwave Extraction		Ultrasound	
	Application	Matrix	Application	Matrix	Application	Matrix	Application	Matrix	Application	Matrix
Transformation	Spray drying	Milk, whey protein powders	Cutting	Carrot	Encapsulation	Jabuticaba skins, rosemary	Drying	Bananas	Cutting	Cheese
	Puffing	Onion chips and apple	Softening	Meat	Fractionation	Oregano, sage, thyme, marigold, grape seed	Baking	Cake batter	Degassing	milk
	Texturation	Apple	Frying	Potato	Textural modification	Corn and potato, wheat flour	Blanching	Brussels sprouts	-------	-------
	Swell drying	Strawberry, Green Moroccan pepper	Fermentation	*Saccharomyces cerevisiae*	Crystallization	Lycopene, shrimp residues	-------	-------	-------	-------
Preservation	UHT decontamination	*Bacillus stearothermophilus*	Inactivation	Carrot puree	Bacteria inactivation	Carrot and dry-cured ham	Pasteurization	Kiwifruit puree	Preservation: Inactivation	Orange juice, milk, phosphate buffer
	Conservation	Waterlogged wood	Freezing or thawing	Apple and spinach	Spore inactivation	Apple juice	Thawing	Strawberries	-------	-------
	Thermal treatment for allergen	Peanuts, lentils, chickpeas, soybean	Osmotic dehydration	Apple and carrot	Enzyme activation	Red beet and orange juice	Sterilization	Palm fruit	-------	-------

Table 3 contd. ....

...*Table 3 contd.*

	Instant Controlled Pressure Drop Technology		Pulsed Electric Field		Supercritical Fluids		Microwave Extraction		Ultrasound	
	Application	Matrix	Application	Matrix	Application	Matrix	Application	Matrix	Application	Matrix
	Protein immunoreactivity	Milk	Convective drying	Carrot and red pepper	Drying	Carrot	-------	-------	-------	-------
Extraction	Diffusion	Olive leaves	Diffusion	Grape pomace and sugar beet	Percolation	Tomato waste and almond oil	MHG,	Grape juice products	Polyphenolic compounds	Citrus peel
	Steam extraction	Oakwood chips	Expression	Apple and grape	Liquid/liquid extraction	Soybean oil and essential oil	SFME	lavender flowers	lycopene	tomatoes
	Deodorization	Rosemary leaves	Filtration	BSA suspension	Pressing	Cocoa and linseeds	-------	-------	vanillin	vanilla pods

and swell-dried products, like tomato, banana, apple, onion and strawberries on an industrial scale.

6.2 Pulsed Electric Field

Pulsed electric field is a non-thermal process where the cell membrane is disrupted by an external electric field. It is also known as electroporation or electropermeabilization (Saulis 2010). Application of controlled electric field creates chare difference in the cell membrane which results in the formation of small hydrophilic spores on the membrane. Increased numbers and size of pores lead to leakage of intracellular compounds. The effectivity of this technique depends on different parameters, such as treatment time, electric field strength, pulse shape and width, temperature, frequency, physicochemical parameters (pH and conductivity), treatment mode (continuous or batch), the configuration of treatment chamber and characteristics of treated cells (shape, size, envelop and membrane structure) (Van den Bosch 2007, Vorobiev and Lebovka 2009). PEF processing is used in food preservation as an alternative to heat treatment as it has a very limited effect on enzymes (Terefe et al. 2015, Roodenburg et al. 2013). Tokusoglu et al. (2014) reported that PEF treated beverages have a higher content of polyphenols, vitamins and carotenoids compared to pasteurized products. The combination of PEF with other treatments provides maximum lethal effect (including spore inactivation) at lower energy and field strength (Siemer et al. 2015, Álvarez and Heinz 2007). Preservation by freezing results in texture deterioration of fruits due to the formation of crystals. But the application of PEF treatment introduces cryoprotectants into biological cells, which prevents crystal formation (Jalté et al. 2009, Parniakov et al. 2016). PEF combined with traditional freeze-drying, osmotic dehydration directs water and solutes migration across the membrane into the food matrix (Wiktor et al. 2014). Extraction of sucrose, polyphenols, colorants and secondary metabolites from plant cells is achieved by the PEF technique (Puértolas et al. 2012). Pre-treatment with PEF weakens the cell membrane prior to the traditional extraction process. Cold extraction with PEF pre-treatment reportedly leads to a higher concentration of sugar and better filterability of juices (Loginova et al. 2011). Besides preservation and extraction, PEF is also used to accelerate the transformation process, skin removal of fruits and tendering of meat (Toepfl 2012). The loss of turgor pressure achieved by the electric treatment gives equal results to steam peeling and also changed the viscoelastic and texture of food (Bekhit et al. 2016).

6.3 Supercritical Fluids

Supercritical fluid is a solvent-free method used mainly in the extraction procedure. When a fluid is heated at its critical temperature (Tc) and pressurized at its critical pressure (Pc), it reaches its supercritical state (Brunner 2005). As their density and viscosity are close to liquid and gases, respectively, and diffusivity is between liquid and gas, SCF has a higher mass transfer rate. SCF-CO_2 is widely used for extraction procedures due to their low critical parameters (Pc: 7.4 MPa and Tc: 31°C), non-inflammability, volatile nature and cheaper cost (Rozzi and Singh 2002, Lumia 2011). SCF extraction from solids is mainly divided into two steps. Extraction, separation and installation are composed of four parts, which are volumetric pump, heat exchanger, extractor, and separator. SCF in combination with extrusion is employed under milder processing conditions to change the texture of the food matrix. It is also used to separate wax from the lipid fraction of oil. High hydrostatic pressure employed in the SCF technique halts microbial growth and inactivates endogenous enzymes. So, this technique is used for preservation by sterilization from bacteria, viruses and spore inactivation (Garcia-Gonzalez et al. 2007, Perrut 2012, Spilimbergo et al. 2002). The effect of SF CO_2 is increased in presence of water and acidic pH, which favours the inactivation of enzymes. Compounds with higher molecular weight (oil) can be extracted by SCF under high pressure (280 bars) through percolation, while aromatic compounds are extracted at lower pressure (70–200 bars) and under moderate conditions (40–60°C) (Reverchon 1997, Sovová 2006).

6.4 Microwave Extraction

Microwave heating is extensively used in different steps of food processing, like preservation and extraction. Drying is one of the preservation techniques where water activity is reduced by removing water by evaporation. The unique characteristic of microwave heating is that heat travels through food and gets absorbed ensuring higher efficiency shorter time and final quality of the product (Fathima et al. 2001). Another positive aspect of microwave drying is that it improves aroma and provides better and faster dehydration (Gowen et al. 2006). Vacuum microwave is successfully applied in the preservation of cranberries, cabbage, fruit gels, mushroom and garlic as it operates in absence of air and retains volatile compounds, colour, texture and nutrients (Giri and Prasad 2006, Gunasekaran 1999). In the case of heat sensitive products, the microwave is used in combination with freeze-drying. Microwave blanching is a pre-treatment process prior to freeze-drying that inactivates enzymes responsible for browning and off-flavour. Besides preventing microbial growth, it consumes less energy and requires less processing time (Brewer and Begum 2003, Lin and Brewer 2005). Microwave thawing uses electromagnetic waves and requires lesser time than the conventional thawing process (Schiffmann 2001). Two advanced techniques, i.e., solvent-free microwave hydro-distillation (SFME) and microwave hydro-diffusion and gravity (HDG) are used in the extraction of essential oil from fruits and plants. It is a combination of distillation at atmospheric pressure and microwave heating without water or any organic solvent (Chemat et al. 2004a, 2004b, Ferhat et al. 2006). MHG is a combined process of microwave heating and gravity at atmospheric pressure and is used in extraction from fruit. Vegetable and aromatic plants (Périno-Issartier et al. 2013).

6.5 Ultrasound-Assisted Food Processing

Low-frequency ultrasound is considered a powerful ultrasound and effectivity depends on temperature, acoustic frequency and pressure. Ultrasonic cleaning bath and ultrasound probe or horn system are two main methods used in food processing. Degassing is an ultrasonic cleaning bath that occurs when rapid vibration brings the gas bubbles together and allows them to grow big enough to rise to the surface against gravity (Tervo et al. 2006). Cutting of food products using ultrasounds requires less maintenance, less energy, reduces waste and increases production but depends on food type and condition (Arnold et al. 2009). Powerful ultrasound is used in meat tenderization rather than the conventional pounding method. Ultrasound acts by either breaking the muscular cells' integrity or enhancing enzymatic activity (Boistier-Marquis et al. 1999). Ultrasound methods are used in the preservation of food by destroying microbes (through thinning of the cell membrane, the release of free radicals and localized heating) and accelerating the sterilization process (Butz and Tauscher 2002, Chisti 2003). Inactivation of *Saccharomyces cerevisiae, Escherichia coli, Staphylococcus aureus, Pseudomonas fluorescens* and *Listeria monocytogenes* is reported when ultra-sonication is applied with heating (Manas et al. 2000). Ultrasound waves interact and change physical and chemical properties, which lead to disruption of the plant cell wall and enable the release of compounds. Extraction of aromatic compounds, pigments, organic and mineral compounds and antioxidants is carried out by ultrasound (Riera et al. 2004).

7. Biorefineries

An alternate and sustainable way of recycling lignocellulosic waste material is to employ the refinery process to transform them into new molecules. The term 'bio-refinery' was coined by (Wyman and Goodman 1993, Cherubini 2010) and is defined as a sustainable transformation of biomass into an array of products and energy (biofuel, heat and electrical power) (Faria et al. 2016). Figure 1 presents an overview on the concept of biorefineries. The most common example of the biorefinery process is the transformation of pentose sugar from sugarcane waste biomass into biogas, *n*-butanol and acetone (Mariano et al. 2013). In biorefineries, under a low supply of oxygen gasification is used to convert organic matter into the air, steam and syngases (CO and H_2) with a little amount of CH_4, CO_2, H_2O and N_2 (Martínez 2012, Pereira et al. 2012). Mohan et al. (2016) has reported that biomass from municipal waste and food waste can be transformed into different biogases; bio-hydrogen (H_2) and bio-hythane (a mixture of CH_4 and H_2). In acidogenesis step of anaerobic digestion, the biomass is converted by fermentation with the help of different enzymes into biohydrogen and bio-hythane and fatty acids in very little amounts. Biohydrogen production is the most cost-effective method as it involves anaerobic digestion and has a higher conversion rate (Waldron 2007). High value-added compounds obtained from biomass conversion are different in comparison to compounds obtained from crude oil. The presence of alcohol, aldehyde, ketone, ester, furans and phenols in these liquids obtained from bio-based feedstock are responsible for their high density, low heating value, chemical, immiscibility with hydrocarbons and thermal corrosiveness (Graça et al. 2013). Ethanol is another important platform chemical with a versatile application as a building block. A range of different products, like synthetic rubber, diethyl ether (a key ingredient in cellulosic plastic production) and acetaldehyde (a key ingredient for several processes) (Villela et al. 2011), is obtained from this process. Low purity glycerol is another by-product obtained from biodiesel production (Amaral et al. 2009). After pre-treatment, glycerol serves as a good

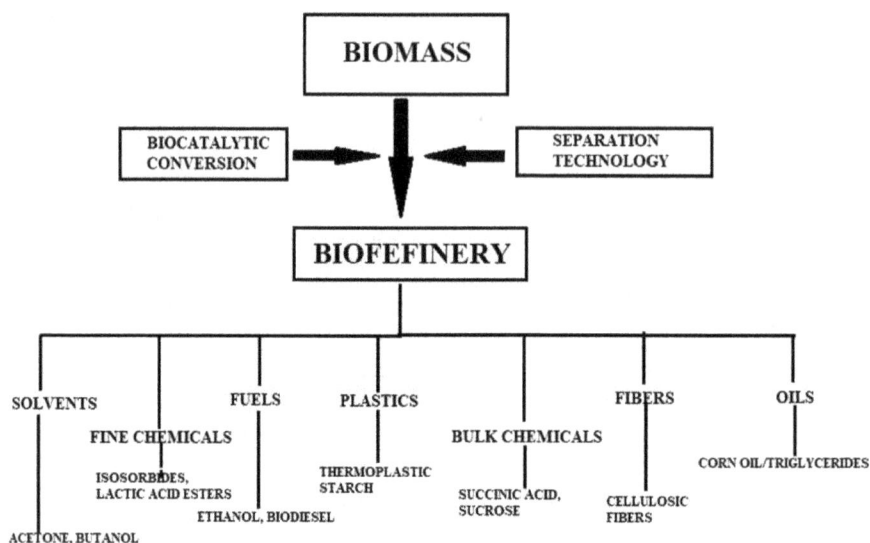

Figure 1. Concept of biorefinery involved in green chemistry (Graça et al. 2013).

carbon source for microbial growth (Quispe et al. 2013, Freitas et al. 2011, Godoy et al. 2009, Godoy et al. 2011, López et al. 2013, Oliveira et al. 2006). Crude bio-oil produced from biomass by pyrolysis contains high added value compounds mixed with several impurities, and they do not mix with other hydrocarbons and need to be deoxygenated to upgrade (Virmond et al. 2013, Graça et al. 2013).

7.1 Reduction of Toxins in Food

The development of toxins during food processing due to high pressure and temperature is another serious issue in the food industry. Heterocyclic amine, nitrosamine, furans, acrylamide, bisphenol and polyaromatic hydrocarbons are some examples of these toxins. Employment of green technologies inhibits the generation of these toxins. High pressure and processing conditions (duration and method of cooking and temperature) greatly affect the generation of toxins (Akhtar 2012). Different approaches, like the selection of correct ingredients and the use of vegetables with lower levels of sugar and asparagine to limit the acrylamide content alternative processing methods, are considered to reduce toxin generation. The material of utensils or containers also plays a role in toxin generation and transfer (Boye and Arcand 2013).

8. Waste Reduction in the Food Supply Chain

As agriculture developed over time, a considerable amount of waste has been generated and most of the agricultural wastes are biodegradable. To meet the requirement of urbanized communities, a global chain of food supply businesses has emerged which produced a considerable amount of waste during production and processing. To maintain the sustainability of the entire food chain, the generated waste must be processed and minimized. Studies have shown that genetic engineering and breeding techniques are being used to create a maximum edible part from a plant but 100% consumption without generating any waste is impossible. Another study on UK's food processing and supply chain showed that 80% of all waste is generated from consumers and 40% of solid waste comes from food processing and packaging. To solve this problem, a multilevel hierarchy process based on the 3RVE strategy of waste reduction. This 3RVE strategy is composed of five steps; reduce, reuse, recycle, valorize and eliminate. Two conventional ways of waste disposal are landfilling and incineration. Landfilling requires a large area of land, and it produces methane (greenhouse gas) due to the anaerobic decomposition of waste by microbes. The incineration technique requires a comparatively smaller area of land but produces toxic gases in the air. The food supply chain generates soluble or suspended waste, which can be treated by different aerobic and anaerobic waste treatments. Pre-treatment like sedimentation and filtration eliminates the particulate matter (Boye and Arcand 2013).

9. Research and Development and Social Perspective of Green Technologies

The food and agriculture industry greatly depend on the research and development section to achieve higher yield, lower inputs, faster pathogen identification, spoilage prevention and improve the health benefits of foods (Boye et al. 2012). Reduction of waste generation and resource management can be achieved by novel approaches, like micro and nano-scale chemistry, the use of innovative ideas and the implication of waste management hierarchy. Micro or nano-scale chemistry operates with small quantities of chemicals maintaining the standard of chemical application while being environmentally safe and pollution-free (Nazzaro et al. 2012). Micro and nanotechnology provide a great opportunity in food quality, food safety, processing, and packaging. Multiplexing is another unique analytical technique that allows concurrent detection and analysis of a large number of compounds in a limited quantity of samples. It offers high specificity and accuracy, low cost and faster analysis, depending on the techniques (DNA sequencing, immunoassays, toxin, allergen, pathogen and antibiotic detection) (Santonen 2012, Henson et al. 2008).

Social perspectives of green technologies include process and product developments. Choices made by consumers have a direct impact on food production, processing and lastly on the economy. Informed choices made by consumers create social pressure on industries to make fundamental changes in their use of technologies (Cardello et al. 2007, Delgado-Gutierrez and Bruhn 2008). The growing interest in the environmental impacts of agricultural and food processing techniques makes consumers choose products that offer maximum safety, quality and nutrition, taste and shelf life (Henson et al. 2008).

10. Current and Future Challenges

Finding the balance between environmental protection and economic profits is required for the long term and continued use of green technologies. Sustainable agriculture and agri-food sectors require to be more associated with carbon sequestration to reduce emissions. The emerging challenges can be solved with new emerging and advanced technologies (Burlingame et al. 2012). Impacts of continuous rise in temperature and sea level, sustainable ways of increasing production and extending shelf life under threatening climate are some of the challenges of the future. Reduction in biodiversity due to current industrial and agricultural practices is another challenge that significantly impacts health and the environment. According to FAO 2010, a sustainable diet is described as a diet that has low environmental impacts and ensures food, nutrition and health security throughout generations. These diets confirm biodiversity, are easily accessible, culturally acceptable, cheaper, nutritionally enriched, safe and healthy. To achieve a sustainable diet, interdisciplinary efforts should be made to solve problems, like malnutrition, erosion of biodiversity and ecosystem, and improved food system and dietary patterns (OCED 2011).

11. Conclusion

There is an evident link between food, environment, health and economy which demands a sustainable balance for the lasting survival of mankind. Productivity in a sustainable way will increase if only high priority is given to research, development, education and innovations in the agri-food system. The above-discussed approaches should be considered to reduce the negative impacts of aggressive agricultural practices and to ensure food security for the growing population. International, national and local collaboration are important factors to continue a functional food supply chain.

References

Akhtar, H. 2012. Reducing process-induced toxins in foods. pp. 571–605. *In*: Boye, J.I. and Arcand, Y. (eds.). Green Technologies in Food Production and Processing. Springer, Boston, MA.

Allaf, K. and Vidal, P. 1989. Feasibility Study of a Process of Drying/Swelling by Instantaneous Decompression Toward Vacuum of in Pieces Vegetables in View of Rapid Re-Hydration Gradient Activity Plotting. University of Technology of Compiègne UTC NoCR/89/103 Industrial SILVALAON Partner.

Allaf, T. and Allaf, K. 2016. Instant Controlled Pressure Drop (DIC) in Food Processing. Springer-Verlag New York.

Alonzo-Macías, M., Cardador-Martínez, A., Mounir, S., Montejano-Gaitán, G. and Allaf, K. 2013. Comparative study of the effects of drying methods on antioxidant activity of dried strawberry (*Fragaria Var. Camarosa*). J. Food. Res. 2(2): 92–107.

Alvarez, I. and Heinz, V. 2007. Hurdle technology and the preservation of food by pulsed electric fields. Food Preservation by Pulsed Electric Fields: From Research to Application. Elsevier. USA.

Amaral, P.F.F., Ferreira, T.F., Fontes, G.C. and Coelho, M.A.Z. 2009. Glycerol valorization: new biotechnological routes. Food. Bioprod. Process. 87(3): 179–186.

Anastas, P.T. and Warner, J.C. 1998. Green chemistry. Frontiers. 640.

Andersen, L.N., Bech, L., Halkier, T. et al. 2003. Transglutaminases from oomycetes. US Patent # 6428993B1.

Aravindan, R., Anbumathi, P. and Viruthagiri, T. 2007. Lipase applications in food industry. Indian. J. Biotechnol. 6(2): 141.

Arcand, Y., Maxime, D. and Zareifard, R. 2012. LCA of processed food. *In*: Boye, J.I. and Arcand, Y. (eds.). Green Technologies in Food Production and Processing. Springer, New York, USA.

Arnold, G., Leiteritz, L., Zahn, S. and Rohm, H. 2009. Ultrasonic cutting of cheese: Composition affects cutting work reduction and energy demand. Int. Dairy. J. 19(5): 314–320.

Bekhit, A.E.D.A., Carne, A., van de Ven, R. and Hopkins, D.L. 2016. Effect of repeated pulsed electric field treatment on the quality of hot-boned beef loins and topsides. Meat. Sci. 111: 139–146.

Berka, R.M. and Cherry, J.R. 2006. Enzyme biotechnology. pp. 477–498. *In*: Ratledge, C. and Kristiansen, B. (eds.). Basic Biotechnology, Cambridge University Press, Cambridge, UK.

Bhandari, S. and Kasana, V. 2018. Application of green chemistry principles in agriculture. Green. Chem. Technol. 4(2): 10–12.

Bharucha, Z. and Pretty, J. 2010. The roles and values of wild foods in agricultural systems. Phil. Trans. R. Soc. B. 365(1554): 2913–2926.

Boistier-Marquis, E., Lagsir-Oulahal, N. and Callard, M. 1999. Applications des ultrasons de puissance en industries alimentaires. Industries Alimentaires Et Agricoles. 116(3): 23–31.

Bond, M., Meacham, T., Bhunnoo, R. and Benton, T. 2013. Food Waste within Global Food Systems. Global Food Security, Swindon, UK.

Boutin, C., Baril, A. and Martin, P.A. 2008. Plant diversity in crop fields and woody hedgerows of organic and conventional farms in contrasting landscapes. Agric. Ecosyst. Environ. 123(1-3): 185–193.

Bouwman, A.F., Boumans, L.J.M. and Batjes, N.H. 2002. Modeling global annual N_2O and NO emissions from fertilized fields. Global Biogeochem Cy. 16(4): 28–31.

Boye, J.I., Maltais, A., Bittner, S. and Arcand, Y. 2012. Greening of research and development. pp. 519–540. *In*: Boye, J.I. and Arcand, Y. (eds.). Green Technologies in Food Production and Processing. Springer, Boston, MA.

Boye, J.I. and Arcand, Y. 2013. Current trends in green technologies in food production and processing. Food Eng. Rev. 5(1): 1–17.

Brentrup, F., Küsters, J., Lammel, J. and Kuhlmann, H. 2000. Methods to estimate on-field nitrogen emissions from crop production as an input to LCA studies in the agricultural sector. Int. J. Life Cycle Assess. 5(6): 349–357.

Brentrup, F. 2012. LCA of crop production. *In*: Boye, J.I. and Arcand, Y. (eds.). Green Technologies in Food Production and Processing. Springer, New York, USA.

Brewer, M.S. and Begum, S. 2003. Effect of microwave power level and time on ascorbic acid content, peroxidase activity and color of selected vegetables. J. Food Process. Preserv. 27(6): 411–426.

Brunner, G. 2005. Supercritical fluids: technology and application to food processing. J. Food. Eng. 67(1-2): 21–33.

Burlingame, B., Charrondiere, U.R., Dernini, S., Stadlmayr, B. and Mondovı, S. 2012. Food biodiversity and sustainable diets: implications of applications for food production and processing. pp. 643–657. *In*: Boye, J.I. and Arcand, Y. (eds.). Green Technologies in Food Production and Processing. Springer, Boston, MA.

Butz, P. and Tauscher, B. 2002. Emerging technologies: chemical aspects. Food. Res. Int. 35(2-3): 279–284.

Cardello, A.V., Schutz, H.G. and Lesher, L.L. 2007. Consumer perceptions of foods processed by innovative and emerging technologies: A conjoint analytic study. Innov. Food. Sci. Emerg. Technol. 8(1): 73–83.

Celestino, K.R.S., Cunha, R.B. and Felix, C.R. 2006. Characterization of a β-glucanase produced by *Rhizopus microsporus var. microsporus,* and its potential for application in the brewing industry. BMC. Biochem. 7(1): 1–9.

Chemat, F., Smadja, J. and Lucchesi, M.E. 2004. Extraction sans solvant assist ée par microondes de produits naturels. EP Pat. 1(439): 218.

Chemat, F., Lucchesi, M. and Smadia, J. 2004. Solvent-free microwave extraction of volatile natural substances. U.S. Patent #10/751,988.

Chemat, F., Rombaut, N., Meullemiestre, A., Turk, M., Perino, S., Fabiano-Tixier, A.S. et al. 2017. Review of green food processing techniques. Preservation, transformation, and extraction. Innov. Food. Sci. Emerg. Technol. 41: 357–377.

Cherubini, F. 2010. The biorefinery concept: using biomass instead of oil for producing energy and chemicals. Energy. Convers. Manag. 51(7): 1412–1421.

Chi, Z., Chi, Z., Zhang, T., Liu, G. and Yue, L. 2009. Inulinase-expressing microorganisms and applications of inulinases. Appl. Micro. Biotechnol. 82(2): 211–220.

Chisti, Y. 2003. Sonobioreactors: using ultrasound for enhanced microbial productivity. TRENDS Biotechnol. 21(2): 89–93.

Conant, R.T. and Paustian, K. 2002. Potential soil carbon sequestration in overgrazed grassland ecosystems. Global. Biogeochem. Cy. 16(4): 90–101.

Cortez, J., Bonner, P.L. and Griffin, M. 2004. Application of transglutaminases in the modification of wool textiles. Enzyme Microb. Technol. 34(1): 64–72.

Couto, S.R. and Herrera, J.L.T. 2006. Industrial and biotechnological applications of laccases: a review. Biotechnol. Adv. 24(5): 500–513.

Das, S.P., Ravindran, R., Ahmed, S., Das, D., Goyal, D., Fontes, C.M. et al. 2012. Bioethanol production involving recombinant *C. thermocellum* hydrolytic hemicellulase and fermentative microbes. Appl. Biochem. Biotechnol. 167(6): 1475–1488.

de Freitas Branco, R., dos Santos, J.C. and da Silva, S.S. 2011. A novel use for sugarcane bagasse hemicellulosic fraction: xylitol enzymatic production. Biomass. Bioenerg. 35(7): 3241–3246.

del Río González, P. 2009. The empirical analysis of the determinants for environmental technological change: A research agenda. Ecol. Econom. 68(3): 861–878.

Delgado-Gutierrez, C. and Bruhn, C.M. 2008. Health professionals' attitudes and educational needs regarding new food processing technologies. J. Food Sci. Educ. 7(4): 78–83.

Diler, G., Chevallier, S., Pöhlmann, I., Guyon, C., Guilloux, M. and Le-Bail, A. 2015. Assessment of amyloglucosidase activity during production and storage of laminated pie dough. Impact on raw dough properties and sweetness after baking. J. Cereal Sci. 61: 63–70.

Doi, R.H. and Kosugi, A. 2004. Cellulosomes: plant-cell-wall-degrading enzyme complexes. Nature. Rev. Microbiol. 2(7): 541–551.

Duan, X., Sun, X. and Wu, J. 2014. Optimization of fermentation conditions of recombinant Pichia pastoris that can produce β-galactosidase. Genom. Appl. Biol. 33: 1288–1293.

El-Batal, A.I., ElKenawy, N.M., Yassin, A.S. and Amin, M.A. 2015. Laccase production by *Pleurotus ostreatus* and its application in synthesis of gold nanoparticles. Biotechnol. Rep. 5: 31–39.

Elkington, J. 1998. Cannibals with Forks: The Triple Bottom Line of 21st Century Business. pp. 407. Gabriola Island, BC/Stony Creek, CT: New Soc.

Espinosa-Ramírez, J., Pérez-Carrillo, E. and Serna-Saldívar, S.O. 2014. Maltose and glucose utilization during fermentation of barley and sorghum lager beers as affected by β-amylase or amyloglucosidase addition. J. Cereal Sci. 60(3): 602–609.

FAO. 2003. World Agriculture: Towards 2015/2030, an FAO Perspective. Earthscan Publications Ltd, London.

FAO. 2009. The State of Food in Agriculture: Livestock in the Balance. Electronic Publishing Policy and Support Branch-Communication Division, FAO, Rome, Italy.

FAO. 2010. FAOSTAT Statistical Database. Rome, Italy.

FAO. 2010. International Scientific Symposium 'Biodiversity and Sustainable Diets', Final Document.

FAO. 2011. Global food losses and food waste–Extent, causes and prevention. SAVE FOOD: An Initiative on Food Loss and Waste Reduction.

FAO. 2013. Food Wastage Footprint. Impacts on Natural Resources. Summary Report.

Faria, D., Machado, G., Eichler, P., Boneberg, B., Fernanda, R., Vilares, M. et al. 2016. Scenarios and perspectives of the main cultures of Rio Grande South in biorefinery processes. UERGS 2(3): 291–306.

Fathima, A., Begum, K. and Rajalakshmi, D. 2001. Microwave drying of selected greens and their sensory characteristics. Plant. Foods. Hum. Nutr. 56(4): 303–311.

Ferguson, A. 2012. Population matters for a sustainable future. OPT J. 12(2): 4–6.

Ferhat, M.A., Meklati, B.Y., Smadja, J. and Chemat, F. 2006. An improved microwave Clevenger apparatus for distillation of essential oils from orange peel. J. Chromatogr. A. 1112(1-2): 121–126.

Fernandes, P. 2010. Enzymes in food processing: a condensed overview on strategies for better biocatalysts. Enzyme Res. 2010: 1–19.

Fernandes, P. 2010. Enzymes in sugar industries. pp. 165–197. *In*: Panesar, P., Marwaha, S.S. and Chopra, H.K. (eds.). Enzymes in Food Processing: Fundamentals and Potential Applications. I.K. International Publishing House, New Delhi, India.

Fernandez-Lafuente, R. 2010. Lipase from *Thermomyces lanuginosus*: uses and prospects as an industrial biocatalyst. J. Mol. Catal. B Enzym. 62(3-4): 197–212.

Ferreira, N.L., Margeot, A., Blanquet, S. et al. 2014. Use of cellulases from *Trichoderma reesei* in the twenty-first century – Part I: Current industrial uses and future applications in the production of second ethanol generation. pp 245–261. *In*: Kubicek, C.P. (ed.). Biotechnology and Biology of Trichoderma. Elsevier, Amsterdam.

Garcia, D.J. and You, F. 2016. The water-energy-food nexus and process systems engineering: A new focus. Comput. Chem. Eng. 91: 49–67.

Garcia, S.M. and Rosenberg, A.A. 2010. Food security and marine capture fisheries: characteristics, trends, drivers and future perspectives. Phil. Trans. R. Soc. B. 365(1554): 2869–2880.

Garcia-Gonzalez, L., Geeraerd, A.H., Spilimbergo, S., Elst, K., Van Ginneken, L., Debevere, J. et al. 2007. High pressure carbon dioxide inactivation of microorganisms in foods: the past, the present and the future. Int. J. Food. Microbiol. 117(1): 1–28.

Gattinger, A., Müller, A., Häni, M., Oehen, B. and Niggli, U. 2010. Klimaleistungen des Biolandbaus - Fakten und Hintergründe Frick.

Gauthier, E.G. 2012. Green food processing technologies: factors affecting consumers' acceptance. pp. 615–41. *In*: Boye, J.I. and Arcand, Y. (eds.). Green Technologies in Food Production and Processing. Springer, Boston, MA.

Gendron, C. and Audet, R. 2012. Key drivers of the food chain. pp. 23–39. *In*: Green Technologies in Food Production and Processing. Springer, Boston, MA.

Giri, S.K. and Prasad, S. 2006. Modeling shrinkage and density changes during microwave-vacuum drying of button mushroom. Int. J. Food. Prop. 9(3): 409–419.

Godfray, H.C.J., Crute, I.R., Haddad, L., Lawrence, D., Muir, J.F., Nisbett, N. et al. 2010. The future of the global food system. Phil. Trans. R. Soc. B. 365: 2769–2777.

Godoy, M.G., Gutarra, M.L., Maciel, F.M., Felix, S.P., Bevilaqua, J.V., Machado et al. 2009. Use of a low-cost methodology for biodetoxification of castor bean waste and lipase production. Enz. Micro. Technol. 44(5): 317–322.

Godoy, M.G., Gutarra, M.L., Castro, A.M., Machado, O.L. and Freire, D.M. 2011. Adding value to a toxic residue from the biodiesel industry: production of two distinct pool of lipases from *Penicillium simplicissimum* in castor bean waste. J. Ind. Microbiol. Biotechnol. 38(8): 945–953.

Golden, J.S. and Handfield, R.B. 2014. Why biobased? Opportunities in the emerging bioeconomy. US Department of Agriculture, Office of Procurement and Property Management. Washington, DC, USA.

Gornall, J., Betts, R., Burke, E., Clark, R., Camp, J. et al. 2010. Implications of climate change for agricultural productivity in the early twenty-first century Phil. Trans. R. Soc. B. 365(1554): 2973–2989.

Goswami, G.K. and Pathak, R.R. 2013. Microbial xylanases and their biomedical applications: a review. Int. J. Basic Clin. Pharmacol. 2(3): 237–246.

Gowen, A., Abu-Ghannam, N., Frias, J. and Oliveira, J. 2006. Optimisation of dehydration and rehydration properties of cooked chickpeas (*Cicer arietinum* L.) undergoing microwave–hot air combination drying. Trends. Food. Sci. Technol. 17(4): 177–183.

Graça, I., Lopes, J.M., Cerqueira, H.S. and Ribeiro, M.F. 2013. Bio-oils upgrading for second generation biofuels. Ind. Eng. Chem. Res. 52(1): 275–287.

Gunasekaran, S. 1999. Pulsed microwave-vacuum drying of food materials. Dry. Technol. 17(3): 395–412.

Gupta, R., Beg, Q. and Lorenz, P. 2002. Bacterial alkaline proteases: molecular approaches and industrial applications. Appl. Microbiol. Biotechnol. 59(1): 15–32.

Haas, G., Wetterich, F. and Köpke, U. 2001. Comparing intensive, extensified and organic grassland farming in southern Germany by process life cycle assessment. Agric. Ecosyst. Environ. 83(1-2): 43–53.

Henson, S., Annou, M., Cranfield, J. and Ryks, J. 2008. Understanding consumer attitudes toward food technologies in Canada. Risk. Anal. 28(6): 1601–1617.

Hjeresen, D.L., Kirchhoff, M.M. and Lankey, R.L. 2002. Green chemistry: Environment, economics, and competitiveness. Corp. Environ. Strategy 9(3): 259–266.

Hoekstra, A.Y. and Chapagain, A.K. 2006. Water footprints of nations: water use by people as a function of their consumption pattern. pp. 35–48. *In*: Integrated Assessment of Water Resources and Global Change. Springer, Dordrecht, Netherlands.

Hunt, R.G., Franklin, W.E. and Hunt, R.G. 1996. LCA—How it came about. Int. J. Life Cycle Assess. 1(1): 4–7.

Jalté, M., Lanoisellé, J.L., Lebovka, N.I. and Vorobiev, E. 2009. Freezing of potato tissue pre-treated by pulsed electric fields. LWT-Food Sci. Technol. 42(2): 576–580.

James, J.A. and Lee, B.H. 1997. Glucoamylases: microbial sources, industrial applications and molecular biology—a review. J. Food. Biochem. 21(6): 1–52.

Juturu, V. and Wu, J.C. 2014. Microbial cellulases: engineering, production and applications. Renewable. Sustainable. Energy. Rev. 33: 188–203.

Kennedy, G.N. and Shetty, P. 2004. Globalization of food systems in developing countries: a synthesis of country case studies. *In*: Globalization of Food Systems in Developing Countries: Impact on Food Security and Nutrition, FAO Food and Nutrition Paper 83. Food and Agriculture Organization of The United Nations Rome, 2004.

Kieliszek, M. and Misiewicz, A. 2014. Microbial transglutaminase and its application in the food industry. A review. Folia. Microbiol. 59(3): 241–250.

Kirk, O., Borchert, T.V. and Fuglsang, C.C. 2002. Industrial enzyme applications. Curr. Opin. Biotechnol. 13(4): 345–351.

Knob, A., Beitel, S.M., Fortkamp, D., Terrasan, C.R.F. and Almeida, A.F.D. 2013. Production, purification, and characterization of a major *Penicillium glabrum* xylanase using brewer's spent grain as substrate. BioMed. Res. Int. 2013: 1–8.

Koepke, U. 2003. August. Conservation agriculture with and without use of agrochemicals. World congress on conservation agriculture "Producing in Harmony with Nature. Brazil 11–15.

Kuddus, M. 2014. Bio-statistical approach for optimization of cold-active α-amylase production by novel psychrotolerant *M. foliorum* GA2 in solid state fermentation. Biocatal. Agric. Biotechnol. 3(2): 175–181.

Kulshrestha, S., Tyagi, P., Sindhi, V. and Yadavilli, K.S. 2013. Invertase and its applications—a brief review. J. Pharm. Res. 7(9): 792–797.

Kumar, D., Kumar, S.S., Kumar, J., Kumar, O., Mishra, S.V., Kumar, R. and Malyan, S.K. 2017. Xylanases and their industrial applications: a review. Biochem. Cell. Arch. 17(1): 353–360.

Kumar, P. and Satyanarayana, T. 2009. Microbial glucoamylases: characteristics and applications. Crit. Rev. Biotechnol. 29(3): 225–255.

Lal, R. 2004. Soil carbon sequestration impacts on global climate change and food security. Science 304(5677): 1623–1627.

Leisola, M., Jokela, J., Pastinen, O., Turunen, O., and Schoemaker, H. 2002. Industrial use of enzymes. pp. 1–25. *In*: Hanninen, O.O.P. and Atalay, M. (eds.). Encyclopedia of Life Support Systems (EOLSS). EOLSS, Oxford, UK.

Lin, S. and Brewer, M.S. 2005. Effects of blanching method on the quality characteristics of frozen peas. J. Food. Qual. 28(4): 350–360.

Loginova, K., Loginov, M., Vorobiev, E. and Lebovka, N.I. 2011. Quality and filtration characteristics of sugar beet juice obtained by "cold" extraction assisted by pulsed electric field. J. Food. Eng. 106(2): 144–151.

López, J.A., da Costa Lázaro, C., dos Reis Castilho, L., Freire, D.M.G. and de Castro, A.M. 2013. Characterization of multienzyme solutions produced by solid-state fermentation of babassu cake, for use in cold hydrolysis of raw biomass. Biochem. Eng. J. 77: 231–239.

Lumia, G. 2011. Extraction par fluides supercritiques. pp. 231–258. *In*: Chemat, F. (ed.). Eco-extraction du végétal. Paris.

Lutz, W. and KC, S. 2010. Dimensions of global population projections: what do we know about future population trends and structures? Phil. Trans. R. Soc. B. 365(1554): 2779–2791.

Macris, B.J. 1981. Production of extracellular lactase from *Fusarium moniliforme*. Eur. J. Appl. Microbiol. 13(3): 161–164.

Mäder, P., Fliessbach, A., Dubois, D., Gunst, L., Fried, P. and Niggli, U. 2002. Soil fertility and biodiversity in organic farming. Science 296(5573): 1694–1697.

Manas, P., Pagan, R., Raso, J., Sala, F.J. and Condon, S. 2000. Inactivation of *Salmonella enteritidis*, *Salmonella typhimurium*, and *Salmonella senftenberg* by ultrasonic waves under pressure. J. Food. Prot. 63(4): 451–456.

Mariano, A.P., Dias, M.O., Junqueira, T.L., Cunha, M.P., Bonomi, A. and Filho, R.M. 2013. Utilization of pentoses from sugarcane biomass: techno-economics of biogas vs. butanol production. Bioresour. Technol. 142: 390–399.

Martínez, J.D., Mahkamov, K., Andrade, R.V. and Lora, E.E.S. 2012. Syngas production in downdraft biomass gasifiers and its application using internal combustion engines. Renew. Energy. 38(1): 1–9.

Mate, D.M. and Alcalde, M. 2015. Laccase engineering: from rational design to directed evolution. Biotechnol. Adv. 33(1): 25–40.

Matharu, A., Melo, E. and Houghton, J.A. 2017. Green chemistry: Opportunities, waste and food supply chains. pp. 457–467. *In*: Routledge Handbook of the Resource Nexus. CRC Press, Florida, USA.

Mena, C. and Stevens, G. 2010. Delivering performance in food supply chains: an introduction. pp. 1–15. *In*: Delivering Performance in Food Supply Chains. Woodhead Publishing.

Mohammadi, M., Habibi, Z., Dezvarei, S., Yousefi, M., Samadi, S. and Ashjari, M. 2014. Improvement of the stability and selectivity of *Rhizomucor miehei* lipase immobilized on silica nanoparticles: Selective hydrolysis of fish oil using immobilized preparations. Proc. Biochem. 49(8): 1314–1323.

Mohan, S.V., Nikhil, G.N., Chiranjeevi, P., Reddy, C.N., Rohit, M.V., Kumar, A.N. and Sarkar, O. 2016. Waste biorefinery models towards sustainable circular bioeconomy: critical review and future perspectives. Bioresour. Technol. 215: 2–12.

Motoki, M. and Seguro, K. 1998. Transglutaminase and its use for food processing. Trends Food Sci. Technol. 9(5): 204–210.

Mounir, S. and Allaf, K. 2008. Three-stage spray drying: new process involving instant controlled pressure drop. Dry. Technol. 26(4): 452–463.

Mulvihill, M.J., Beach, E.S., Zimmerman, J.B. and Anastas, P.T. 2011. Green chemistry and green engineering: a framework for sustainable technology development. Annu. Rev. Environ. Resour. 36: 271–293.

Nakkharat, P. and Haltrich, D. 2006. Purification and characterisation of an intracellular enzyme with β-glucosidase and β-galactosidase activity from the thermophilic fungus *Talaromyces thermophilus* CBS 236.58. J. Biotechnol. 123(3): 304–313.

Nazzaro, F., Fratianni, F. and Coppola, R. 2012. Nano and micro analyses. pp. 471–494. *In*: Boye, J.I. and Arcand, Y. (eds.). Green Technologies in Food Production and Processing. Springer, Boston, MA.

Nemecek, T., Huguenin-Elie, O., Dubois, D. and Gaillard, G. 2005. Life cycle assessment of cultivation systems in Swiss arable and forage production. FAL Series of Publications, Agroscope FAL Reckenholz, Zurich, Switzerland.

Niggli, U., Schader, C. and Stolze, M. 2010. Organic Farming-An efficient and integrated system approach responding to pressing challenges. Organic food and farming-A system approach to meet the sustainability challenge, The New EU Regulation for Organic Food and Farming pp. 17–20.

Noda, S., Miyazaki, T., Tanaka, T., Chiaki, O. and Kondo, A. 2013. High-level production of mature active-form *Streptomyces mobaraensis* transglutaminase via pro-transglutaminase processing using *Streptomyces lividans* as a host. Biochem. Eng. J. 74: 76–80.

Norton, L., Johnson, P., Joys, A., Stuart, R., Chamberlain, D. and Feber, R. 2009. Consequences of organic and non-organic farming practices for field, farm and landscape complexity. Agric. Ecosyst. Environ. 129(1-3): 221–227.

Norus, J. 2006. Building sustainable competitive advantage from knowledge in the region: the industrial enzymes industry. Eur. Plan. Stud. 14(5): 681–696.

OECD. 2011. A Green Growth Strategy for Food and Agriculture: Preliminary Report. OECD 2011.

OECD. 2013. Material Resources, Productivity and the Environment: Key Findings. OECD Green Growth Studies. OECD Publishing.

Oliveira, L.A., Porto, A.L. and Tambourgi, E.B. 2006. Production of xylanase and protease by *Penicillium janthinellum* CRC 87M-115 from different agricultural wastes. Bioresour. Technol. 97(6): 862–867.

Osterburg, B. and Runge, T. 2007. Maßnahmen zur reduzierung von tickstoffeinträgen in gewässer— eine wasserschutzorientierte landwirtschaft zur umsetzung der wasserrahmenrichtline, vol. 307. Braunschweig: Bundesforschungsanstalt für Landwirtschaft (FAL), Landbauforschung Völkenrode—FAL Agricultural Research.

Parniakov, O., Bals, O., Lebovka, N. and Vorobiev, E. 2016. Effects of pulsed electric fields assisted osmotic dehydration on freezing-thawing and texture of apple tissue. J. Food. Eng. 183: 32–38.

Pereira, E.G., Da Silva, J.N., de Oliveira, J.L. and Machado, C.S. 2012. Sustainable energy: a review of gasification technologies. Renewable. Sustainable. Energy. Rev. 16(7): 4753–4762.

Périno-Issartier, S., Ginies, C., Cravotto, G. and Chemat, F. 2013. A comparison of essential oils obtained from lavandin via different extraction processes: Ultrasound, microwave, turbohydrodistillation, steam and hydrodistillation. J. Chromatogr. A. 1305: 41–47.

Perrut, M. 2012. Sterilization and virus inactivation by supercritical fluids (a review). J. Supercrit. Fluids 66: 359–371.

Pillai, P., Mandge, S. and Archana, G. 2011. Statistical optimization of production and tannery applications of a keratinolytic serine protease from *Bacillus subtilis* P13. Proc. Biochem. 46(5): 1110–1117.

Pimentel, D., Harvey, C., Resosudarmo, P., Sinclair, K., Kurz, D. and McNair, M. 1995. Environmental and economic costs of soil erosion and conservation benefits. Science 267(5201): 1117–1123.

Pimentel, D. and Pimentel, M. 2003. Sustainability of meat-based and plant-based diets and the environment. Am. J. Clin. Nutr. 78(3): 660S–663S.

Popkin, B.M. 1998. The nutrition transition and its health implications in lower-income countries. Public Health Nutr. 1(1): 5–21.

Poulsen, P.B. and Buchholz, H.K. 2003. History of Enzymology with Emphasis on Food Production. Handbook of Food Enzymology, CRC Press. London.

Prakash, B., Vidyasagar, M., Madhukumar, M.S., Muralikrishna, G. and Sreeramulu, K. 2009. Production, purification, and characterization of two extremely halotolerant, thermostable, and alkali-stable α-amylases from *Chromohalobacter* sp. TVSP 101. Proc Biochem. 44(2): 210–215.

Prasad, M.P. and Manjunath, K. 2011. Comparative study on biodegradation of lipid-rich wastewater using lipase producing bacterial species. Indian. J. Biotechnol. 10: 121–124.

Puértolas, E., Luengo, E., Álvarez, I. and Raso, J. 2012. Improving mass transfer to soften tissues by pulsed electric fields: Fundamentals and applications. Annu. Rev. Food. Sci. Technol. 3: 263–282.

Quispe, C.A., Coronado, C.J. and Carvalho Jr, J.A. 2013. Glycerol: Production, consumption, prices, characterization and new trends in combustion. Renewable. Sustainable. Energy. Rev. 27: 475–493.

Radha, S., Nithya, V.J., Himakiran Babu, R., Sridevi, A., Prasad, N. and Narasimha, G. 2011. Production and optimization of acid protease by *Aspergillus* spp. under submerged fermentation. Arch. Appl. Sci. Res. 3(2): 155–163.

Ravindran, R. and Jaiswal, A.K. 2018. Enzymes in bioconversion and food processing. pp. 19–40. *In*: Enzymes in Food Technology. Springer, Singapore.

Reverchon, E. 1997. Supercritical fluid extraction and fractionation of essential oils and related products. J. Supercrit. Fluids 10(1): 1–37.

Riera, E., Golas, Y., Blanco, A., Gallego, J.A., Blasco, M. and Mulet, A. 2004. Mass transfer enhancement in supercritical fluids extraction by means of power ultrasound. Ultrason. Sonochem. 11(3-4): 241–244.

Roodenburg, B., De Haan, S.W.H., FERREIRA, J.A., Coronel, P., Wouters, P.C. and Hatt, V. 2013. Toward 6 log10 pulsed electric field inactivation with conductive plastic packaging material. J. Food Process Eng. 36(1): 77–86.

Rozzi, N.L. and Singh, R.K. 2002. Supercritical fluids and the food industry. Compr. Rev. Food Sci. Food Saf. 1(1): 33–44.

Sahnoun, M., Kriaa, M., Elgharbi, F., Ayadi, D.Z., Bejar, S. and Kammoun, R. 2015. *Aspergillus oryzae* S2 alpha-amylase production under solid state fermentation: optimization of culture conditions. Int. J. Biol. Macromol. 75: 73–80.

Santonen, T. 2012. Massidea.org—a greener way to innovate. pp. 541–568. *In*: Boye, J.I. and Arcand, Y. (eds.). Green Technologies in Food Production and Processing. Springer, Boston, MA.

Satterthwaite, D., McGranahan, G. and Tacoli, C. 2010. Urbanization and its implications for food and farming. Phil. Trans. R. Soc. B. 365(1554): 2809–2820.

Saulis, G. 2010. Electroporation of cell membranes: the fundamental effects of pulsed electric fields in food processing. Food. Eng. Rev. 2(2): 52–73.

Sawant, R. and Nagendran, S. 2014. Protease: an enzyme with multiple industrial applications. World. J. Pharm. Sci. 3: 568–579.

Schader, C. 2009. Cost-effectiveness of organic farming for achieving environmental policy targets in Switzerland. Ph.D Thesis, Institute of Biological, Environmental and Rural Sciences, Aberystwyth, Aberystwyth University, Wales. Research Institute of Organic Agriculture (FiBL), Frick, Switzerland.

Schader, C., Pfiffner, L., Schlatter, C. and Stolze, M. 2009. Umsetzung von Agrarumweltmassnahmen auf Bio-und konventionellen Betrieben der Schweiz. pp. 11–13. *In*: Mayer, J., Alföldi, T., Leiber, F. et al.

(eds.). Werte—Wege—Wirkungen: Biolandbau im Spannungsfeld zwischen Ernährungssicherung, Markt und Klimawandel. Beiträge zur 10. Wissenschaftstagung Ökologischer Landbau.

Schader, C., Stolze, M. and Gattinger, A. 2012. Environmental performance of organic farming. pp. 183–210. *In*: Green Technologies in Food Production and Processing. Springer, Boston, MA.

Schafer, T., Kirk, O., Borchert, T.V. et al. 2005. Enzymes for technical applications. pp. 557–617. *In*: Polysaccharides and Polyamides in the Food Industry: Properties, Production, and Patents, Biopolymers. Wiley-VCH, Weinheim, Germany.

Schäfer, T., Borchert, T.W., Nielsen, V.S., Skagerlind, P., Gibson, K. and Wenger, K. 2006. Industrial enzymes. White biotechnology. pp. 59–131. *In*: Adv. Biochem. Engin/Biotechnol. Spinger. Berlin. Germany.

Schiffmann, R.F. 2001. Microwave processes for the food industry. pp. 299. *In*: Datta, A.K. and Anantheswaran, R.C. (eds.). Handbook of Microwave Technology for Food Applications. New York, USA.

Sen, S.K., Dora, T.K., Bandyopadhyay, B., Mohapatra, P.K.D. and Raut, S. 2014. Thermostable alpha-amylase enzyme production from hot spring isolates *Alcaligenes faecalis* SSB17–statistical optimization. Biocatal. Agric. Biotechnol. 3(4): 218–226.

Seyis, I. and Aksoz, N. 2004. Production of lactase by *Trichoderma* sp. Food. Technol. Biotechnol. 42(2): 121–124.

Shekher, R., Sehgal, S., Kamthania, M. and Kumar, A. 2011. Laccase: microbial sources, production, purification, and potential biotechnological applications. Enz. Res. 2011: 1–11.

Shepherd, M.A., Harrison, R. and Webb, J. 2002. Managing soil organic matter–implications for soil structure on organic farms. Soil Use Manag. 18: 284–292.

Siemer, A.K., Toepfl, S. and Heinz, V. 2015. Application of pulsed electric fields in food. pp. 645–672. *In*: Bhattacharya, S. (ed.). Conventional and Advanced Food Processing Technologies, Food Science and Technology. John Wiley and Sons. New York, USA.

Singhania, R.R., Saini, J.K., Saini, R., Adsul, M., Mathur, A. and Gupta, R. 2014. Bioethanol production from wheat straw via enzymatic route employing *Penicillium janthinellum* cellulases. Biores. Technol. 169: 490–495.

Smith, P., Martino, D., Cai, Z., Gwary, D., Janzen, H. and Kumar, P. et al. 2007. Agriculture. *In*: Metz, B., Davidson, O.R., Bosch, P.R., Dave, R. and Meyer, L.A. (eds.). Climate Change 2007: Mitigation. Contribution of Working Group III to the Fourth Assessment Report of the Intergovernmental Panel on Climate Change. Cambridge University Press, Cambridge, UK.

Smith, P., Martino, D., Cai, Z., Gwary, H., Janzen, H. and Kumar, P. et al. 2007. Agriculture, Climate Change 2007: Mitigation. Cambridge: Contribution of Working Group III to the Fourth Assessment Report of the Intergovernmental Panel on Climate Change. Cambridge University Press. London, UK.

Sovová, H., Aleksovski, S.A., Bocevska, M. and Stateva, R.P. 2006. Supercritical fluid extraction of essential oils: results of joint research. Chem. Ind. Chem. Eng. Q. 12(3): 168–174.

Spilimbergo, S., Elvassore, N. and Bertucco, A. 2002. Microbial inactivation by high-pressure. J. Supercrit. Fluids 22(1): 55–63.

Steinfeld, H., Gerber, P., Wassenaar, T.D., Castel, V., Rosales, M. et al. 2006. Livestock's long shadow: environmental issues and options. Food and Agriculture Org.

Stolze, M., Piorr, A., Häring, A.M. and Dabbert, S. 2000. Environmental Impacts of Organic Farming in Europe. Universität Hohenheim, Stuttgart-Hohenheim. Germany.

Sundarram, A. and Murthy, T.P.K. 2014. α-amylase production and applications: a review. Appl. Environ. Microbiol. 2(4): 166–175.

Tang, X.J., He, G.Q., Chen, Q.H., Zhang, X.Y. and Ali, M.A. 2004. Medium optimization for the production of thermal stable β-glucanase by *Bacillus subtilis* ZJF-1A5 using response surface methodology. Biores. Technol. 93(2): 175–181.

Terefe, N.S., Buckow, R. and Versteeg, C. 2015. Quality-related enzymes in plant-based products: effects of novel food processing technologies part 2: pulsed electric field processing. Crit. Rev. Food Sci. Nutr. 55(1): 1–15.

Tervo, J.T., Mettin, R. and Lauterborn, W. 2006. Bubble cluster dynamics in acoustic cavitation. Acta. Acust. United. Ac. 92(1): 178–180.

Thomassen, M.A., van Calker, K.J., Smits, M.C., Iepema, G.L. and de Boer, I.J. 2008. Life cycle assessment of conventional and organic milk production in the Netherlands. Agric Syst. 96(1-3): 95–107.

Toepfl, S. 2012. Pulsed electric field food processing—industrial equipment design and commercial applications. Stewart. Postharvest. Rev. 8: 1–7.

Tokusoglu, Ö., Odriozola-Serrano, I. and Martín-Belloso, O. 2014. Quality, Safety, and Shelf-Life Improvement in Fruit. Improving Food Quality with Novel Food Processing Technologies, CRC Press, USA.

Van den Bosch, H.F.M. 2007. Chamber design and process conditions for pulsed electric field treatment of food. pp. 70–93. *In*: Lelieveld, H.L., Notermans, S. and De Haan, S.W.H. (eds.). Food Preservation by Pulsed Electric Fields: From Research to Application. Elsevier. USA.

Van, C. 2010. Mastering the Technological Aptitude of Oilseeds by Structural Modification: Applications to *In-Situ* Extraction and Transesterification Operations. Ph.D. Thesis, University of La Rochelle, France.

Várnai, A., Mäkelä, M.R., Djajadi, D.T. et al. 2014. Carbohydrate-binding modules of fungal cellulases: occurrence in nature, function, and relevance in industrial biomass conversion. pp. 103–165. *In*: Sima, S. and Geoffrey Michael, G. (eds.). Advances in Applied Microbiology. Academic/Elsevier, Amsterdam.

Vasic-Racki, D. 2006. History of industrial biotransformations—dreams and realities. pp. 1–35. *In*: Liese, A., Seelbach, K. and Wandrey, C. (eds.). Industrial Biotransformations. WileyVCH, Weinheim, Germany.

Vayssie`res, J.M.C. 2012. Management of agricultural inputs, waste and farm outputs: present and future best management practices. *In*: Boye, J.I. and Arcand, Y. (eds.). Green Technologies in Food Production and Processing. Springer, New York, USA.

Verge, X.P.C., Worth, D.E., Desjardins, R.L., McConkey, B.G. and Dyer, J.A. 2012. LCA of animal production. *In*: Boye, J.I. and Arcand, Y. (eds.). Green Technologies in Food Production and Processing. Springer, New York, USA.

Vijayaraghavan, K., Yamini, D., Ambika, V. and Sowdamini, N.S. 2009. Trends in inulinase production—a review. Crit. Rev. Biotechnol. 29(1): 67–77.

Vilariño, M.V., Franco, C. and Quarrington, C. 2017. Food loss and waste reduction as an integral part of a circular economy. Front. Environ. Sci. 5: 21.

Villela Filho, M., Araujo, C., Bonfá, A. and Porto, W. 2011. Chemistry based on renewable raw materials: perspectives for a sugar cane-based biorefinery. Enzyme. Res. 2011: 1–8.

Virmond, E., Rocha, J.D., Moreira, R.F.P.M. and José, H.J. 2013. Valorization of agroindustrial solid residues and residues from biofuel production chains by thermochemical conversion: a review, citing Brazil as a case study. Braz. J. Chem. Eng. 30(2): 197–230.

Vorobiev, E. and Lebovka, N. 2008. Electrotechnologies for Extraction from Food Plants and Biomaterials. Food Engineering Series. Springer, New York, USA.

Waldron, K. 2007. Waste minimization, management and co-product recovery in food processing: An introduction. pp. 3–20. *In*: Handbook of Waste Management and Co-Product Recovery in Food Processing. Woodhead Publishing, Cambridge, UK.

WHO/FAO. 2003. Diet, Nutrition and the Prevention of Chronic Diseases. WHO/FAO Expert Consultation, Geneva, Switzerland.

Wiktor, A., Śledź, M., Nowacka, M., Chudoba, T. and Witrowa-Rajchert, D. 2014. Pulsed electric field pretreatment for osmotic dehydration of apple tissue: Experimental and mathematical modeling studies. Dry. Technol. 32(4): 408–417.

Wilson, M.P. and Schwarzman, M.R. 2009. Toward a new US chemicals policy: rebuilding the foundation to advance new science, green chemistry, and environmental health. Environ. Health Perspect. 117(8): 1202–1209.

Woodhouse, A., Davis, J., Pénicaud, C. and Östergren, K. 2018. Sustainability checklist in support of the design of food processing. Sustain. Prod. Consum. 16: 110–120.

Wyman, C.E. and Goodman, B.J. 1993. Biotechnology for production of fuels, chemicals, and materials from biomass. Appl. Biochem. Biotechnol. 39(1): 41–59.

CHAPTER 10

Green Technologies for Reduction of Toxins in Food Production and Processing

Neha Kumari,[1] *Ankit Srivastava*[2,3] and *Saurabh Bansal*[1,*]

1. Introduction

New global challenges for the food industry are to ensure food safety while maintaining a good nutritional status of food functionality. Modern industrialization has resulted in the overuse of various chemicals in food to enhance the longevity, aroma, texture, and shelf-life of food. The chemicals pose major health hazards as they are carcinogenic. Moreover, the conventional approaches utilize a very high temperature, leading to a chemical reaction between food ingredients, and consequently forming undesirable toxic compounds, like acrylamide, nitrosamine, furans, and heterocyclic aromatic amines.

Green technologies proved to be the best alternative as it overcomes all the hurdles associated with conventional approaches which utilize thermal methodologies and carcinogenic chemicals. Globally, 25% of agricultural food is contaminated due to these mycotoxins, imposing more economic pressure to meet the annual food demand (Marin et al. 2013). Mycotoxins are secondary metabolites produced by several fungal species and introduced into food as raw materials, like grains and nuts, are contaminated with their spores or added up during food processing, storage, packaging, and distribution. The major mycotoxin-producing fungi are *Aspergillus, Fusarium,* and *Penicillium,* which result in major health concerns and safety issues

[1] Department of Biotechnology and Bioinformatics, Jaypee University of Information Technology, Waknaghat, District. Solan, Himachal Pradesh, India.
[2] Kusuma School of Biological Sciences, IIT Delhi, Hauz Khas, India.
[3] Rocky Mountain Laboratories, National Institute of Allergy and Infectious Diseases, NIH, Hamilton, MT, USA.
* Corresponding author: saurab.bansal02@gmail.com

due to their unpredictable health issues. Till now, 450 mycotoxins have been discovered so far, among them aflatoxin (AF), ochratoxin (OT), deoxynivalenol (DON), nivalenol (NIV), fumonisin (FM), zearalenone (ZEN), and patulin (PAT) have proved to be the most potent agents for causing major multifaceted human health diseases (Alshannaq and Yu 2017). Most mycotoxins are chemically inert for any kind of sterilization, thermal destruction, and pasteurization process. Most of these mycotoxins directly enter by consuming contaminated food ingredients and indirectly enter the human food chain through contaminated animal feed (Richard 2007, Kaushik 2015). The robust chemical nature of mycotoxins contributed to their resistance to physical, chemical, and biological traditional methods. Therefore, currently, there is a huge demand for versatile novel approaches to overcome the shortcomings associated with classical methodologies.

The novel green techniques are more ecofriendly and non-thermal with minor effects on functional attributes of the food matrix and have proven to be suitable for heat liable food ingredients. The underlying mechanism behind their action is that the presence of water helps in the generation of reactive species, which results in oxidative stress. Reactive species cause disruption, transformation, and modification in functional groups in mycotoxins structure (Figure 1) (Adebo et al. 2021). The major factors influencing the efficacy of these novel technologies are the availability of moisture, initial concentration, type of mycotoxins, additive effect with other food ingredients, and other physio-biochemical attributes of the food matrix. The current chapter aims to critically underlie the applications of novel non-thermal approaches for mycotoxins decontamination.

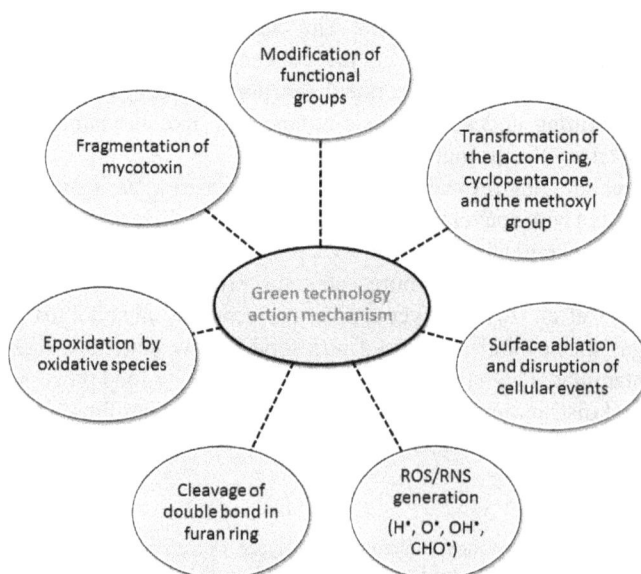

Figure 1. A different mechanism of action of green technology for mycotoxin detoxification.

2. Major Kind of Mycotoxins

Among various mycotoxins, 20 classes have been classified, each varying in its occurrence and toxicological potency. Maize is the most contaminated crop, while rice is the least one (Chulze 2010). Since a large number of cereals, fruits, and nuts are being contaminated with mycotoxins, food organizations like the World Health Organization (WHO), Food and Agricultural Organization (FAO), US Food and Drug Administration (FDA), and European Food Safety Authority (EFSA) have set up their minimal permissible limit in food and feed (Lee and Ryu 2017). Aflatoxin is one of the most dangerous toxins produced by *Aspergillus* species. The most contaminated food of these aflatoxins is peanut butter, millet, sorghum, corn, rice, chillies, cottonseeds, peanut oils, and dairy products (Kong et al. 2012, Teniola et al. 2005). Ochratoxin A1 (OTA1) is mainly produced by *Aspergillus* and *Penicillin* species and is detected primarily in cereals, wine, pork, nuts, spices, and beer (Péteri et al. 2007, Duarte et al. 2010, Chang et al. 2015). Deoxynivalenol is mainly produced from *Fusarium* species and wheat, maize, and oats-derived food, and feed is mainly contaminated (Zhu et al. 2018, He et al. 2016). Zearalenone is an estrogenic mycotoxin produced by *Fusarium* species and mainly grains are the most infected ones (Kowalska et al. 2016). Patulin is the most commonly contaminated mycotoxin produced by *Penicillium, Aspergillus,* and *Byssochlamys* (Assatarakul et al. 2012). Table 1 shows the significant mycotoxins causing major health concerns.

3. Approaches for Reduction of Toxins in Food Production and Processing

Since, the presence of mycotoxins in food preparation may cause mild to fatal symptoms to the consumers, therefore, the mycotoxins removal from the food preparation is necessary to avoid the clinical consequences. Various novel non-thermal approaches including, Irradiation, High pressure processing (HPP), Cold plasma treatment, Pulse Electric Field (PET) treatment, Pulse Light (PL) Treatment, Ultrasound (US) treatment, Enzymatic treatment and treatment with nanomaterials have been developed for the mycotoxin decontamination.

3.1 Irradiation

Irradiation is a dual physical approach applied to get rid of pathogenic microbes and detoxify mycotoxins. Both ionizing (Gamma radiation) and nonionizing (solar, UV, and microwave) radiations are used and evidenced to be competent for industrial applications (Figure 2). Solar radiation has been proved as the most economical mode of irradiation. Unrefined groundnut oil contaminated with AF when exposed for 15 minutes to bright sunlight resulted in 99% toxin destruction (Shantha and Murthy 1977). AFB1 contaminated groundnut cake partially (50%) destroyed by sunlight, whereas 83% reduced load was observed in casein (Shantha and Murthy 1981). The exposure to sunlight for an optimal time resulted in a significantly decreased total AF concentration in the artificially spiked feed. AF content in sheep feed was disinfected (83.7%) when exposed to direct sunlight for

Table 1. Significant types and characteristics of potential mycotoxins.

Type of Mycotoxins	Producing Fungi	Abbreviations	Chemical Component/ Derivative	Health Hazard/Toxicity	References
Aflatoxin B1, B2, G1, G2	*Aspergillus flavus*, *A. parasiticus*	AFB1, AFB2, AFG1, AFG2	Difuranocoumarin derivatives	Carcinogenic, teratogenic, hepatotoxic, mutagenic, and immunosuppressive	(Bennett and Klich 2003, Wilson et al. 2002, Liu and Wu 2010)
Aflatoxin M1	Due to metabolism of Aflatoxin B1	AFM1		Carcinogenicity, mutagenicity, genotoxicity, teratogenicity, and immunosuppression	(Prandini et al. 2009, Min et al. 2021)
Ochratoxin A1	*A. ochraceus* *Penicillium verrucosum* *A. carbonarius*	OTA1	Polyketide derived modified isocoumarin	Balkan Endemic Nephropathy, Nephrotoxic, hepatotoxic, immunogenicity, genocity, neurotoxic, embryotoxic, and teratogenicity	(Bennett and Klich 2003, Duarte et al. 2010)
Deoxynivalenol	*Fusarium graminearum* *F. culmorum*	DON	Type B trichothecene mycotoxins	Carcinogenic, immunosuppressive, Nausea, Vomiting, diarrhoea, abdominal pain, headache, dizziness, and fever	(Bottalico and Perrone 2002, He et al. 2016, Zhu et al. 2018)
Zearalenone	*F. graminearum* *F. culmorum*	ZEN	Macrocyclic β-resorcylic acid lactone	Infertility and hyperestrogenism	(Bennett and Klich 2003, Kowalska et al. 2016)
Fumonisin B1, B2, B3	*F. verticillioides* *F. proliferatum*	FMB1, FMB2, FMB3	1, 2, 3-propanetricar-boxylic acid	Carcinogenic	(Rheeder et al. 2002)
Patulin	*Penicillium expansum*	PAT	4-hydroxy-4H-furo (3,2-c)-pyran-2-(6H)-one	Nausea, vomiting, ulceration, and hemorrhage	(Drusch and Ragab 2003, Assatarakul et al. 2012)
Ergot alkaloids	*Claviceps purpurea*	EA		Acute neurotoxicity, chronic toxicity, and ergotism	(EFSA Panel on Contaminants in the Food Chain (CONTAM) 2012)

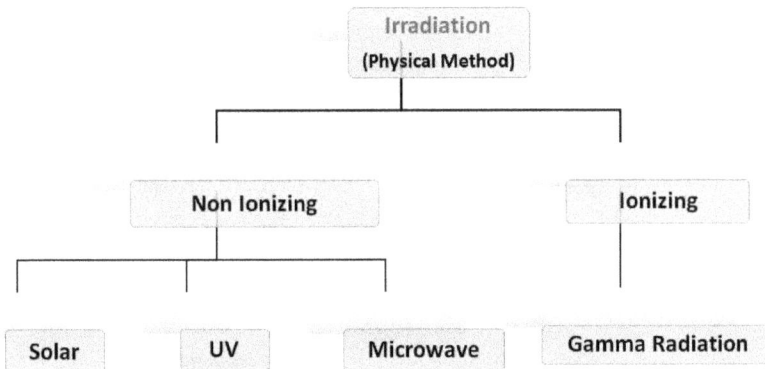

Figure 2. Mycotoxins mitigation by different types of irradiation.

14 hours at ambient temperature (25/37°C) (Gowda et al. 2007). Solar irradiation proved to be more efficient than gamma radiation (25 KGy) and microwave heat (10 minutes) treatment (Herzallah et al. 2008). AFB1 was reduced to 74.85% in broken rice when exposed to sunlight for 16 hours (Sapcota et al. 2011). One of the limiting factors associated with solar radiation is penetration power in the food matrix.

3.1.1 Gamma Radiation Treatment

Gamma radiation causes mycotoxin fragmentation and causes no other toxicological or nutritional alterations. Food (peanut, pistachio, rice, and corn) and feed artificially contaminated with AFB1 were successfully decontaminated with 10 kGy (Ghanem et al. 2008). Another study conducted for decontaminating almonds showed that the maximum decline observed was 19.25%, 10.99%, 21.11%, 16.62%, and 23.90% for AFB1, AFB2, AFG1, AFG2, and OTA, respectively, that were found at 15 kGy. However, treatment was not effective for complete detoxification (Di Stefano and Pitonzo 2014). Even earlier studies carried out for detoxification of black pepper also showed similar findings that AF and OTA are more resistant to gamma radiation (Jalili et al. 2010). Most recently, peanuts were irradiated with 20 kGy showed complete eradiation of OTA and ZEN and a 59% reduction recorded for AFB1 (Abdel-Rahman et al. 2021).

3.1.2 Microwave (MW) Treatment

The mechanism of detoxification uses microwave functions by denaturing inter/intracellular proteins, extruding the cellular contents of the microbial cells. MW disrupts enzyme structure (oxidase) and aflatoxin structure by a non-thermal heating system. One of the major disadvantages of this approach is its non-uniform distribution of heating and low penetration depth to the food matrix (Li et al. 2016, Lee et al. 2017). Kim et al. studied the effect of MW treatment on chilli sauce and elucidated that low sugar and dielectric properties positively enhance the impact of treatment without affecting its physical appearance (Kim et al. 2018) (Figure 3). Synergistic effects of radiation (7 kGy) and MW at 600 W resulted in 95% reduced AFB1 load and 85% reduced mutagenicity in artificially spiked peanuts (Patil et al.

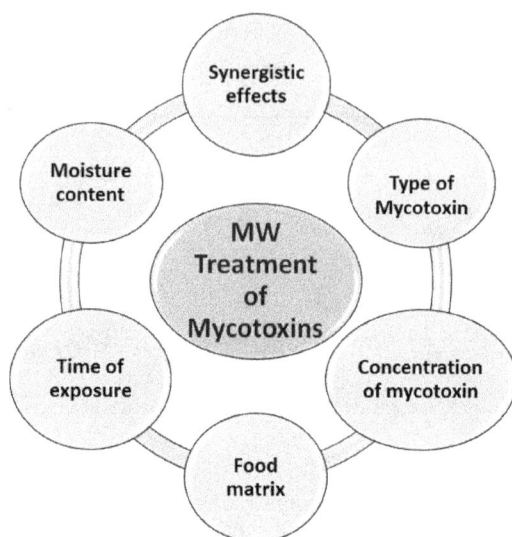

Figure 3. Factors affecting the effectiveness of Microwave (MW) treatment for mycotoxin degradation.

2019). Water-assisted MW treatment to mitigate AFB1 load in corn with negligible effect on its color. Water helps maintain a low temperature to avoid the damage caused by heat-assisted kernels of corn (Zhang et al. 2020). MW treatment of wheat flour with artificially spiked AFB1 and OTA showed that initial mycotoxin concentration and exposure time play an imperative role in detoxification (Alkadi and Altal 2019). Figure 3 shows the various factors affecting microwave treatment.

3.1.3 Ultraviolet Light (UV) Treatment

The UV treatment method is the most appropriate, inexpensive, efficient, and easy to use non-thermal way of reducing mycotoxins and mycotoxins producing fungal strains from food and feed matrix (He et al. 2021). The mechanism of this method relies on the efficacy of chemical contaminants that undergo photodegradation by absorbing photons when exposed to UV radiation that falls under a wavelength of 100–400 nm. Commercially for UV sources, mercury lamps, collimated beam systems, ciderSure UV systems, and 3D UV-C chambers have been used for mycotoxins degradations (Assatarakul et al. 2012, Popović et al. 2018, Xiang et al. 2014, Koutchma 2019). The feasibility of this treatment depends upon the nature of the photo degraded produced formed at last. It should be either nontoxic or less toxic without interfering with the organoleptic characteristics of food (Chandra et al. 2017, Shanakhat et al. 2019). The effectiveness and efficacy of mycotoxins degradation by the UV treatment method rely on the wavelength of the UV absorbance spectrum of mycotoxins and the food matrix (Zhu et al. 2014, Chandra et al. 2017). Various research teams have studied the kinetic models for mycotoxin photodegradation by UV treatment.

Most liquid media have been decontaminated with zero, first, and second-order kinetic models. When apple cider and apple juices were treated with UV-C radiation

Table 2. Mycotoxins mitigation on various food matrices by UV (Ultraviolet) treatment.

Food matrix	Mycotoxin	UV treatment	References
Animal feed	ZEN and DON	intensity 0.1 mW cm⁻² at 254 nm UV-C for 60 min	(Murata et al. 2008)
Apple cider, apple juice	PAT	UV exposure, ranging from 14.2 mJ/cm² (one pass) to 99.4 mJ/cm² (seven passes)	(Assatarakul et al. 2012)
Apple juice and apple cider	PAT	UV exposure at 253.7 nm for 40 minutes results in first-order kinetics for PAT degradation	(Zhu et al. 2013)
Apple juice and apple cider	PAT	Far UVC (222 nm) proved to be most efficient with UV fluences of 19.6 mJ/cm²	(Zhu et al. 2014)
Apple juice	PAT	UV-C dose of 400 mJ/cm²	(Chandra et al. 2017)
Peanut oil	AFB1	40 minutes exposure time	(Xiang et al. 2014)
Maize and Wheat	DON, ZEN, and OTA	UV-C dose of 15,000 mJ cm⁻²	(Popovic et al. 2018)
Semolina	OTA1, DON, and Enniatin B (ENB)	UVC irradiation for 120 minutes at a distance of 15 cm	(Shanakhat et al. 2019)

(UVC) for seven cycles at 253.7 nm, it results in 99.4 mJ/cm² fluence (Assatarakul et al. 2012). Studies have concluded that UV exposure resulted in more efficient PAT reduction in apple juice than apple cider. The latter has more insoluble solids (pectin, cellulose, and hemicellulose) (Assatarakul et al. 2012, Zhu et al. 2014). Another explanation for enhanced PAT reduction in apple juice compared to apple cider is the presence of ascorbic acid having a maximum absorbance spectrum at 254 nm, rendering instability in PAT conformation (Zhu et al. 2013, El-Hajj Assaf et al. 2019).

Various food matrixes have been studied to be effectively treated with UV treatment listed in Table 2. One of the most undesirable effects of UV treatment is the loss of nutritional status and the organoleptic character of the food matrix (Sew et al. 2014). OTA photodegradation showed first-order kinetics by multi-UV wavelength in a pH and time-dependent manner (Ibarz et al. 2015). Recent studies have found that maize and wheat kernels infected with DON, ZEN, and OTA showed reduced mycotoxin load by UV 3D sources (Popovic et al. 2018). Significant shortcomings associated with UV treatment are low penetration potency and transformed end products may still be toxic. Nowadays, much success is achieved by combining the UV treatment with other oxidizing agents, like ozone and hydrogen peroxide (Shen and Singh 2021).

3.2 High-Pressure Processing (HPP)

The HPP treatment works by applying high pressure, which results in breakage of the plasma membrane, alters genetic material, and disrupts the chemical structure of mycotoxins. The primary critical limiting factor of HPP is pressure, temperature, and exposure time (Woldemariam and Emire 2019). Olives contaminated with mycotoxin citrinin produced by *Penicillium* citrinum were treated at 250 MPa at 35°C for 1 minute exposure time (Tokusoglu et al. 2010). Avsaroglu et al. (2015) first evidenced that pulsed HPP results in 62.11 reduced PAT in apple juice (Avsaroglu et al. 2015). Blended juice of fruits and vegetables was successfully treated with 600 MPa for 300 s at 11°C for PAT mitigation (Hao et al. 2016). A study showed complete eradication of DON and ZEN from contaminated maize after 550 MPa pressure was applied for 20 minutes at 45°C (Kalagatur et al. 2018). HPP acts uniformly on a wide variety of food matrices without altering its flavor, odor, and physical appearance (Barba et al. 2012). The only shortcoming linked with the process is that it acts in batch and needs standardization of optimum treatment time. Some studies formulated that the combinatorial approach is much more fascinating for detoxification (Wu et al. 2021). Evelyn and Silva formulated that strawberry puree contaminated with *Byssochlamys nivea* was treated more effectively by combining HPP at 600 MPa, beside ultrasound processing (24 kHz) at 75°C than alone treatment (Evelyn and Silva 2015).

3.3 Cold Plasma (CP) Treatment

The term 'plasma' has been used since 1928, which is described as the region of ionized ions and atoms and is technically termed the fourth state of matter (Langmuir 1928). Different sources are utilized for plasma generation, like dry air, high atmospheric pressure (Cold Atmospheric Pressure Plasma, CAPP), and high voltage. Microwave-induced argon gas plasma resulted in 100% exclusion of AFB1, DON, and NIV after 5 seconds of treatment. The cytotoxic assessment showed no toxic effects on mouse macrophage RAW264.7 cells (Park et al. 2007). CP-treated date palm fruits showed 100% efficacy against OTA after a 7.5 minute exposure time (Ouf et al. 2015). CAPP resulted in a 70% reduced AFB1 in treated hazelnuts while maintaining its organoleptic characteristics (Siciliano et al. 2016). A study showed that CAPP results in 68% deduced AFB1 and FMB1 in maize. Degraded products formed after AFB1 decontaminations showed no cytotoxicity against human hepatocellular cells (HCC) (Wielogorska et al. 2019). Plasma gas and exposure time determine the efficacy of CAPP treatment (Durek et al. 2018). CAPP treatment effectively mitigates mycotoxins as compared to UV treatment. Mycotoxin reduction observed after 8 minutes of treatment was AF by 93%, trichothecenes by 90%, FM by 93%, and ZEN by 100% (Hojnik et al. 2019). High voltage treatment either acts directly on mycotoxins or decontaminates the mycotoxins producing fungi from the food matrix.

High Voltage Atmospheric Cold Plasma (HVACP) inactivated the fungal spores and toxins by using air for a generation of reactive oxygen (ROS) and nitrogen species (RNS) (O_3, NO_2, NO_3, N_2O_4, and N_2O_5). Firstly, the plasma species interact with the food matrix and penetrate it. Then, these ROS and RNS either result in oxidative stress leading to surface ablation and membrane degradation of fungal spores or transform toxic metabolites into less toxic forms (Shi et al. 2017b, Ott et al. 2021). CP treatment is much more effective as long-lived molecular reactive species, like ozone, are produced during the process showing synergistic effects for decontamination. The O and OH· species are short-lived and directly act on

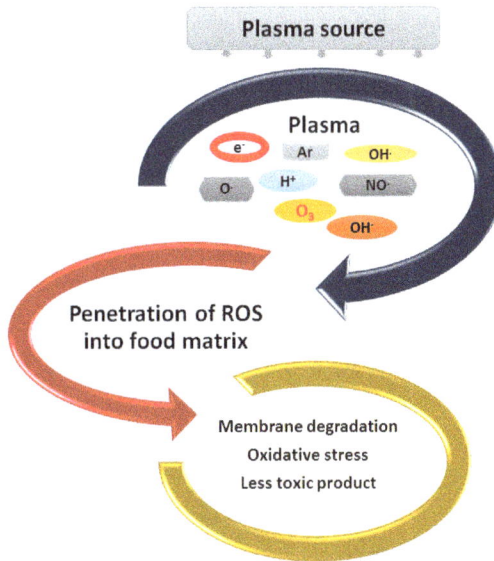

Figure 4. Mycotoxin decontamination by cold plasma (CP) treatment.

membranes causing oxidation of lipids. These antimicrobial oxidative species cause significant intracellular damage (Laurita et al. 2015, Hojnik et al. 2019) (Figure 4). The therapy showed effective results in a dose-dependent manner. Previous studies have shown that cold plasma treatment is much more efficient than UV treatment for decontaminating aflatoxin and zearalenone.

HVACP treatment for 5 minutes resulted in 76% degradation of AFB1. The mechanism behind mycotoxins degradation involves two pathways. One approach suggests that free radicals (H$^{\cdot}$, OH$^{\cdot}$, and CHO$^{\cdot}$) formed during treatment consequently destabilize the mycotoxins structures. Other pathways showed that epoxidation by HO$_2$$^{\cdot}$ radicals and other oxidative species OH$^{\cdot}$, H$_2$O$_2$, and O$_3$ results in the vanishing of the double bond in the furfural ring and the transformation of the lactone ring, cyclopentanone, and the methoxyl group results in degradation of AFB1 (Shi et al. 2017a,b). HVACP treatment for 1 minute exposure resulted in 50% of spores destructed in *Aspergillus flavus*, whereas 99% destruction of DON mycotoxins (Ott et al. 2021). Moreover, cytotoxic assessment using HepG2 cell lines also suggests that 20 minutes HVACP treatment is safe for AFB1 decontamination (Nishimwe et al. 2021).

Despite having various advantages, a major drawback of the CP approach is that plasma species can interact with biological biomolecules (Carbohydrates, proteins, and lipids). ROS species interact with the amino acids and cause disruption/ modification in the native structure of the protein (Attri et al. 2015). Plasma induces the breakage of glycosidic bonds in polysaccharides. Lipids are most sensitive to CP treatment owing to primary and secondary oxidative end products causing off-odor in high-fat foods products. Precise control of airflow, voltage, and long-lived plasma species is another safety concern associated with CP (Niemira 2012, Misra et al. 2011, 2016). Table 3 shows different matrix studies for mycotoxin detoxification

Table 3. Various mycotoxin detoxification studies by CP treatment.

Matrix	Mycotoxin	Plasma Source	Mycotoxin Reduction	Reference
Slide glasses	AFB1, DON, and NIV.	MW induced argon gas.	100% exclusion of AFB1, DON, NIV after 5 seconds of treatment	(Park et al. 2007)
Hazelnut	AFB1, AFB2, AFG1, and AFG2.	Low-pressure CP.	50% reduction achieved at 20 minutes	(Basaran et al. 2008)
Hazelnut	AFB1, AFG1, and AFG2.	Dielectric barrier discharge CP.	AFB1 and AFG2 were more sensitive to treatment 70% AFB1 reduced	(Siciliano et al. 2016)
Thin layer	DON, ZEN, and FMB1.	CCAP.	Destroyed after 60 s.	(Ten Bosch et al. 2017)
Glass slide	AFB1	HVACP.	76% reduction observed after 5 minutes.	(Shi et al. 2017a)
Hazelnut	AFB1, AFB2, AFG1, and AFG2.	CP and Cold low pressure.	72–73% AFB1 detoxify at 1.5 minutes exposure time.	(Sen et al. 2019)
Maize	AFB1 and FMB1.	CAPP.	68% deduction in AFB1 and FMB1.	(Wielogorska et al. 2019)
Glass coverslip	AFB1, AFB2, AFG1, and AFG2.	CAPP.	83% reduction in 8 minutes.	(Hojnik et al. 2019)
Peanut	AF	Atmospheric pressure plasma jet (APPJ).	23% and 39% reduction observed after 2 and 5 minutes exposure, respectively.	(Iqdiam et al. 2020)
Agar plates and pure DON	DON	HVACP.	1 minute exposure resulted in 50% spores destructed in *Aspergillus flavus*, whereas 99% destruction of DON mycotoxins.	(Ott et al. 2021)

by CP treatment. CP technology is novel and yet standardization needs to be done to apply it on a commercial scale. Ingredients of food matrix-like vitamin C, E, and phenolic content have a scavenging capacity to remove ROS and RNS. Complete elucidation behind the CP mechanism needs to be confirmed through more elaborate studies and urgently arrested for being used in the commercial-scale food industry.

3.4 Pulsed Electric Field (PEF) Treatment

This technology utilizes small electrical pulses for the inactivation of microorganisms and imposes minimum harmful influence on the food quality. This treatment provides the electric field of various power ranges from 1–40 kV/cm for a specific period, and the product is positioned amid two electrodes (Misra et al. 2018). PEF maintained the freshness, nutrition, color, and essence of food. This processing structure comprises a high voltage, chamber, capacitor, and other switches (Singh and Kumar 2011, Butz

Table 4. Pulsed electric field treatment for mycotoxins reduction.

Treatment Conditions	Toxins	Percentage Reduction	Reference
Pulse width- 10–26 µs pH- 4–10 Output Voltage- 20–65%	AFB1	77–97%	(Vijayalakshmi et al. 2018)
Pulse field energy- 0.97 to 17.28 J	AFG1, AFG2, AFB1, and AFB2	94.7%, 92.7%, 86.9%, and 98.7%, respectively.	(Bulut et al. 2020)
High hydrostatic pressure and pulsed electric field	AFB2 and AFG1	72% and 84%, respectively	(Pallares et al. 2021)
Voltage-30 kV Field strength-3 kV/cm Max. temp.-75°C Specific energy-500 kJ/kg Pulse-238	Different types of Enniatins (ENA, ENA1, ENB, ENB1) and Beauvericin (BEA)	43%–70%	(Pallares et al. 2020)
Time-12 hours PEF-100–800 Hz	AFG1 and AFB1	83%	(Eisa et al. 2003)

and Tauscher 2002). Mycotoxins are generally present in feed and food products and are produced from the derived metabolism processes of fungal strains (Carballo et al. 2018). This pulse-field treatment has evolved as one of the non-thermal processing methods. Also, it has applications in fungal growth (Evrendilek et al. 2019) and toxin inhibition (Singh and Kumar 2011). This technique also effortlessly treats the seeds, liquid foods, and grains (Evrendilek et al. 2017, Evrendilek et al. 2019). In food treatment, the PEF treatment takes into account value-added products, microbes' inactivation, perking up the osmotic, drying, and freezing processes (Koubaa et al. 2016, Gabric et al. 2018, Putnik et al. 2018) (Table 4). This treatment process did not degrade the quality of food parameters compared with other conventional methods. Therefore, the recommendation of response surface methodology (RSM) and other optimizing processes assisted in designing the process for better outcomes (Gavahian et al. 2020).

3.5 Pulsed Light (PL) Treatment

The pulse light treatment has also emerged as a non-thermal technology to decontaminate food and feed products and has a fluence limit of 12 J cm^{-2}. This treatment is endorsed by the food and drug administration to produce and facilitate the processing. This equipment generates high strength light with emission ranges from 100–1,100 nm, including visible, infrared, and UV light. The intensity of emitted light at sea level is higher and more potent than the sunlight (Dunn et al. 1995, Oms-Oliu et al. 2008). This methodology relies upon a wide variety of light treatments. It can trigger the chemical and biochemical processes in food, making it promising to attain certain co-products through its mechanism. This treatment benefited in reducing the overall equipped cost through inactivating the fungal species, microbial community, and endospores in a short period (Garvey et al. 2015, Kramer et al. 2017). Table 5 shows different PL treatment effects on various mycotoxins.

Table 5. PL treatment effects on various mycotoxins reductions.

Treatment Conditions	Toxins	Reduction %	Reference
Light flux- 1 J cm^{-2}	AFB1, OTA, ZEN, and DON	92.7%, 98.1%, 84.5% and 72.5%, respectively	(Moreau et al. 2013)
Pulse light- 0.4 J cm^{-1}	AF	91%	(Wang et al. 2016)
Pulse Light- 0.52 J cm^{-1} Time- 15 second	AFB1 and AFB2	90.3% and 86.7%	(Abuagela et al. 2018)
Fluence-0.4 J cm^{-2} per pulse Number of pulses- 10–80 Supplemented with glutathione and Fe^{2+}	PAT	Maximum reduction-98.5%	(Rodríguez-Bencomo et al. 2020)
Pulse light-35.8 J cm^{-2}	PAT	22%–51% reduction	(Funes et al. 2013)
Pulse rate-180 pulse for 1 minute	DON	35.5%	(Chen et al. 2019)

3.6 Ultrasound (US) Treatment

Ultrasound (US) is an emerging cheap, energy-saving treatment that causes cavitation and removes toxigenic fungi and mycotoxin without producing any toxic end product. AFB1, DON, ZEN, and OTA were attained at a duty cycle of 25% and reduced up to 96.5, 60.8, 95.9, and 91.6%, respectively, by ultrasonic intensity of 2.2–11 W/cm^3 and sonication time range from 10 to 50 minutes. DON was found to be less sensitive to US treatment. However, the US treatment is affected by preliminary mycotoxin concentrations, US power intensity, time duration, and duty cycle (Liu et al. 2019). The US treatment resulted in an 85.1% reduced AFB1 after 80 minutes of ultrasound exposure. US waves cause cavitation in an aqueous solution until bubbles outburst and form H• and OH• radicals for detoxification. Another mechanism underlying H$_2$O$_2$ epoxidation eventually leads to the oxidation of AFB1. The US detoxified AFB1 by breaking the C8–C9 double bond in the furan ring and amending the lactone ring and methoxy group (Liu et al. 2019). Thermo-US treatment at a rate of 20 kHz and 95% amplitude for 10–15 minutes of exposure efficiently reduces the AFM1 in milk without hampering its microbiological and physiological properties (Hernández-Falcón et al. 2018). The US is detrimental to being used on a commercial scale as it denatures proteins and can cause the oxidation of fats (Adebo et al. 2021).

3.7 Enzymatic Treatment

Enzymes are attractive food processing tools for mycotoxin reduction as they are very specific. Acrylamide, a potent toxic metabolite, was removed from bread by recombinant purified L-asparaginase of *Thermococcus kodakarensis* (Hong et al. 2014). AFB1 was reduced to 87.34% when treated with a pure laccase enzyme isolated from *Trametes versicolor* (1 U/ml) (Alberts et al. 2009). AFB1 was transformed to the less toxic form AFB1-8,9-dihydrodiol with the application of 48 hours exposure of manganese peroxidase (5 nkat) from *Phanerochaete sordida* YK-624. The 86% abolition of AFB1 was detected after 48 hours treatment (Wang et al. 2011). Orotate phosphoribosyltransferase (0.15 g/l) isolated from *Rhodotorula mucilaginosa* effectively degrades 80% PAT-contaminated apple in 18 hours at 25°C (Tang et al. 2019). One of the major shortcomings concomitant with enzymatic treatment is its chemical nature as they are proteins and can cause allergic reactions. Since then, no enzyme has been assigned by the EU to be used as a detoxifying agent on a commercial scale (Karlovsky et al. 2016).

3.8 Treatment with Nanomaterials

Treatment with nanomaterials is a recently introduced novel method to mitigate mycotoxin contamination while ameliorating food's functional and nutritional properties. This nanomaterial acts as an absorbent for mycotoxin mitigation (Horky et al. 2018). Magnetic noncomposite prepared from maize straw results in a 90% reduced AFB1 (Zahoor and Khan 2016). Zeolite samples showed anti-mycotoxins activity against AFB1 produced by *A. flavus* (Savi et al. 2017). Magnetic graphene oxide nanocomposites treatment at pH 6.2 for 5.2 h at 40.6°C resulted in 69.57 and 67.28% reduced DON and ZEN levels, respectively, in *Fusarium* mycotoxin. Chitosan-coated Fe_3O_4 particles effectively removed PAT from fruit juices (Luo et al. 2017). But further detailed toxicological studies need to be done to implement this approach on a commercial scale.

4. Conclusion

Novel non-thermal green technologies help overcome all the hurdles allied with traditional detoxification procedures and thus could be fruitfully executed in commercial food processing. This novel technology either removes or transforms mycotoxins into a less toxic form without affecting the physical, chemical, and biological attributes of the food matrix. Green techniques are more fascinating as heat liable food products undergo detoxifying mycotoxins more efficiently than conventional approaches. More process parameters optimization, standardization, interpretation of mechanism for detoxification, and further elaborated future research must be done. More extensive future research is compulsory for accessing the safety concerns associated with byproducts generated after applying novel technologies to the food matrices. Although complete eradication cannot be achieved with the solo method, combining one or more modes can be explored to accomplish 100% efficacy.

References

Abdel-Rahman, Gomaa N., Yousef Y. Sultan, Salah H. Salem and May M. Amer. 2021. Identify the natural levels of mycotoxins in egyptian roasted peanuts and the destructive effect of gamma radiation. Journal of Microbiology, Biotechnology and Food Sciences 10: 1174–77.

Abuagela, Manal O., Basheer M. Iqdiam, Hussein Mostafa, Liwei Gu, Matthew E. Smith and Paul J. Sarnoski. 2018. Assessing pulsed light treatment on the reduction of aflatoxins in peanuts with and without skin. International Journal of Food Science & Technology 53(11): 2567–75. https://doi.org/10.1111/ijfs.13851.

Adebo, Oluwafemi Ayodeji, Tumisi Molelekoa, Rhulani Makhuvele, Janet Adeyinka Adebiyi, Ajibola Bamikole Oyedeji, Sefater Gbashi, Martins Ajibade Adefisoye, Opeoluwa Mayowa Ogundele and Patrick Berka Njobeh. 2021. A review on novel non-thermal food processing techniques for mycotoxin reduction. International Journal of Food Science & Technology 56(1): 13–27. https://doi.org/10.1111/ijfs.14734.

Alkadi, Hourieh and Jihad Altal. 2019. Effect of microwave oven processing treatments on reduction of aflatoxin b1 and ochratoxin a in maize flour. European Journal of Chemistry 10(3): 224–27. https://doi.org/10.5155/eurjchem.10.3.224-227.1840.

Alshannaq, Ahmad and Jae-Hyuk Yu. 2017. Occurrence, toxicity, and analysis of major mycotoxins in food. International Journal of Environmental Research and Public Health 14(6). https://doi.org/10.3390/ijerph14060632.

Alberts, J.f., Gelderblom, W.c., Botha, A. and van Zyl, Wh. 2009. Degradation of aflatoxin B(1) by fungal laccase enzymes. International Journal of Food Microbiology 135(1). https://doi.org/10.1016/j.ijfoodmicro.2009.07.022.

Assatarakul, Kitipong, John J. churey, David C. Manns and Randy W. Worobo. 2012. Patulin reduction in apple juice from concentrate by UV radiation and comparison of kinetic degradation models between apple juice and apple cider. Journal of Food Protection 75(4): 717–24. https://doi.org/10.4315/0362-028X.JFP-11-429.

Attri, P., Sarinont, T., Kim, M., Amano, T., Koga, K., Cho Ae, Choi Eh and Shiratani, M. 2015. Influence of ionic liquid and ionic salt on protein against the reactive species generated using dielectric barrier discharge plasma. Scientific Reports 5(December): 17781–17781. https://doi.org/10.1038/srep17781.

Avsaroglu, M.D., Bozoglu, F., Alpas, H., Largeteau, A. and Demazeau, G. 2015. use of pulsed-high hydrostatic pressure treatment to decrease patulin in apple juice. High Pressure Research 35(2): 214–22. https://doi.org/10.1080/08957959.2015.1027700.

Barba, F.J., Esteve, M.J. and Frígola, A. 2012. High pressure treatment effect on physicochemical and nutritional properties of fluid foods during storage: a review. Comprehensive Reviews in Food Science and Food Safety 11(3): 307–22.

Basaran, Pervin, Nese Basaran-Akgul and Lutfi Oksuz. 2008. Elimination of *Aspergillus parasiticus* from nut surface with low pressure cold plasma (LPCP) treatment. Food Microbiology 25(4): 626–32. https://doi.org/10.1016/j.fm.2007.12.005.

Bennett, J.W. and Klich, M. 2003. Mycotoxins. Clinical Microbiology Reviews 16(3): 497–516. https://doi.org/10.1128/CMR.16.3.497-516.2003.

Bottalico, Antonio and Giancarlo Perrone. 2002. Toxigenic *Fusarium* species and mycotoxins associated with head blight in small-grain cereals in Europe. European Journal of Plant Pathology 108(7): 611–24. https://doi.org/10.1023/A:1020635214971.

Bulut, Nurullah, Bahar Atmaca, Gülsün Akdemir Evrendilek and Sibel Uzuner. 2020. Potential of pulsed electric field to control *Aspergillus parasiticus*, aflatoxin and mutagenicity levels: sesame seed quality. Journal of Food Safety 40(6): e12855. https://doi.org/10.1111/jfs.12855.

Butz, P. and Tauscher, B. 2002. Emerging technologies: chemical aspects. Food Research International 35(2): 279–84. https://doi.org/10.1016/S0963-9969(01)00197-1.

Carballo, D., Pinheiro-Fernandes-Vieira, P., Tolosa, J., Font, G., Berrada, H. and Ferrer, E. 2018. Dietary exposure to mycotoxins through fruits juice consumption. Rev. Toxicol. 2–6.

Chandra, Sharath, Ankit Patras, Bharat Pokharel, Rishipal R. Bansode, Afroza Begum and Michael Sasges. 2017. Patulin degradation and cytotoxicity evaluation of UV irradiated apple juice using human peripheral blood mononuclear cells. Journal of Food Process Engineering 40(6): e12586. https://doi.org/10.1111/jfpe.12586.

Chang, Xiaojiao, Zidan Wu, Songling Wu, Yanshi Dai and Changpo Sun. 2015. Degradation of ochratoxin A by *Bacillus amyloliquefaciens* ASAG1. Food Additives & Contaminants: Part A 32(4): 564–71. https://doi.org/10.1080/19440049.2014.991948.

Chen, Dongjie, Paul Chen, Yanling Cheng, Peng Peng, Juer Liu, Yiwei Ma, Yuhuan Liu and Roger Ruan. 2019. Deoxynivalenol decontamination in raw and germinating barley treated by plasma-activated water and intense pulsed light. Food and Bioprocess Technology 12(2): 246–54. https://doi.org/10.1007/s11947-018-2206-2.

Chulze, S.N. 2010. Strategies to reduce mycotoxin levels in maize during storage: a review. Food Additives & Contaminants: Part A 27(5): 651–57. https://doi.org/10.1080/19440040903573032.

Di Stefano, Vita and Rosa Pitonzo. 2014. Effect of gamma irradiation on aflatoxins and ochratoxin a reduction in almond samples. Journal of Food Research 3(May): 113. https://doi.org/10.5539/jfr.v3n4p113.

Drusch, S. and Ragab, W. 2003. Mycotoxins in fruits, fruit juices, and dried fruits. Journal of Food Protection 66(8): 1514–27. https://doi.org/10.4315/0362-028X-66.8.1514.

Duarte, S.C., Pena, A. and Lino, C.M. 2010. A review on ochratoxin a occurrence and effects of processing of cereal and cereal derived food products. Food Microbiology 27(2): 187–98. https://doi.org/10.1016/j.fm.2009.11.016.

Dunn, J., OTT, T. and Clark, W. 1995. Pulsed-light treatment of food and packaging. Pulsed-Light Treatment of Food and Packaging 49(9): 95–98.

Durek, Julia, Oliver Schlüter, Anne Roscher, Pawel Durek and Antje Fröhling. 2018. Inhibition or stimulation of ochratoxin a synthesis on inoculated barley triggered by diffuse coplanar surface barrier discharge plasma. Frontiers in Microbiology 9. https://doi.org/10.3389/fmicb.2018.02782.

EFSA Panel on Contaminants in the Food Chain (CONTAM). 2012. Scientific opinion on ergot alkaloids in food and feed. EFSA Journal 10(7). https://doi.org/10.2903/j.efsa.2012.2798.

Eisa, N.A., Ali, F.M., El-Habbaa, G.M., Abdel-Reheem, S.K. and Abou-El-Ella, M.F. 2003. Pulsed electric field technology for checking aflatoxin production in cultures and corn grains. Egypt J. Phytopathol. 31(1–2): 75–86.

El Hajj Assaf, Christelle, Nikki De Clercq, Christof Van Poucke, Geertrui Vlaemynck, Els Van Coillie and Els Van Pamel. 2019. Effects of ascorbic acid on patulin in aqueous solution and in cloudy apple juice. Mycotoxin Research 35(4): 341–51. https://doi.org/10.1007/s12550-019-00354-y.

Evelyn and Silva, F.V.M. 2015. Inactivation of *Byssochlamys nivea* ascospores in strawberry puree by high pressure, power ultrasound and thermal processing. International Journal of Food Microbiology 214(December): 129–36. https://doi.org/10.1016/j.ijfoodmicro.2015.07.031.

Evrendilek, Akdemir, G. and Tanasov, I. 2017. Configuring pulsed electric fields to treat seeds: an innovative method of seed disinfection. Seed Science and Technology 45(1): 72–80. https://doi.org/10.15258/sst.2017.45.1.13.

Evrendilek, Gulsun A., Berna Karatas, Sibel Uzuner and Igor Tanasov. 2019. Design and effectiveness of pulsed electric fields towards seed disinfection. Journal of the Science of Food and Agriculture 99(7): 3475–80. https://doi.org/10.1002/jsfa.9566.

Funes, Gustavo J., Paula L. Gómez, Silvia L. Resnik and Stella M. Alzamora. 2013. Application of pulsed light to patulin reduction in mcilvaine buffer and apple products. Food Control 30(2): 405–10. https://doi.org/10.1016/j.foodcont.2012.09.001.

Gabrić, Domagoj, Francisco Barba, Shahin Roohinejad, Seyed Mohammad Taghi Gharibzahedi, Milivoj Radojčin, Predrag Putnik and Danijela Bursać Kovačević. 2018. Pulsed electric fields as an alternative to thermal processing for preservation of nutritive and physicochemical properties of beverages: a review. Journal of Food Process Engineering 41(1): e12638. https://doi.org/10.1111/jfpe.12638.

Garvey, Mary, Joao Paulo Andrade Fernandes and Neil Rowan. 2015. Pulsed light for the inactivation of fungal biofilms of clinically important pathogenic candida species. Yeast 32(7): 533–40. https://doi.org/10.1002/yea.3077.

Gavahian, Mohsen, Noelia Pallares, Fadila Al Khawli, Emilia Ferrer and Francisco J. Barba. 2020. Recent advances in the application of innovative food processing technologies for mycotoxins and pesticide reduction in foods. Trends in Food Science & Technology 106(December): 209–18. https://doi.org/10.1016/j.tifs.2020.09.018.

Ghanem, I., Orfi, M. and Shamma, M. 2008. Effect of gamma radiation on the inactivation of aflatoxin b1 in food and feed crops. Brazilian Journal of Microbiology 39(4): 787–91. https://doi.org/10.1590/S1517-83822008000400035.

Gowda, N.K.S., Suganthi, R.U., Malathi, V. and Raghavendra, A. 2007. Efficacy of heat treatment and sun drying of aflatoxin-contaminated feed for reducing the harmful biological effects in sheep. Animal Feed Science and Technology, Feed Safety 133(1): 167–75. https://doi.org/10.1016/j.anifeedsci.2006.08.009.

Hao, Heying, Ting Zhou, Tatiana Koutchma, Fan Wu and Keith Warriner. 2016. High hydrostatic pressure assisted degradation of patulin in fruit and vegetable juice blends. Food Control 62(April): 237–42. https://doi.org/10.1016/j.foodcont.2015.10.042.

He, Jian Wei, Yousef I. Hassan, Norma Perilla, Xiu-Zhen Li, Greg J. Boland and Ting Zhou. 2016. Bacterial epimerization as a route for deoxynivalenol detoxification: the influence of growth and environmental conditions. Frontiers in Microbiology 7. https://doi.org/10.3389/fmicb.2016.00572.

He, Jiang, Natasha Marie Evans, Huaizhi Liu, Yan Zhu, Ting Zhou and Suqin Shao. 2021. UV treatment for degradation of chemical contaminants in food: a review. Comprehensive Reviews in Food Science and Food Safety 20(2): 1857–86. https://doi.org/10.1111/1541-4337.12698.

Hernández-Falcón, Tania Atzimba, Araceli Monter-Arciniega, Nelly del Socorro Cruz-Cansino, Ernesto Alanís-García, Gabriela Mariana Rodríguez-Serrano, Araceli Castañeda-Ovando, Mariano García-Garibay, Esther Ramírez-Moreno and Judith Jaimez-Ordaz. 2018. Effect of thermoultrasound on aflatoxin m1 levels, physicochemical and microbiological properties of milk during storage. Ultrasonics Sonochemistry 48(November): 396–403. https://doi.org/10.1016/j.ultsonch.2018.06.018.

Herzallah, S., Alshawabkeh, K. and AL Fataftah, A. 2008. Aflatoxin decontamination of artificially contaminated feeds by sunlight, γ-radiation, and microwave heating. Journal of Applied Poultry Research 17(4): 515–21. https://doi.org/10.3382/japr.2007-00107.

Hojnik, Nataša, Martina Modic, Gabrijela Tavčar-Kalcher, Janja Babič, James L. Walsh and Uroš Cvelbar. 2019. Mycotoxin decontamination efficacy of atmospheric pressure air plasma. Toxins 11(4): 219. https://doi.org/10.3390/toxins11040219.

Hong, Sung-Jun, Yun-Ha Lee, Abdur Rahim Khan, Ihsan Ullah, Changhee Lee, Choi Kyu Park and Jae-Ho Shin. 2014. Cloning, expression, and characterization of thermophilic L-asparaginase from *Thermococcus kodakarensis* KOD1. Journal of Basic Microbiology 54(6): 500–508. https://doi.org/10.1002/jobm.201300741.

Horky, Pavel, Sylvie Skalickova, Daria Baholet and Jiri Skladanka. 2018. Nanoparticles as a solution for eliminating the risk of mycotoxins. Nanomaterials 8(9): 727. https://doi.org/10.3390/nano8090727.

Ibarz, Raquel, Alfonso Garvín, Ebner Azuara and Albert Ibarz. 2015. Modelling of ochratoxin a photo-degradation by a UV multi-wavelength emitting lamp. LWT - Food Science and Technology 61(2): 385–92. https://doi.org/10.1016/j.lwt.2014.12.017.

Iqdiam, Basheer M., Manal O. Abuagela, Ziynet Boz, Sara M. Marshall, Renee Goodrich-Schneider, Charles A. Sims, Maurice R. Marshall, Andrew J. MacIntosh and Bruce A. Welt. 2020. Effects of atmospheric pressure plasma jet treatment on aflatoxin level, physiochemical quality, and sensory attributes of peanuts. Journal of Food Processing and Preservation 44(1): e14305. https://doi.org/10.1111/jfpp.14305.

Jalili, M., Jinap, S. and Noranizan, A. 2010. Effect of gamma radiation on reduction of mycotoxins in black pepper. Food Control 21(10): 1388–93. https://doi.org/10.1016/j.foodcont.2010.04.012.

Kalagatur, Naveen Kumar, Jalarama Reddy Kamasani, Venkataramana Mudili, Kadirvelu Krishna, Om Prakash Chauhan and Murali Harishchandra Sreepathi. 2018. Effect of high pressure processing on growth and mycotoxin production of *Fusarium graminearum* in maize. Food Bioscience 21(February): 53–59. https://doi.org/10.1016/j.fbio.2017.11.005.

Karlovsky, Petr, Michele Suman, Franz Berthiller, Johan De Meester, Gerhard Eisenbrand, Irène Perrin, Isabelle P. Oswald et al. 2016. Impact of food processing and detoxification treatments on mycotoxin contamination. Mycotoxin Research 32(4): 179–205. https://doi.org/10.1007/s12550-016-0257-7.

Kaushik, Geetanjali. 2015. Effect of processing on mycotoxin content in grains. Critical Reviews in Food Science and Nutrition 55(12): 1672–83. https://doi.org/10.1080/10408398.2012.701254.

Kim, Woo-Ju, Sang-Hyun Park and Dong-Hyun Kang. 2018. Inactivation of foodborne pathogens influenced by dielectric properties, relevant to sugar contents, in chili sauce by 915 MHz microwaves. Lebensmittel-Wissenschaft Technologie. https://agris.fao.org/agris-search/search.do?recordID=US201800399426.

Kong, Qing, Cuiping Zhai, Bin Guan, Chunjuan Li, Shihua Shan and Jiujiang Yu. 2012. Mathematic modeling for optimum conditions on aflatoxin b1 degradation by the aerobic bacterium *Rhodococcus erythropolis*. Toxins 4(11): 1181–95. https://doi.org/10.3390/toxins4111181.

Koubaa, Mohamed, Francisco J. Barba, Nabil Grimi, Houcine Mhemdi, Wael Koubaa, Nadia Boussetta and Eugène Vorobiev. 2016. Recovery of colorants from red prickly pear peels and pulps enhanced by pulsed electric field and ultrasound. Innovative Food Science & Emerging Technologies 37(October): 336–44. https://doi.org/10.1016/j.ifset.2016.04.015.

Koutchma, Tatiana. 2019. Ultraviolet Light in Food Technology: Principles and Applications. CRC Press.

Kowalska, Karolina, Dominika Ewa Habrowska-Górczyńska and Agnieszka Wanda Piastowska-Ciesielska. 2016. Zearalenone as an endocrine disruptor in humans. Environmental Toxicology and Pharmacology 48(December): 141–49. https://doi.org/10.1016/j.etap.2016.10.015.

Kramer, B., Wunderlich, J. and Muranyi, P. 2017. Recent findings in pulsed light disinfection. Journal of Applied Microbiology 122(4): 830–56. https://doi.org/10.1111/jam.13389.

Langmuir, Irving. 1928. Oscillations in ionized gases. Proceedings of the National Academy of Sciences of the United States of America 14(8): 627–37.

Laurita, R., Barbieri, D., Gherardi, M., Colombo, V. and Lukes, P. 2015. Chemical analysis of reactive species and antimicrobial activity of water treated by nanosecond pulsed DBD air plasma. Clinical Plasma Medicine, Plasma-Liquid Interactions 3(2): 53–61. https://doi.org/10.1016/j.cpme.2015.10.001.

Lee, Hyun Jung and Dojin Ryu. 2017. Worldwide occurrence of mycotoxins in cereals and cereal-derived food products: public health perspectives of their co-occurrence. Journal of Agricultural and Food Chemistry 65(33): 7034–51. https://doi.org/10.1021/acs.jafc.6b04847.

Lee, Seung-Hun, Shin Young Park, Kye-Hwan Byun, Hyang Sook Chun and Sang-Do Ha. 2017. Effects of microwaves on the reduction of *Aspergillus flavus* and *Aspergillus parasiticus* on brown rice (*Oryza sativa* L.) and barley (*Hordeum vulgare* L.). Food Additives & Contaminants. Part A, Chemistry, Analysis, Control, Exposure & Risk Assessment 34(7): 1193–1200. https://doi.org/10.1080/19440049.2017.1319072.

Li, Hongqiang, Yongshui Qu, Yongqing Yang, Senlin Chang and Jian Xu. 2016. Microwave irradiation—a green and efficient way to pretreat biomass. Bioresource Technology 199(January): 34–41. https://doi.org/10.1016/j.biortech.2015.08.099.

Liu, Y., Li, M., Liu, Y., Bai, F. and Bian, K. 2019. Effects of pulsed ultrasound at 20 kHz on the sonochemical degradation of mycotoxins. World Mycotoxin Journal 12(4): 357–66. https://doi.org/10.3920/WMJ2018.2431.

Liu Yan and Wu Felicia. 2010. Global burden of aflatoxin-induced hepatocellular carcinoma: a risk assessment. Environmental Health Perspectives 118(6): 818–24. https://doi.org/10.1289/ehp.0901388.

Liu, Yuanfang, Mengmeng Li, Yuanxiao Liu and Ke Bian. 2019. Structures of reaction products and degradation pathways of aflatoxin b1 by ultrasound treatment. Toxins 11(9): 526. https://doi.org/10.3390/toxins11090526.

Luo, Ying, Zhengkun Zhou and Tianli Yue. 2017. Synthesis and characterization of nontoxic chitosan-coated Fe₃O₄ particles for patulin adsorption in a juice-PH simulation aqueous. Food Chemistry 221(April): 317–23. https://doi.org/10.1016/j.foodchem.2016.09.008.

Marin, S., Ramos, A.J., Cano-Sancho, G. and Sanchis, V. 2013. Mycotoxins: occurrence, toxicology, and exposure assessment. Food and Chemical Toxicology 60(October): 218–37. https://doi.org/10.1016/j.fct.2013.07.047.

Min, L., Johanna Fink-Gremmels, Dagang Li, Xiong Tong, Jing Tang, Xuemei Nan, Zhongtang Yu, Weidong Chen and Gang Wang. 2021. An overview of aflatoxin b1 biotransformation and aflatoxin m1 secretion in lactating dairy cows. Animal Nutrition, January. https://doi.org/10.1016/j.aninu.2020.11.002.

Misra, N.N., Tiwari, B.K., Raghavarao, K.S.M.S. and Cullen, P.J. 2011. Non-thermal plasma inactivation of food-borne pathogens. Food Engineering Reviews 3(3): 159–70. https://doi.org/10.1007/s12393-011-9041-9.

Misra, N.N., Pankaj, S.K., Annalisa Segat and Kenji Ishikawa. 2016. Cold plasma interactions with enzymes in foods and model systems. Trends in Food Science & Technology 55(September): 39–47. https://doi.org/10.1016/j.tifs.2016.07.001.

Misra, N.N., Alex Martynenko, Farid Chemat, Larysa Paniwnyk, Francisco J. Barba and Anet Režek Jambrak. 2018. Thermodynamics, transport phenomena, and electrochemistry of external field-assisted nonthermal food technologies. Critical Reviews in Food Science and Nutrition 58(11): 1832–63. https://doi.org/10.1080/10408398.2017.1287660.

Moreau, Morgane, Geoffroy Lescure, Adrien Agoulon, Pascal Svinareff, Nicole Orange and Marc Feuilloley. 2013. Application of the pulsed light technology to mycotoxin degradation and inactivation. Journal of Applied Toxicology 33(5): 357–63. https://doi.org/10.1002/jat.1749.

Murata, H., Mitsumatsu, M. and Shimada, N. 2008. Reduction of feed-contaminating mycotoxins by ultraviolet irradiation: an *in vitro* study. Food Additives & Contaminants: Part A 25(9): 1107–10. https://doi.org/10.1080/02652030802057343.

Niemira Ba. 2012. Cold plasma decontamination of foods. Annual Review of Food Science and Technology 3. https://doi.org/10.1146/annurev-food-022811-101132.

Nishimwe, Kizito, Isaac Agbemafle, Manju B. Reddy, Kevin Keener and Dirk E. Maier. 2021. Cytotoxicity assessment of aflatoxin b1 after high voltage atmospheric cold plasma treatment. Toxicon 194(April): 17–22. https://doi.org/10.1016/j.toxicon.2021.02.008.

Oms-Oliu, Gemma, Olga Martín-Belloso and Robert Soliva-Fortuny. 2008. Pulsed light treatments for food preservation. a review. Food and Bioprocess Technology 3(1): 13. https://doi.org/10.1007/s11947-008-0147-x.

Ott, Logan C., Holly J. Appleton, Hu Shi, Kevin Keener and Melha Mellata. 2021. High voltage atmospheric cold plasma treatment inactivates *Aspergillus flavus* spores and deoxynivalenol toxin. *Food Microbiology* 95(May): 103669. https://doi.org/10.1016/j.fm.2020.103669.

Ouf Sa, Basher Ah and Mohamed Aa. 2015. Inhibitory effect of double atmospheric pressure argon cold plasma on spores and mycotoxin production of *Aspergillus niger* contaminating date palm fruits. Journal of the Science of Food and Agriculture 95(15). https://doi.org/10.1002/jsfa.7060.

Pallarés, Noelia, Francisco J. Barba, Houda Berrada, Josefa Tolosa and Emilia Ferrer. 2020. Pulsed electric fields (PEF) to mitigate emerging mycotoxins in juices and smoothies. Applied Sciences 10(19): 6989. https://doi.org/10.3390/app10196989.

Pallarés, Noelia, Houda Berrada, Josefa Tolosa and Emilia Ferrer. 2021. Effect of high hydrostatic pressure (HPP) and pulsed electric field (PEF) technologies on reduction of aflatoxins in fruit juices. LWT 142(May): 111000. https://doi.org/10.1016/j.lwt.2021.111000.

Park, Bong Joo, Kosuke Takatori, Yoshiko Sugita-Konishi, Ik Hwi Kim, Mi Hee Lee, Dong Wook Han, Kie Hyung Chung, Soon O. Hyun and Jong Chul Park. 2007. Degradation of mycotoxins using microwave-induced argon plasma at atmospheric pressure. Surface and Coatings Technology 201(9-11 SPEC. ISS.): 5733–37. https://doi.org/10.1016/j.surfcoat.2006.07.092.

Patil, H., Shah, N.g., Hajare, S.n., Gautam, S. and Kumar, G. 2019. Combination of microwave and gamma irradiation for reduction of aflatoxin b1 and microbiological contamination in peanuts (*Arachis hypogaea* L.). World Mycotoxin Journal 12(3): 269–80. https://doi.org/10.3920/WMJ2018.2384.

Péteri, Z., Téren, J., Vágvölgyi, C. and Varga, J. 2007. Ochratoxin degradation and adsorption caused by astaxanthin-producing yeasts. Food Microbiology 24(3): 205–10. https://doi.org/10.1016/j.fm.2006.06.003.

Popović, Vladimir, Nicholas Fairbanks, Jacob Pierscianowski, Michael Biancaniello, Ting Zhou and Tatiana Koutchma. 2018. Feasibility of 3D UV-C treatment to reduce fungal growth and mycotoxin loads on maize and wheat kernels. Mycotoxin Research 34(3): 211–21. https://doi.org/10.1007/s12550-018-0316-3.

Prandini, A., Tansini, G., Sigolo, S., Filippi, L., Laporta, M. and Piva, G. 2009. On the occurrence of aflatoxin m1 in milk and dairy products. Food and Chemical Toxicology, Early Awareness of Emerging Risks to Food and Feed Safety 47(5): 984–91. https://doi.org/10.1016/j.fct.2007.10.005.

Putnik, Predrag, Jose M. Lorenzo, Francisco J. Barba, Shahin Roohinejad, Anet Režek Jambrak, Daniel Granato, Domenico Montesano and Danijela Bursać Kovačević. 2018. Novel food processing and extraction technologies of high-added value compounds from plant materials. Foods 7(7): 106. https://doi.org/10.3390/foods7070106.

Rheeder, John P., Walter F.O. Marasas and Hester F. Vismer. 2002. Production of fumonisin analogs by *Fusarium* species. Applied and Environmental Microbiology 68(5): 2101–5. https://doi.org/10.1128/AEM.68.5.2101-2105.2002.

Richard, John L. 2007. Some major mycotoxins and their mycotoxicoses—an overview. International Journal of Food Microbiology, Mycotoxins from the Field to the Table 119(1): 3–10. https://doi.org/10.1016/j.ijfoodmicro.2007.07.019.

Rodríguez-Bencomo, Juan José, Vicente Sanchis, Inmaculada Viñas, Olga Martín-Belloso and Robert Soliva-Fortuny. 2020. Formation of patulin-glutathione conjugates induced by pulsed light: a tentative strategy for patulin degradation in apple juices. Food Chemistry 315(June): 126283. https://doi.org/10.1016/j.foodchem.2020.126283.

Sapcota, Deben, Rafiqul Islam and Borah, M. 2011. Effect of light and duration of exposure in counteracting aflatoxin b1 in broken rice. Indian Journal of Animal Sciences 81(April): 380–81.

Savi, Geovana D., Willian A. Cardoso, Bianca G. Furtado, Tiago Bortolotto, Luciana O.V. Da Agostin, Janaína Nones, Elton Torres Zanoni, Oscar R.K. Montedo and Elidio Angioletto. 2017. New ion-exchanged zeolite derivatives: antifungal and antimycotoxin properties against *Aspergillus flavus* and aflatoxin b1. Materials Research Express 4(August): 085401. https://doi.org/10.1088/2053-1591/aa84a5.

Sen, Yasin, Baran Onal-Ulusoy and Mehmet Mutlu. 2019. Detoxification of hazelnuts by different cold plasmas and gamma irradiation treatments. Innovative Food Science & Emerging Technologies. https://agris.fao.org/agris-search/search.do?recordID=US201900362268.

Sew, Chang Chew, Hasanah Mohd Ghazali, Olga Martín-Belloso and Mohd Adzahan Noranizan. 2014. Effects of combining ultraviolet and mild heat treatments on enzymatic activities and total phenolic contents in pineapple juice. Innovative Food Science & Emerging Technologies 26(December): 511–16. https://doi.org/10.1016/j.ifset.2014.05.008.

Shanakhat, Hina, Angela Sorrentino, Assunta Raiola, Massimo Reverberi, Manuel Salustri, Paolo Masi and Silvana Cavella. 2019. Technological properties of durum wheat semolina treated by heating and UV irradiation for reduction of mycotoxin content. Journal of Food Process Engineering 42(3): e13006. https://doi.org/10.1111/jfpe.13006.

Shantha, T. and Sreenivasa Murthy, V. 1977. Photo-destruction of aflatoxin in groundnut oil. Indian Journal of Technology 15(10): 453–54, 6.

Shantha, Thimmappaji and Venkateshaiah Sreenivasa Murthy. 1981. Use of sunlight to partially detoxify groundnut (peanut) cake flour and casein contaminated with aflatoxin b1. Journal of Association of Official Analytical Chemists 64(2): 291–93. https://doi.org/10.1093/jaoac/64.2.291.

Shen, Ming-Hsun and Rakesh K. Singh. 2021. Detoxification of aflatoxins in foods by ultraviolet irradiation, hydrogen peroxide, and their combination—a review. LWT 142(May): 110986. https://doi.org/10.1016/j.lwt.2021.110986.

Shi, Hu, Bruce Cooper, Richard L. Stroshine, Klein E. Ileleji and Kevin M. Keener. 2017a. Structures of degradation products and degradation pathways of aflatoxin b1 by high-voltage atmospheric cold plasma (HVACP) treatment. Journal of Agricultural and Food Chemistry 65(30): 6222–30. https://doi.org/10.1021/acs.jafc.7b01604.

Shi, Hu, Klein Ileleji, Richard L. Stroshine, Kevin Keener and Jeanette L. Jensen. 2017b. Reduction of aflatoxin in corn by high voltage atmospheric cold plasma. Food and Bioprocess Technology 6(10): 1042–52. https://doi.org/10.1007/s11947-017-1873-8.

Siciliano, Ilenia, Davide Spadaro, Ambra Prelle, Dario Vallauri, Maria Chiara Cavallero, Angelo Garibaldi and Maria Lodovica Gullino. 2016. Use of cold atmospheric plasma to detoxify hazelnuts from aflatoxins. Toxins 8(5). https://doi.org/10.3390/toxins8050125.

Singh, Ranjeet and Ashok Kumar. 2011. Pulsed electric fields, processing and application in food industry. European Journal of Nutrition & Food Safety 71–93.

Tang, Hui, Xiaohong Li, Fang Zhang, Xianghong Meng and Bingjie Liu. 2019. Biodegradation of the mycotoxin patulin in apple juice by orotate phosphoribosyltransferase from *Rhodotorula mucilaginosa*. Food Control 100(June): 158–64. https://doi.org/10.1016/j.foodcont.2019.01.020.

Ten Bosch, Lars, Katharina Pfohl, Georg Avramidis, Stephan Wieneke, Wolfgang Viöl and Petr Karlovsky. 2017. Plasma-based degradation of mycotoxins produced by *Fusarium*, *Aspergillus* and *Alternaria* species. Toxins 9(3): 97. https://doi.org/10.3390/toxins9030097.

Teniola, O.D., Addo, P.A., Brost, I.M., Färber, P., Jany, K.-D., Alberts, J.F., van Zyl, W.H., Steyn, P.S. and Holzapfel, W.H. 2005. Degradation of aflatoxin b(1) by cell-free extracts of *Rhodococcus erythropolis* and mycobacterium *Fluoranthenivorans* sp. Nov. DSM44556(T). International Journal of Food Microbiology 105(2): 111–17. https://doi.org/10.1016/j.ijfoodmicro.2005.05.004.

Tokuşoğlu, Özlem, Hami Alpas and Faruk Bozoğlu. 2010. High hydrostatic pressure effects on mold flora, citrinin mycotoxin, hydroxytyrosol, oleuropein phenolics and antioxidant activity of black table olives. Innovative Food Science & Emerging Technologies 11(2): 250–58. https://doi.org/10.1016/j.ifset.2009.11.005.

Vijayalakshmi, Subramanian, Shanmugam Nadanasabhapathi, Ranganathan Kumar and Sunny Kumar, S. 2018. Effect of pH and pulsed electric field process parameters on the aflatoxin reduction in model system using response surface methodology. Journal of Food Science and Technology 55(3): 868–78. https://doi.org/10.1007/s13197-017-2939-3.

Wang, Bei, Noreen E. Mahoney, Zhongli Pan, Ragab Khir, Bengang Wu, Haile Ma and Liming Zhao. 2016. Effectiveness of pulsed light treatment for degradation and detoxification of aflatoxin b1 and b2 in rough rice and rice bran. Food Control 59(January): 461–67. https://doi.org/10.1016/j.foodcont.2015.06.030.

Wang, J., Ogata, M., Hirai, H. and Kawagishi, H. 2011. Detoxification of aflatoxin b1 by manganese peroxidase from the white-rot fungus *Phanerochaete sordida* YK-624. FEMS Microbiology Letters 314(2). https://doi.org/10.1111/j.1574-6968.2010.02158.x.

Wielogorska, Ewa, Yusuf Ahmed, Julie Meneely, William G. Graham, Christopher T. Elliott and Brendan F. Gilmore. 2019. A holistic study to understand the detoxification of mycotoxins in maize and impact on its molecular integrity using cold atmospheric plasma treatment. Food Chemistry 301(December): 125281. https://doi.org/10.1016/j.foodchem.2019.125281.

Wilson, David M., Wellington Mubatanhema and Zeljko Jurjevic. 2002. Biology and ecology of mycotoxigenic *Aspergillus* species as related to economic and health concerns. Advances in Experimental Medicine and Biology 504: 3–17. https://doi.org/10.1007/978-1-4615-0629-4_2.

Woldemariam, Henock Woldemichael and Shimelis Admassu Emire. 2019. High pressure processing of foods for microbial and mycotoxins control: current trends and future prospects. Cogent Food & Agriculture, June. https://www.tandfonline.com/doi/abs/10.1080/23311932.2019.1622184.

Wu, Yue, Jun-Hu Cheng and Da-Wen Sun. 2021. Blocking and degradation of aflatoxins by cold plasma treatments: applications and mechanisms. Trends in Food Science & Technology 109(March): 647–61. https://doi.org/10.1016/j.tifs.2021.01.053.

Xiang Zhen, Shen, Diao EnJie, Zhang Zheng, Ji Ning, Ma WenWen and Hai-Zhou. 2014. Effects of UV-irradiation detoxification in a photodegradation reactor on quality of peanut oil. International Food Research Journal 21(6): 2311–14.

Zahoor, Muhammad and Farhat Ali Khan. 2016. Aflatoxin b1 detoxification by magnetic carbon nanostructures prepared from maize straw. Desalination and Water Treatment 57(25): 11893–903. https://doi.org/10.1080/19443994.2015.1046147.

Zhang, Yaolei, Mengmeng Li, Yuanxiao Liu, Erqi Guan and Ke Bian. 2020. Reduction of aflatoxin b1 in corn by water-assisted microwaves treatment and its effects on corn quality. Toxins 12(9): 605. https://doi.org/10.3390/toxins12090605.

Zhu, Yan, Tatiana Koutchma, Keith Warriner, Suqin Shao and Ting Zhou. 2013. Kinetics of patulin degradation in model solution, apple cider and apple juice by ultraviolet radiation. Food Science and Technology International Ciencia Y Tecnologia De Los Alimentos Internacional 19(4): 291–303. https://doi.org/10.1177/1082013212452414.

Zhu, Yan, Tatiana Koutchma, Keith Warriner and Ting Zhou. 2014. Reduction of patulin in apple juice products by UV light of different wavelengths in the UVC range. Journal of Food Protection 77(6): 963–71. https://doi.org/10.4315/0362-028X.JFP-13-429.

Zhu, Yan, Yousef I. Hassan, Suqin Shao and Ting Zhou. 2018. Employing immuno-affinity for the analysis of various microbial metabolites of the mycotoxin deoxynivalenol. Journal of Chromatography A 1556(June): 81–87. https://doi.org/10.1016/j.chroma.2018.04.067.

CHAPTER 11

Green Technologies for Food Analysis

Ana Paula Rebellato,[#] Joyce Grazielle Siqueira Silva,[#]
Elem Tamirys dos Santos Caramês, José Luan da Paixão Teixeira
and *Juliana Azevedo Lima Pallone**

1. Introduction

The main objective of food analysis is the determination of specific or multiple components. The traditional and/or official method usually investigates one analyte at a time and requires several stages for sample preparation. They require toxic reagents, which may cause risk to the operator, are time-consuming, have excessive energy consumption and are harmful to the environment (Tobiszewski and Namieśnik 2017). The Green Analytical Chemistry (GAC) concept considers the environmental impact of analytical methods and safe practice. Since 1990, new analytical methods have been developed for different types of samples, aiming to implement GAC (Alberto and Machado 2018). An important challenge has been the replacement of traditional methods with techniques that generate less toxic waste (Gałuszka et al. 2013).

Miniaturization and automation of methods, decontamination of in-line waste, use of alternative (non-toxic) reagents as well as the reduction in energy consumption are the main challenges in the use of green chemistry (Płotka-Wasylka et al. 2018). In addition, the determination of low concentration analytes, complex matrices, samples with high water concentration and the substitution of organic solvents by green (less toxic) solvents are considered to be the challenges of GAC in food analysis (Pallone et al. 2018). Green methods examine beyond the traditional parameters of analytical quality and have robust figures of merit and high efficiency in the detection and quantification of compounds using the fundamentals of GAC. It replaces the use of

Department of Food Science and Nutrition, School of Food Engineering, University of Campinas, Monteiro Lobato Street, 80, Zip Code: 13083-862, Campinas, São Paulo, Brazil.
* Corresponding author: jpallone@unicamp.br
# These authors contributed equally to the work.

toxic and dangerous solvents with green reagents or with alternative methods that do not need sample preparation (Mabood et al. 2017a).

Hence, analytical techniques based on vibrational spectroscopy, such as near-infrared spectroscopy (NIR), mid-infrared spectroscopy (MIR), Raman, methods using hyperspectral images and RGB image system, along with magnetic resonance spectroscopy comply with the GAC. However, data obtained through vibrational spectroscopies and imaging techniques are complex and numerous, so interpretation of the same must be done using chemometric tools, to allow for the maximum acquisition of chemical information (Lobato et al. 2018).

These methodologies are increasingly present in several sectors of the food production chain as they can be used directly in the production line or in quality control of both raw materials and/or the final product (Valderrama et al. 2015). These techniques have already been used successfully for determining fish firmness, ethanol content in beer, wheat bran composition, geographical origin of wines and the physical characterization of meat, cheese, grains, spices, fruits and vegetables (Cheng et al. 2014, Grassi et al. 2014, Santos et al. 2015, Mandrile et al. 2016, Hell et al. 2016).

Therefore, the use of techniques considered 'green' with chemometric tools requires little to no sample preparation, rapid delivery of quantitative results, simultaneous data analysis (chemometrics) and no chemical reagents at low cost (Tsuchikawa and Kobori 2015, Valderrama et al. 2015, Hein et al. 2016, Hwang et al. 2016). This association of spectroscopic techniques with chemometric tools can boost the modernization and advance food analysis while contributing to the establishment of new 'green' methodologies with more efficiency in comparison to conventional methods.

2. GAC Principles and Challenges for Food Analysis

The movement associated with the development of GAC or 'Green Chemistry' began in the early 1990s mainly in the United States, England and Italy. It introduced new concepts and values for several fundamental activities of chemistry as well as for various sectors of industrial and economic activity (Santos and Royer 2004).

The GAC is based on 12 principles, i.e., (1) applying analytical techniques requiring little or no sample preparation; (2) reducing sample size; (3) performing *in situ* measurements; (4) integrating processes; (5) using automated and miniaturized analytical methods; (6) avoiding analytical derivatization; (7) reducing the volume of analytical residues generated by chemical analyzes; (8) prioritizing multi-analytical or multi-parameter methods when compared to the methods to which they only analyze one analyte at a time; (9) minimizing energy use; (10) using reagents from renewable sources; (11) reducing the use of toxic reagents; (12) and reducing risk to the operator. These principles serve as a basis for guiding both the scientific community in activity and instilling a new generation of researchers with critical and constructive thoughts in the quest for environmentally friendly and analytical chemistry that respects environmental preservation (Gałuszka et al. 2013).

The main objectives of GAC are to eliminate or reduce the use of chemicals (solvents, reagents, and preservatives), minimize energy consumption, manage waste generated during analysis as well as increase operator safety (Zhang 2018). Traditional chemical analysis techniques applied to food analysis, comprise several steps involving at least one chemical reagent, leading to chemical residue production (Haq et al. 2017). In choosing an appropriate method for food analysis, it is necessary to consider the relationship between the method and the analyte by observing various criteria. These include accuracy, limits of detection and quantification, and the food matrix, which can have different interferences and concentrations for the same analyte (Płotka-Wasylka et al. 2018).

Based on the GAC principles, several studies have been carried out to verify the efficiency of the use of green techniques, such as vibrations (NIR, MIR, Raman, hyperspectral and digital imaging) in the control of micronutrients, macronutrients, volatile bioactive, microbiological parameters, physical attributes and authenticity of food (Bázár et al. 2016a, Alamprese et al. 2016a, Basri et al. 2017a, Mabood et al. 2017a, Pallone et al. 2018).

NIR and MIR have already been used for the evaluation of physical-chemical parameters in several food matrices, such as quinoa, proteins in milk powder, soybean moisture and starch in black beans (Santiago-Ramos et al. 2018). Raman, for the same purpose, has also been used in the determination of proteins and lipids in milk, antioxidant capacity and phenolic compounds in Chinese rice wine, control of *Escherichia coli* in meat products and addition of maltodextrin in milk powder (Wu et al. 2015, Lohumi et al. 2015a, Rodrigues Júnior et al. 2016, Silveira et al. 2016).

Digital image systems, such as RGB, were used in the determination of several analytes in food matrices, such as honey (Dominguez and Centurión 2015), milk (Kucheryavskiy et al. 2014) and alcoholic beverages (Lima et al. 2015). At the same time, hyperspectral images have already been used in the determination of fungi in maize and the determination of origin in seeds of *Jatropha curcas* L. (Del Fiore et al. 2010, Gao et al. 2013) and nuclear magnetic resonance used to identify compounds in starch and various foodstuffs (Larsen et al. 2008).

Although we may find several applications in food analysis emphasizing the importance of using 'green' techniques, the challenges are still enormous and vary among the techniques cited. The near and medium infrared spectra generally face challenges as a large amount of noise, baseline displacement, as well as wide absorption bands with little definition (Cortés et al. 2019). The main disadvantage of Raman spectroscopy is its sensitivity to light and background signals (Harting and Kleinebudde 2019). The use of hyperspectral images used in food analysis requires a lot of work in data processing (Foster and Amano 2019), whereas the Nuclear Magnetic Resonance (NMR) technique presents difficulties in the signal-to-noise ratio, requiring a magnetic field when compared to other methods (Colnago et al. 2014). These difficulties can be circumvented through the use of chemometric tools and in the analysis of foods as each sample behaves in a particular way.

Thus, 'green' techniques for the food industry are achievable; for use in the determination of compounds present in foods and evaluation of product quality. They represent an evolution in terms of decreasing the use of toxic reagents, the amount of sample used and of operational costs.

3. Green Techniques in Food Analysis

Nowadays, several alternative analytical techniques, such as NIR and MIR spectroscopy, coupled or not to Hyperspectral Images, RGB imaging systems, Raman spectroscopy and Nuclear Magnetic Resonance (NMR), are currently being used in food industries, quality control laboratories and in related areas.

Spectroscopic methods are based on measuring the amount of radiation produced or absorbed by molecules or by the atomic species and classified according to the region of the electromagnetic spectrum involved in the measurement. These techniques are in accordance with the green chemistry principles, as they are considered non-destructive, require little or no sample preparation, do not generate toxic residues, and consequently avoid risks to human health and the environment; besides that, they are fast and low-cost techniques.

3.1 Near Infrared Spectroscopy (NIR) and Medium Infrared Spectroscopy (MIR)

The electromagnetic spectrum range between 780 and 100,000 nm comprises the infrared (IR) region, which is further subdivided into near-infrared (750 to 2,500 nm), medium infrared (2,500 to 50,000 nm) and far-infrared (50,000 to 100,000 nm); the latter is not normally used in analytical procedures (Sun 2008, Burns and Ciurczak 2009).

In NIR and MIR spectroscopy, the absorption of radiation by matter occurs through the interaction between the radiation source and the chemical bonds of a given compound. Radiation in the infrared region does not have enough energy to excite electrons but makes the atoms or groups of atoms of the compounds vibrate more rapidly and with better amplitude around the bonds that bind them. This is called 'vibrational spectroscopy'. In NIR spectroscopy, overtones or combinations of fundamental vibrations are observed. In MIR spectroscopy the fundamental vibrations are quantized and when they occur, the compounds absorb IR energy in certain regions of the electromagnetic spectrum (Pavia et al. 2008, Sun 2008).

NIR has been used since 1960 for food analysis. The interactions of this electromagnetic radiation with organic molecules obtain a spectrum with information that is qualitative and quantitative to the constituents of a given matrix. Thus, NIR spectroscopy has become a widely used method in the quality control of food industries (Osborne 1999, Lee 2004).

Basically, the principles involved in NIR spectroscopy are the production, recording and interpretation of spectra resulting from the interaction of electromagnetic radiation emitted by a source with matter (Sun 2009).

To absorb radiation in the infrared region, a molecule must have a change in the dipole moment, as a consequence of the vibrational or rotational movement in the vibrations. The NIR spectra present wide and superposed bands, corresponding to overtones and fundamental vibration combinations. NIR absorption bands are weaker than the fundamental infrared medium bands (Skoog et al. 2009, Sun 2009).

The spectrum obtained in the NIR region, as it does not present characteristic bands for each functional group, cannot be interpreted in the same way as the spectrum obtained in the medium infrared, which presents sharp and narrow bands in this region. Thus, in order to obtain important chemical information in the NIR region, it is essential to use chemometric tools (Soares 2006).

Several compounds may present overtones that correspond to their fundamental vibrational modes. Thus, the NIR spectra are generated by clustering, involving simple chemical bonds that have strong interatomic bonds, such as CH, OH and NH, which are of interest. Since the first overtones of other types of vibration, such as CO, still occur in the region of the average infrared and thus only present the second overtone in the NIR region. For food components, there are regions of the spectrum in which overtones or combinations of chemical bonds appear, which are characteristic of proteins, lipids, and carbohydrates, among others. For example, in NIR spectral data, peaks at 1,450 nm and 1,920 nm (6,900 and 5,200 cm^{-1}) are characteristic of OH binding in foods, mostly from water, whereas lipids are generally correlated with the supertons near 1,041 nm, 1,200 nm, 1,740–1,800 nm and 2,440 nm (9,600 cm^{-1}, 8,300 cm^{-1}, 5,730–5,560 cm^{-1} and 4,100 cm^{-1}). In contrast to MIR, NH amide bonds are found at 3,350 cm^{-1} and the region known as fingerprint between 1,800–900 cm^{-1} is correlated with fats, carbohydrates, proteins, bioactive compounds and water (Alamprese et al. 2013).

Normally, the spectra in the NIR for liquid samples are obtained by transmittance or transflectance, according to the configuration of the equipment; whereas for solid matrices, the spectrum can be obtained through direct measurement in whole or ground samples using diffuse reflectance (Ferrão 2000).

Diffuse reflectance is obtained when the infrared radiation penetrates the surface of the sample (about 1 micrometer), which interacts with the matrix through partial absorption and multiple scattering before returning to the surface. Diffuse reflectance results from a combination of reflection, refraction and diffraction phenomena and is a characteristic of each material. The ratio of reflectance, refraction, transmittance and absorbance that occurs is dependent on the chemical and physical properties of the material (Ferrão 2000, Sun 2009).

In diffuse reflectance, the incident radiation comes into contact several times with particles of the samples; in this way, the radiation that undergoes diffuse reflectance provides both qualitative and quantitative information on the composition of the sample. In addition, it is interesting to note that the radiation that returns from the diffuse form of a matrix is usually smaller in magnitude than the incident radiation. Thus, most diffuse reflectance accessories have optical components that aim to converge the radiation, so that it can be oriented over the detection system of the equipment (Ferrão 2000, Sun 2009).

Benchtop equipment is available on the market that operates in the NIR region, with Fourier Transform (FT-NIR) or dispersive, in addition to portable instruments with different configurations, giving smaller bands of electromagnetic radiation (915–1,700 nm) and still the possibility to use NIR on-line sensors installed on production lines.

Although the differences in NIR spectra are often not perceptible to the naked eye, the true value of NIR spectroscopy as an analytical tool depends on the use of statistical and mathematical tools for use on the spectral data (chemometrics).

The MIR comprises the 4,000 to 400 cm^{-1} (2,500 to 50,000 nm) range in the electromagnetic spectrum, where the fundamental vibrations of chemical bonds are observed. The MIR region, as mentioned above, is associated with fundamental vibrations, therefore it provides information on functional groups that absorb radiation of the same intensity and the same wavelength from different classes of compounds. The spectra obtained in the MIR allow for the identification of organic molecules and the characterization of functional groups; besides the possibility to work in a similar way to the spectra obtained in the NIR for qualitative and quantitative analysis (Pavia et al. 2008, Skoog et al. 2009).

In practice, to analyze a sample using MIR, a beam of radiation in the MIR region is passed through the sample and the energy absorbed at each wavelength is recorded. This process can be performed in two different ways. The first, by scanning the spectrum with a monochromatic beam that changes the wavelength over time; or a Fourier transform system (FT-MIR) can be applied to measure all wavelengths at the same time. In this case, all the effects of the different functional groups are considered. As a result, an absorbance or transmittance spectrum is obtained, which demonstrates at what wavelengths of infrared radiation the sample absorbs and consequently the interpretation of the chemical bonds (Sun 2008).

Since the spectra obtained in the MIR are quite complex, the possibility of two different compounds having the same spectrum is extremely small. Thus, the MIR spectrum is known as the fingerprint of the molecule and therefore can be used to confirm the identity of a given sample when compared to a standard (Soares 2006).

For food analysis, it is possible to carry out qualitative analyzes (classification of samples as to the origin and authenticity among others) and quantitative evaluations using whole or ground samples. For solid samples, attenuated total reflectance (ATR) is recommended. After collecting the spectra of a representative number of samples, it is possible to obtain multivariate calibration models for the quantification of compounds or to perform the classification. For some applications, MIR has the advantage that fundamental vibrations can generate regions in the fingerprint spectrum of samples of the same food with different compositions, and this may facilitate the necessary chemometric treatment. On the other hand, portable equipment is rarer and generally has a narrower range of electromagnetic radiation (Soares 2006, Sun 2008).

Most of the quality control laboratories that use infrared spectroscopy use the NIR, mainly due to the good range of equipment arrangements on the online market, portability, and its low cost among others. This fact is also observed in studies related to food matrices since most foods contain functional groups, such as C-H, N-H, and O-H. These groups are closely linked with vibrational (energy) stretching and rotational combinations in the NIR region. In addition, the technique can be used to measure different quality attributes simultaneously and the centesimal composition of different cultivars, fresh and processed foods among others (Sun 2008, Alamprese et al. 2013).

3.2 Raman Spectroscopy

Special interest in Raman spectroscopy has increased in different fields, such as quality control, agricultural, pharmaceutical, petrochemical, and food process control due to its ease of use, minimal sample preparation, and high sensitivity for molecular structure evaluation and formation of biological substances (Almeida et al. 2010)

Raman spectroscopy is based on the Raman scattering phenomenon. When irradiating a sample with a visible monochromatic UV beam or NIR beam generated by a laser, the resulting excitation of the molecules is found at the same wavelength as the incident radiation and this phenomenon is called elastic or 'Rayleigh scattering'. However, a small fraction of scattered radiation could be found at wavelengths that are different to the incident; this 'inelastic' dispersion process is known as the Raman scattering effect. Raman scattering can produce photons with energy higher or less than the incident photon. When the energy of the photon decreases, the vibration of the molecule occurs and this effect is called 'Stokes' (Sun 2008, Skoog et al. 2009).

However, when the photon energy increases, losing the vibration energy of the molecule, the process is called 'anti-Stokes'. In this scattering, small frequency variations are observed, both above and below the Rayleigh scattering, which makes it possible to obtain information about the vibrational energy bands of the molecules. Thus, a plot of the intensities of the incident radiation dispersion versus the deviations in the frequency between the incident and dispersed radiation can be constructed and the Raman spectrum obtained. Interpretation of Raman spectra bands is performed in the same way as for MIR signals (Sun 2008, Skoog et al. 2009).

Although Raman scattering is a weak phenomenon, analyses of trace elements in food matrices can also be studied successfully. In addition, the development of the Fourier Transform significantly improved the use of Raman Spectroscopy since it was possible to obtain spectra in a shorter time, besides reducing the photodecomposition of complex molecules (Sun 2008).

Raman spectroscopy can be used for both qualitative and quantitative analysis of inorganic, organic and biological systems in liquid, solid and gaseous samples. It is a technique considered selective in the detection of apolar molecules, ring structures and structures of the double or triple bond. This spectroscopic technique provides information on the structure of molecules and on the interaction with neighbouring molecules, which can be used to identify molecules present in a given compound. This information is obtained from the interpretation of the absorption bands and peaks and/or emission spectra of the atoms and/or molecules of the matrix under analysis (Schrader and Bougeard 1995, Sun 2008, Skoog et al. 2009).

In food analysis, Raman spectroscopy provides useful information on the chemical nature of food molecules that can be used for food quality assessment, including authentication or adulteration of food products, and may assist in the determination of changes during the production process. With the combination of the data obtained in Raman spectroscopy and the use of chemometric analysis, it is possible to discriminate between different origins or to classify different food matrices, as well as identify food adulterations (Sun 2009, Almeida et al. 2010).

3.3 Red-Green-Blue (RGB) Image

Different instruments can be used to acquire images, such as cell phones, webcams, desktop scanners, and digital cameras. The most common way to obtain digital image information is to decompose it into a color system called Red-Green-Blue (RGB) image. The RGB system consists of a set of colors that uses the combination of the colors Red (R), Green (G) and Blue (B) to form a wide variety of tones (Godinho et al. 2010, Iqbal and Bjorklund 2011, Santos et al. 2012).

Each pixel, the basic forming unit of a digital image, is composed of a combination of these three colors. The intensity of each color in the RGB system is measured in channels (256 for each color). Channel 0 refers to the absence of color, while channel 255 corresponds to that same color but at its maximum intensity. The combination of RGB system channels provides 2,563 possible tonalities.

To obtain the frequency histogram, it is necessary to perform the decomposition of all pixels in the image and calculate the frequency of each channel. In this way, the histogram can be treated in a similar way to the spectral data used in the development of chemometric models. The RGB variables can also be combined with other color parameters, such as saturation, intensity, and hue, resulting in colorgrams (Santos et al. 2012, Oliveira et al. 2013).

The samples decomposed into the RGB system generate OOV (pixel × pixel × channel) arrangements, where 'O' represents an object and 'V' is a variable. If the images of several samples are grouped, a fourth dimension arrangement is obtained (OOOV). For simplicity and agility in computational processing, RGB variables can be converted to grayscale by reducing one dimension of the data and thus are condensed to a three-dimensional (OOO) arrangement (sample × pixel × pixel). An arrangement of type 'OOO' is not congruent, so it is considered non-trilinear and consequently cannot be directly dealt with by higher-order methods. Thus, the use of domain transformation techniques, such as the Fast Fourier Transform (FFT) or the wavelet transform, is required. These techniques are responsible for converting arrangements of type OOO into OVV arrangements (sample × frequency component × frequency component). This technique is applied when you want to insert information related to the texture of the sample into the model. When the analyzed images are homogeneous and do not present texture variations, it is more appropriate to use the direct modelling of color frequency histograms since it is a simpler and more economical technique (Huang et al. 2003, Oliveira et al. 2013).

The application of digital imaging systems has been employed in food studies, mainly in the determination of color by obtaining the color space CIE L*a*b*. One of the most important aspects of the use of the digital imaging system is the image processing stage where digital cameras obtain values expressed in RGB (red, green, and blue), which need to be transformed into the CIE L*a*b* (Wu and Sun 2013).

3.4 Hyperspectral Imaging System (HIS)

Hyperspectral imaging system (HIS) or image spectroscopy is a technique that integrates spectroscopy and images to obtain spectral and spatial information about the sample. In this way, the technique expands the capacity of spectroscopy, adding a new spatial dimension through the use of conventional images in the same system to simultaneously provide spatial and spectral information. Since each pixel in a hyperspectral image has its spectrum, the attribute concentration (chemical composition/quality) can be calculated on each pixel of the sample to generate distribution maps. Thus, the technique can be used to know which chemical compositions, contents and the place where they are located in the analysis matrix and also allows for reaching smaller limits of detection and quantification of compounds (Shieber 2008).

The hyperspectral imaging technology depends on the hardware to acquire the images and the software to perform the image processing. The main components of a HIS system are a source of radiation (illumination or excitation), the detector, the wavelength dispersing device and a computer compatible with the software for the processing of the images (Shieber 2008).

Most HIS systems are designed in reflectance mode. Thus, the image by hyperspectral reflectance detects external quality characteristics (color, size, shape and surface defects) with greater efficiency than the internal characteristics (Wu and Sun 2013).

The equipment used for the acquisition of hyperspectral images, also known as spectral cameras, has three basic camera configurations for hyperspectral images. Each configuration is associated with the form of image acquisition; a point scan is where a complete spectrum of reflectance or transmittance is obtained at a single point of the sample and after scanning of the whole sample when the hyperspectral image is obtained; a line scan, on the other hand, enables spectra in a dotted line and the hyperspectral image is created while a platform containing the sample moves; a plane scan is where the detection system is parallel to the sample, and there is no sample or detector movement and all samples come into contact with the radiation and after measurements at all wavelengths as the hyperspectral image is obtained. Therefore, modes differ in lighting and detection settings, resulting in different effects for the same sample. The appropriate mode of an acquisition depends on the type of sample, constituent, and/or property being evaluated.

Hyperspectral image analysis is a technique with good potential because it integrates spectroscopy and image techniques with the objective of obtaining spatial and spectral information of each pixel of the image (Elmasry et al. 2012). Hyperspectral imaging is a non-destructive, robust, and flexible technique and thus presents the potential for food quality control. The technique is feasible because it is possible to evaluate a wide spectral region in each pixel of the image and flexible because it is able to act in several spectral regions and according to the desired modality (Xu and Lam 2010).

3.5 Nuclear Magnetic Resonance Spectroscopy (NMR)

NMR spectroscopy is a powerful tool for the study of molecular systems since it provides a large amount of information regarding both nuclear properties and electron effects in regions close to magnetic nuclei. This allows the application of NMR in several branches of chemistry, physics and biology, ranging from studies of conformational analysis of small molecules to the structural and dynamic elucidation of macromolecules (Simpson 2008).

Several nuclei can be studied by the NMR technique, but the most used are hydrogen (1H) and carbon (13C). In NMR the information obtained is on the number of magnetically distinct atoms of the isotope under analysis.

By placing a compound with 1H, 13C, 19F, and 31P atoms among others in an intense magnetic field, the nuclei can absorb energy from certain electromagnetic radiation. This energy is quantized and produces a characteristic spectrum of the compound. Energy absorption will occur only if the magnetic field strength and the electromagnetic radiation frequency are well defined. Thus, NMR spectroscopy evaluates the frequency corresponding to energy absorption (resonance frequency) and the time at which nuclei, after irradiation, return to room temperature. This cooling period is called the 'relaxation time'. As molecular motion and interactions with magnetic environments are specific to each molecule, this makes the relaxation time between nuclei distinct. This fact about the relaxation period provides information about the chemical composition and properties of the sample (Solomons et al. 1996).

The fundamental characteristics for the differentiation of the high and low field NMR techniques are the resolution of the measurements of the equipment and the sensitivity, which is lower in the low field NMR. Low-field or low-resolution NMR equipment is one that operates at less than 1T (Tesla). Normally, the equipment works with the signal in the time domain in low-field NMR; so instead of studying the chemical shifts, the relaxation phenomena and the constants related to the physical-chemical properties of the samples are evaluated.

The chemical shifts and proton coupling studies are usually employed in high field NMR, where they are performed through the application of the Fourier Transform, where the signal is worked in the frequency domain. High field NMR is less employed due to the high cost of equipment, the need for specialized personnel and the sensitivity of the equipment for industrial use.

Low-field NMR has been widely used in academia in several areas due to the low cost of equipment and maintenance. The low field analysis recognizes different physical states in the same sample since the magnetic field gradients resulting from differences in magnetic susceptibility in the sample do not affect the signal, a negative characteristic in high field equipment. Problems of sample homogeneity are solved in the low field NMR since the volume is evaluated instead of only evaluating the surface. It is a non-invasive and non-destructive technique and requires little or no pre-treatment of the sample (Silva 2009).

Studies that use relaxation time under the field (relaxometry) are the most used as an analytical tool for quality control, mainly in the food industry, which is basically based on the difference of proton relaxation in the different neighborhoods in which the water molecules, organic matter and/or viscosity of the analyzed fluid (Silva 2009).

NMR spectroscopy is able to simultaneously provide qualitative and quantitative information, even in complex samples that contain several chemical compounds present (Colnago et al. 1996, Silva 2009).

In this context, the mentioned spectroscopic techniques (NIR, MIR, RAMAN, RGB, HSI, and NMR) have been used with success in the food area. In general, for quantitative evaluations using spectra collected in spectroscopic techniques, the analyte of interest is only obtained after establishing a multivariate calibration model.

Thus, the use of chemometrics, which aims at the optimization of experiments, exploratory analysis, classification, and calibration among others is an essential tool for the data treatment obtained by spectroscopic techniques. The data treatment is relevant to the effective use of the results and can be used for different purposes, such as identification, classification, quantification and discrimination of the samples (Elmasry et al. 2012).

4. Chemometrics and Examples of Applied Techniques

Chemometrics is a multivariate statistical method used to analyze complex and large datasets and allow a visual approach to the data (Granato et al. 2018, Kumar and Sharma 2018). The common methods of dataset analysis require a lot of time to be analyzed and the manual evaluation is susceptible to false-positive results. Despite that, with chemometrics, it is possible to obtain important and accurate results in less time (Kumar and Sharma 2018). After the results are obtained they can be analyzed by some numerical and statistical software, such as MATLAB, SPSS, R, Excel Stat, Unscrambler x, Pirouette and Minitab (Kumar and Sharma 2018).

Thus, chemometrics has good applicability when there are multiple results to be analyzed and interpreted, such as a large number of samples, variables and responses (Granato et al. 2018). This includes spectral data obtained in NIR, MIR, Raman, HIS or RMN spectroscopy. The traditional methods of sample evaluation are time-consuming, destructive and use a large number of toxic reagents, whereas chemometrics is associated as a green method since spectroscopy could be considered a green technology (Kumar and Sharma 2018).

According to the requested approach, the analysis of data sets could be evaluated by recognition of patterns using unsupervised and supervised methods (Ferreira 2015, Kumar and Sharma 2018). The unsupervised methods include exploratory analysis, and the supervised methods include classification and calibration analysis (Granato et al. 2018, Kumar and Sharma 2018). The methods of exploratory analysis are Principal Component Analysis (PCA) and Hierarchical Cluster Analysis (HCA) (Ferreira 2015, Kumar and Sharma 2018).

The use of the PCA is interesting when there are a large number of independent variables or when there is a high correlation between the variables (positively or negatively). It is used to evaluate the relationship between variables and the total variation of the data by the reduction of the data dimension and expression of the variables in the form of a linear combination (Alkarkhi et al. 2019). It is composed of principal components (PC), the first PC constitutes the maximum variance of the data (Oliveri and Simonetti 2016).

The Hierarchical Cluster Analysis (HCA) is a method that presents the results in a dendrogram, a graphic that shows the relation between the samples in a tree form and makes the evaluation of the similarities and differences of the samples (intra- and inter-group) possible. It is a method based on the sample organization in clusters and groups with a hierarchy (Caesar et al. 2018, Granato et al. 2018).

4.1 Classification Methods

The principal classification methods are k-nearest neighbor (k-NN), Soft Independent Modeling of Class Analogy (SIMCA), Linear Discriminant Analysis (LDA), and Partial Least Squares Discriminant Analysis (PLS-DA).

The k-nearest neighbor (k-NN) is based on the distance between the unknown object and the training objects. It provides good results when there is a high correlation between the variables (Kumar and Sharma 2018). The KNN is commonly used in pattern recognition but when there are small sample sizes and outliers the sensitivity of the neighborhood size degrades the classification performance (Gou et al. 2019).

Soft Independent Modeling of Class Analogy (SIMCA) is a distance-based technique, and it performs PCA on the samples to be modeled and identifies the significant ones. The PCs define an inner space that contains important information. According to the number of PCs, it has different shapes segment, rectangle, parallelepiped or hyper-parallelepiped when it has one, two, three or more than three PCs, respectively (Oliveri and Simonetti 2016).

Linear discriminant analysis (LDA) is a probability-based discriminant method, and it considers the same dispersion for all the classes (Oliveri and Simonetti 2016). It is considered for the reduction method but instead of PCA, it selects a direction to obtain the maximum separation among the known classes. This method maximizes the ratio of between-class variance and minimizes the variance into class (Wang et al. 2019a).

Partial least squares discriminant analysis (PLS-DA) is a method that joins the properties of the PLS regression and the discrimination provided by classification techniques. PLS-DA try to correlate latent variables (LVs) and Y-variables with high correlation according to the PLS regression algorithm. It provides a graphical visualization of the data, used in LV scores and loadings. The scores represent the samples and the loadings of the variables (Wang et al. 2019b).

The main multivariate regression methods (calibration methods) are included Classical Least Squares (CLS), Multiple Linear Regression (MLR), Principal Component Regression (PCR) and Partial Least Squares (PLS) (Ferreira 2015, Kumar and Sharma 2018).

Among the calibration methods, the PLS is the most popular method in the regression methods. This method uses factor analysis, which is a denomination of latent variables (LV) or PLS components. In this method, the aim is to decrease the space of the original measurements (Ferreira 2015).

4.2 Calibration Methods

The calibration or regression methods are used to create mathematical expressions (models) to evaluate some property of interest. The models are built with the spectra data set of samples (X) and with the values of the property of interest (Y) obtained from a reference method. It is then possible to obtain a numerical expression to correlate the spectrum and the property of interest. With this numerical expression, it is possible to predict the property of interest for new samples according to their spectrum obtained under the same conditions as the samples used in the model (Ferreira 2015).

For the development of the calibration method, the inclusion of all the data measured by the spectral range is not necessary. Instead, some frequencies or wavelengths might be selected to be applied in PLS or PCR models. Some spectral information may be excluded from the calibration due to spectral interferences, such as high spectral noise or nonlinear spectrometer response. However, for some physical or performance properties, the inclusion of all spectral data available could be required (ASTM).

After the use of the pretreatments and the development of the classification and/or calibration models, it is possible to evaluate their quality with statistical parameters. The parameters generally considered are Standard Error of Calibration (SEC), Standard Error of Cross Validation (SECV), and the Standard Error of Prediction (SEP) (Ferreira 2015).

The SEC compares the values measured by the reference method and the values estimated by the calibration models and evaluates how much they are in agreement. The SEC is defined as the following formula:

$$SEC = \sqrt{\frac{e'e}{d}} \qquad (1)$$

where e is the calibration error vector and d is the degrees of freedom in the calibration model.

The SECV is obtained from the errors of cross validation procedures, which are used to estimate the best number of variables that will be included in a model. In the cross validation process, the data related to a sample is removed from the data matrix and the reference values, and the model is built with the other samples. This process is made for all the samples. The best model will be the one with the smallest SECV and with fewer variables. SECV could be calculated as the following expression (2). The SEP evaluate the errors related to the prediction (ASTM).

$$SECV = \sqrt{\frac{PRESS}{n}} \qquad (2)$$

where PRESS implies Predicted Residual Error Sum of Squares.

The calibrated model should be tested by cross validation and by external validation; the figures of merit should be calculated and the best model is chosen to do predictions. The evaluation of quality and confidence of the models obtained should be analyzed by some figures of merit. The common figures of merit used are limit of detection (LD), the limit of quantification (LQ), sensibility (SEN), selectivity (SEL), standard deviation (RSD), repeatability, and accuracy (Ferreira 2015).

Thus, to establish a method using these spectroscopic techniques, it is necessary to evaluate a large number of samples, which present good variation of both the reference method and the spectroscopic technique (Ferreira 2015).

4.3 Pretreatments

The spectroscopy method challenges are the huge amounts of data, unfavourable spectral variation (caused by multiplicative errors of light scattering), spectral noise, variations of particle size, viscosity and temperature of the samples, and all of these factors affect the modelling process (Porep et al. 2015).

Against this context, after the acquisition of the spectra data (NIR, MIR, Raman and RMN spectroscopy, HIS, or RGB image), it is generally necessary to do some mathematical pretreatment(s) to improve the quality of signals to remove noise and unnecessary information, which are common in spectrometry results (Oliveri and Simonetti 2016). This is followed by the evaluation of the data set through analysis using unsupervised and/or supervised chemometric methods. When the pretreatments are applied to the samples, they are called transformations and when they are applied to the variables, they are called pre-processing (Ferreira 2015).

The pretreatments to reduce noise are moving average, Savitzky-Golay smoothing and Fourier transformation. Derivatives are commonly used to correct baseline shifts and/or drifts. The first derivative allows a correction for baseline shifts; a derivation of the first derivative, and the second derivative provides a correction for both baseline shifts and drifts. When there are high-frequency slope variations in the signals, the smoothing, as Savitzky and Golay, could be used instead of the derivatives because they generally cause an amplification of the noise (Oliveri and Simonetti 2016).

Other pretreatments are used to eliminate or minimize systematic interferences or effects as Standard Normal Variate (SNV) transformations, logarithm, normalization, Kubelka-Munk transformation, Multiplicative Scatter Correction (MSC), Standard Normal Variate (SNV) and Ortogonal Signal Correction (OSC) (Ferreira 2015).

The mean center, variance escalation, and auto-scaling are the most common pretreatments applied to the variables (Ferreira 2015). The data obtained from spectroscopy analysis generates a large number of variables. Therefore, the use of chemometrics tools allows the acquisition of more information about the data available. The pretreatments could be used separately or in combination to improve the quality of the data (Oliveri and Simonetti 2016).

5. Applications of Green Technologies in Food Analysis

The chemometrics associated with vibrational spectroscopy (NIR, MIR, Raman, HIS or RGB image, and RMN spectroscopy) are classified as non-destructive methods, have minimum or no sample preparation, do not offer risks for the operator, and could provide a rapid and cost-effective evaluation of samples (Ríos-Reina et al. 2018, Pallone et al. 2018).

Chemometrics has been associated with methods, like vibrational spectroscopy such as NIR and MIR, Raman spectroscopy, HIS, RGB image, and NMR. Also to quantify macronutrients (proteins, lipids, sugar, moisture, and fibers), micronutrients (phenolic compounds, volatile, anthocyanin, and physical parameters), firmness and viscosity in food (Pallone et al. 2018) and for classification, such as food fraud detection, food safety, discrimination, and traceability (Roberts et al. 2018, Medina et al. 2019, Rodríguez et al. 2019).

5.1 Near-Infrared Spectroscopy (NIR)

NIR associated with chemometrics was used to differentiate types of wine vinegar by PLS-DA (Ríos-Reina et al. 2018). Despite that, the NIR spectroscopy has also been evaluated in automated processes in food industries (Grassi and Alamprese 2018). NIR and chemometrics were used to evaluate quality parameters (acidity, total sugar, and soluble solids) and ascorbic acid in cashew apple and guava nectar; and also to determine micro compounds, such as bioactive in grape juice, vitamin C, total polyphenol and sugar content in apples (Pissard et al. 2013, Caramês et al. 2017a).

Physical parameters, like texture, maturity in tomatoes, firmness in apples and color in nectarine, can also be measured using NIR spectral data (Cortés et al. 2019).

NIR has been applied to several cases of food authentication as in cases of addition of fraudulence substances in black rice, honey adulterated with syrups, melamine, freeze-dried açai and also in cases of origin fraud as in white wines and white asparagus (Lohumi et al. 2015b, Chen et al. 2018, Lobato et al. 2018, Richter et al. 2019).

5.2 Mid Infrared Spectroscopy (MIR)

MIR associated with multivariate chemometrics was applied in the detection of adulterated quinoa flour. The research evaluated the presence of soybean, maize, and wheat flour in quinoa flour. The classification of adulterated and pure samples was evaluated using the PLS-DA and SIMCA models. To do the classification models, pure and adulterated samples were used (Rodríguez et al. 2019).

MIR and PLS methods were applied in the quantification of adulterants (soybean oil and sunflower oil) in extra virgin flaxseed oil (de Souza et al. 2015). MIR and NIR were also applied in the evaluation of butter oil adulteration with soybean oil. PCA discriminated the spectrums and PLS models were used to predict the percentage of adulteration (Pereira et al. 2019).

MIR has also been applied to detect and quantify bioactive, such as carotenoids in passion fruit and antioxidant capacity and phenolic compounds in chocolate (de Oliveira et al. 2014b, Hu et al. 2016).

Microbiological spoilage can also be detected and quantified by MIR; for example, *Escherichia coli* in ground beef and *Salmonella enterica* in chicken breast (Lee et al. 2019, Nolasco-Perez et al. 2019a).

5.3 Raman

Surface-enhanced Raman Spectroscopy (SERS) was used to identify artificial colorants in foods. Raman was used as a means to obtain the spectral fingerprints of the different compounds, and PCA was the chemometric method used to differentiate the colorants. Differentiation was possible even when the colorants had similar visual or chemical structures (Gukowsky et al. 2018).

Raman was used to evaluating the carotenoid content in fruit pulp. The data were evaluated by PCA to evaluate the reproducibility, and PLS regression was used for modeling the total carotenoids. Antioxidant capacity and phenolic compounds in Chinese rice wine were determined using Raman spectra and the PLS regression method (Wu et al. 2015).

Cases of maltodextrin addition in milk powder and whey addition in fluid milk can also be resolved using Raman spectra. Origin frauds in wine and honey were also detected by Raman (Mandrile et al. 2016, Li et al. 2016).

5.4 Nuclear Magnetic Resonance Spectroscopy (NMR)

NMR combined with chemometrics was used to discriminate lager beer from two different classes with PLS-DA (da Silva et al. 2019) and to discriminate between the varietal and geographical classification of wines (Magdas et al. 2019). NMR was also applied in the quantification of cations in mineral water using PLS and in the differentiation of organic and conventional coffee using PLS-DA (Monakhova et al. 2017, Consonni et al. 2018).

NMR was also applied in the quantification of food ingredients, such as polysaccharides; it also applied in the evaluation of classes of non-polar compounds, like fatty acids in roasted coffee beans, coffee beverages and spent coffee grounds (Merkx et al. 2018, Williamson and Hatzakis 2019).

5.5 Hyperspectral Imaging (HIS)

Hyperspectral imaging (HIS) has also been applied to foods analysis. It offers the advantage to obtain multiple images and the wavelengths in NIR, MIR and Raman with the possibility to obtain spatial, spectral, and multi-constituent information about the sample. On the other hand, to evaluate all the data generated and create models, it is necessary to combine multivariate data analysis and algorithms (Liu et al. 2017, Roberts et al. 2018).

Hyperspectral imaging associated with near-infrared (NIR) and chemometrics can provide an online monitoring system and have been used to monitor food processes, such as drying, cooking, freezing, salting, dehydrating, and process analytical technology applications (Cheng et al. 2015, Xie et al. 2016, Liu et al. 2017, Ma et al. 2017, 2019, Qu et al. 2017, Achata et al. 2018). Hyperspectral imaging with NIR was applied to honey analysis (Noviyanto and Abdulla 2019) and for detecting citrus disease in different seasons and cultivars (Weng et al. 2018).

Near-infrared hyperspectral imaging (NIR-HSI) and Raman hyperspectral imaging (R-HSI) in combination with chemometrics were used to evaluate binary mixtures of food powders in different proportions (Achata et al. 2018). HIS was used to evaluate protein content in meat (Ma et al. 2019). Near-infrared hyperspectral imaging (NIR-HIS) was used in the evaluation of adulterated samples of berry black pepper with berry papaya seeds (Orrillo et al. 2019).

5.6 Digital Image – Red, Green and Blue (RGB)

RGB was applied to the selection and classification of fruits, volume estimation and evaluation of the number of assets per each kind of fruit (Méndez Perez et al. 2017). It was also used in the evaluation of potato volume (Long et al. 2018).

RGB was used in the detection of defective hazelnuts and the discrimination of two types of defects, the rotten and the pest-affected kernels (Giraudo et al. 2018). RGB with multivariate calibration models was used to evaluate the quantification of color-related properties of food, such as pesto sauce, and to predict visual attributes and various pigments. The results obtained with this method were reliable and could be also used to monitor raw materials, the different stages of the manufacture and end products (Foca et al. 2011).

Table 1. Summary of applicability of green methods in food analysis.

Technique	Attribute evaluated	Food matrix	Reference
NIR	Macro and micronutrients content, volatile and bioactive content, microbiological parameters, physical attributes and authenticity; geographical origin; fiber; quality and origin; adulterants; nutrients, heavy metals and pesticide composition; different types; automated processes; quality parameters (acidity, total sugar, soluble solids) and ascorbic acid.	Green coffee beans; semolina; rice; grape nectars; rice varieties; wine vinegar; food industries; cashew apple and guava nectar.	(Alamprese et al. 2016b, Bázár et al. 2016b, Basri et al. 2017b, Caramês et al. 2017b, Mabood et al. 2017b, Ríos-Reina et al. 2018, Giraudo et al. 2018, Grassi and Alamprese 2018, Miaw et al. 2018, Pallone et al. 2018, Badaró et al. 2019, Teye et al. 2019, David et al. 2019)
MIR	Adulterants (soybean oil and sunflower oil); adulteration.	Extra virgin flaxseed oil; quinoa flour.	(De Souza et al. 2015, Rodríguez et al. 2019)
NIR and MIR	Physical-chemical parameters; sugars, organic acids and carotenoids; adulteration.	Quinoa flour; protein in milk powder; moisture in soy and starch in black beans; passion fruit; butter oil.	(de Oliveira et al. 2014a, Ferreira et al. 2015, Chen et al. 2018, Santiago-Ramos et al. 2018, Pereira et al. 2019).
Raman and NIR	Food fraud	Hazelnut	(Mir-Marqués et al. 2016)
Raman	Online control; proteins and lipids; antioxidant capacity and phenolic compounds; control of *Escherichia coli*; addition of maltodextrin; artificial colorants; carotenoids.	glucose fermentation by *Saccharomyces cerevisiae*; milk; rice wine chines; meat products and powdered milk; foods; fruit pulp.	(Ávila et al. 2012, Wu et al. 2015, Lohumi et al. 2015a, Rodrigues Júnior et al. 2016, Silveira et al. 2016, Gukowsky et al. 2018)
RGB	Multiple parameters	Honey; milk and alcoholic beverage.	(Kucheryavskiy et al. 2014, Dominguez and Centurión 2015)
RGB and hyperspectral imaging	Grading and color evolution	Apples	(Garrido-Novell et al. 2012)
NIR hyperspectral imaging	Adulterants; binary mixtures; adulterated samples; online monitoring.	Milk; food powders; berry black pepper; monitoring food process as drying, cooking, freezing, salting, dehydrating and process analytical technology applications.	(Cheng et al. 2015, Xie et al. 2016, Forchetti and Poppi 2017, Liu et al. 2017, Ma et al. 2017, 2019, Achata et al. 2018, Orrillo et al. 2019, Su et al. 2019).

Table 1 contd. ...

...Table 1 contd.

Technique	Attribute evaluated	Food matrix	Reference
Hyperspectral imaging	Fermentation index, polyphenol content and antioxidant activity; measuring ripening; food quality and safety; fungi; origin; analyses; detection disease; protein content.	Single cocoa beans; tomatoes; foods undergoing processing; corn; *Jatropha curcas* L. seed; honey; citrus; meat.	(Del Fiore et al. 2010, Polder and Heijden 2010, Gao et al. 2013, Liu et al. 2017, Caporaso et al. 2018, Weng et al. 2018, Ma et al. 2019, Noviyanto and Abdulla 2019)
RMN	Origin; identification of metabolites; physical characterization; classification of different classes; evaluate discrimination, varietal and geographical classification.	Lamb meat; wheat; starch; larger beer; wine.	(Larsen et al. 2008, Santos et al. 2015, Magdas et al. 2019)

Classification of ground meat from chicken, pork, and beef could also be achieved with the RGB image systems using the PLS-DA method of discrimination. Fresh-sliced tissues of pork and salmon could also be evaluated by RGB digital images proving the efficiency of this type of analysis (Nolasco-Perez et al. 2019b, Verdú et al. 2019).

A summary of the main applicability of green methods in the food matrix is presented in Table 1.

References

Achata, E.M., Esquerre, C., Gowen, A.A. and O'Donnell, C.P. 2018. Feasibility of near infrared and Raman hyperspectral imaging combined with multivariate analysis to assess binary mixtures of food powders. Powder Technol. 336: 555–566. https://doi.org/10.1016/j.powtec.2018.06.025.

Alamprese, C., Casale, M., Sinelli, N. et al. 2013. Detection of minced beef adulteration with turkey meat by UV-vis, NIR and MIR spectroscopy. LWT - Food Sci. Technol. 53: 225–232. https://doi.org/10.1016/j.lwt.2013.01.027.

Alamprese, C., Amigo, J.M., Casiraghi, E. and Engelsen, S.B. 2016a. Identification and quantification of turkey meat adulteration in fresh, frozen-thawed and cooked minced beef by FT-NIR spectroscopy and chemometrics. Meat Sci. 121: 175–181. https://doi.org/10.1016/j.meatsci.2016.06.018.

Alamprese, C., Amigo, J.M., Casiraghi, E. and Engelsen, S.B. 2016b. Identification and quantification of turkey meat adulteration in fresh, frozen-thawed and cooked minced beef by FT-NIR spectroscopy and chemometrics. Meat Sci. 121: 175–181. https://doi.org/10.1016/j.meatsci.2016.06.018.

Alberto, C. and Machado, A.A.S.C. 2018. Una visión sobre propuestas de enseñanza de la Química Verde. Rev. Electrónica Enseñanza las Ciencias 17: 19–43.

Alkarkhi, A.F.M., Alqaraghuli, W.A.A., Alkarkhi, A.F.M. and Alqaraghuli, W.A.A. 2019. Principal components analysis. Easy Stat. Food Sci. with R 125–141. https://doi.org/10.1016/B978-0-12-814262-2.00008-X.

Almeida, M.R., Alves, R.S., Nascimbem, L.B. et al. 2010. Determination of amylose content in starch using Raman spectroscopy and multivariate calibration analysis. Anal. Bioanal. Chem. 397: 2693–2701.

Ávila, T.C., Poppi, R.J., Lunardi, I. et al. 2012. Raman spectroscopy and chemometrics for *on-line* control of glucose fermentation by *Saccharomyces cerevisiae*. Biotechnol. Prog. 28: 1598–1604. https://doi.org/10.1002/btpr.1615.

Badaró, A.T., Morimitsu, F.L., Ferreira, A.R. et al. 2019. Identification of fiber added to semolina by near infrared (NIR) spectral techniques. Food Chem. 289: 195–203. https://doi.org/10.1016/j.foodchem.2019.03.057.

Basri, K.N., Hussain, M.N., Bakar, J. et al. 2017a. Classification and quantification of palm oil adulteration via portable NIR spectroscopy. Spectrochim Acta Part A Mol. Biomol. Spectrosc. 173: 335–342. https://doi.org/10.1016/j.saa.2016.09.028.

Basri, K.N., Hussain, M.N., Bakar, J. et al. 2017b. Classification and quantification of palm oil adulteration via portable NIR spectroscopy. Spectrochim Acta - Part A Mol. Biomol. Spectrosc. 173: 335–342. https://doi.org/10.1016/j.saa.2016.09.028.

Bázár, G., Romvári, R., Szabó, A. et al. 2016a. NIR detection of honey adulteration reveals differences in water spectral pattern. Food Chem. 194: 873–880. https://doi.org/10.1016/j.foodchem.2015.08.092.

Bázár, G., Romvári, R., Szabó, A. et al. 2016b. NIR detection of honey adulteration reveals differences in water spectral pattern. Food Chem. 194: 873–880. https://doi.org/10.1016/j.foodchem.2015.08.092.

Burns, D.A. and Ciurczak, E.W. 2009. Handbook of Near-Infrared Analysis. 3rd ed. Anal. Bioanal. Chem. 393: 1387–1389.

Caesar, L.K., Kvalheim, O.M. and Cech, N.B. 2018. Hierarchical cluster analysis of technical replicates to identify interferents in untargeted mass spectrometry metabolomics. Anal. Chim. Acta 1021: 69–77. https://doi.org/10.1016/j.aca.2018.03.013.

Caporaso, N., Whitworth, M.B., Fowler, M.S. and Fisk, I.D. 2018. Hyperspectral imaging for non-destructive prediction of fermentation index, polyphenol content and antioxidant activity in single cocoa beans. Food Chem. 258: 343–351. https://doi.org/10.1016/j.foodchem.2018.03.039.

Caramês, E.T.S., Alamar, P.D., Poppi, R.J. and Pallone, J.A.L. 2017a. Rapid assessment of total phenolic and anthocyanin contents in grape juice using infrared spectroscopy and multivariate calibration. Food Anal. Methods 10: 1609–1615. https://doi.org/10.1007/s12161-016-0721-1.

Caramês, E.T.S., Alamar, P.D., Poppi, R.J. and Pallone, J.A.L. 2017b. Quality control of cashew apple and guava nectar by near infrared spectroscopy. J. Food Compos Anal. 56: 41–46. https://doi.org/10.1016/j.jfca.2016.12.002.

Chen, H., Tan, C. and Lin, Z. 2018. Authenticity detection of black rice by near-infrared spectroscopy and support vector data description. Int. J. Anal. Chem. 2018: 1–8. https://doi.org/10.1155/2018/8032831.

Cheng, J.H., Qu, J.H., Sun, D.W. and Zeng, X.A. 2014. Visible/near-infrared hyperspectral imaging prediction of textural firmness of grass carp (*Ctenopharyngodon idella*) as affected by frozen storage. Food Res. Int. 56: 190–198. https://doi.org/10.1016/j.foodres.2013.12.009.

Cheng, J.H., Sun, D.W., Pu, H., Bin et al. 2015. Integration of classifiers analysis and hyperspectral imaging for rapid discrimination of fresh from cold-stored and frozen-thawed fish fillets. J. Food Eng. 161: 33–39. https://doi.org/10.1016/j.jfoodeng.2015.03.011.

Colnago, L.A., Andrade, F.D., Souza, A.A. et al. 2014. Why is inline NMR rarely used as industrial sensor? challenges and opportunities. Chem. Eng. Technol. 37: 191–203. https://doi.org/10.1002/ceat.201300380.

Colnago, L.S., Torre Neto, A., Ferrazini, J. and Oste, R. 1996. Espectrometro de RMN para Análise Quantitativa. INPI.

Consonni, R., Polla, D. and Cagliani, L.R. 2018. Organic and conventional coffee differentiation by NMR spectroscopy. Food Control 94: 284–288. https://doi.org/10.1016/j.foodcont.2018.07.013.

Cortés, V., Blasco, J., Aleixos, N. et al. 2019. Monitoring strategies for quality control of agricultural products using visible and near-infrared spectroscopy: A review. Trends Food Sci. Technol. 85: 138–148. https://doi.org/10.1016/j.tifs.2019.01.015.

da Silva, L.A., Flumignan, D.L., Tininis, A.G. et al. 2019. Discrimination of Brazilian lager beer by 1H NMR spectroscopy combined with chemometrics. Food Chem. 272: 488–493. https://doi.org/10.1016/j.foodchem.2018.08.077.

David, E., Eleazu, C., Igweibor, N. et al. 2019. Comparative study on the nutrients, heavy metals and pesticide composition of some locally produced and marketed rice varieties in Nigeria. Food Chem. 278: 617–624. https://doi.org/10.1016/j.foodchem.2018.11.100.

de Oliveira, G.A., de Castilhos, F., Renard, C.M.-G.C. and Bureau, S. 2014a. Comparison of NIR and MIR spectroscopic methods for determination of individual sugars, organic acids and carotenoids in passion fruit. Food Res. Int. 60: 154–162. https://doi.org/10.1016/j.foodres.2013.10.051.

de Oliveira, G.A., de Castilhos, F., Renard, C.M.G.C. and Bureau, S. 2014b. Comparison of NIR and MIR spectroscopic methods for determination of individual sugars, organic acids and carotenoids in passion fruit. Food Res. Int. 60: 154–162. https://doi.org/10.1016/j.foodres.2013.10.051.

de Souza, L.M., de Santana, F.B., Gontijo, L.C. et al. 2015. Quantification of adulterations in extra virgin flaxseed oil using MIR and PLS. Food Chem. 182: 35–40. https://doi.org/10.1016/j.foodchem.2015.02.081.

De Souza, L.M., De Santana, F.B., Gontijo, L.C. et al. 2015. Quantification of adulterations in extra virgin flaxseed oil using MIR and PLS. Food Chem. 182: 35–40. https://doi.org/10.1016/j.foodchem.2015.02.081.

Del Fiore, A., Reverberi, M., Ricelli, A. et al. 2010. Early detection of toxigenic fungi on maize by hyperspectral imaging analysis. Int. J. Food Microbiol. 144: 64–71. https://doi.org/10.1016/j.ijfoodmicro.2010.08.001.

Dominguez, M.A. and Centurión, M.E. 2015. Application of digital images to determine color in honey samples from Argentina. Microchem. J. 118: 110–114. https://doi.org/10.1016/j.microc.2014.08.002.

Elmasry, G., Barbin, D.F., Sun, D. and Allen, P. 2012. Meat quality evaluation by hyperspectral imaging technique: an overview. Food Sci. Nutr. 52: 689–711.

Ferrão, M.F. 2000. Aplicações de técnicas Espectroscópicas de reflexão o infravermelho no controle de qualidade de farinha de trigo. Unicamp.

Ferreira, D.S., Pallone, J.A.L. and Poppi, R.J. 2015. Direct analysis of the main chemical constituents in Chenopodium quinoa grain using Fourier transform near-infrared spectroscopy. Food Control 48: 91–95. https://doi.org/10.1016/j.foodcont.2014.04.016.

Ferreira, M.M.C. 2015. Quimiometria-conceitos, métodos e aplicações, 1 st. Editora UNICAMP, Campinas, SP.

Foca, G., Masino, F., Antonelli, A. and Ulrici, A. 2011. Prediction of compositional and sensory characteristics using RGB digital images and multivariate calibration techniques. Anal. Chim. Acta 706: 238–245. https://doi.org/10.1016/J.ACA.2011.08.046.

Forchetti, D.A.P. and Poppi, R.J. 2017. Use of NIR hyperspectral imaging and multivariate curve resolution (MCR) for detection and quantification of adulterants in milk powder. LWT - Food Sci. Technol. 76: 337–343. https://doi.org/10.1016/j.lwt.2016.06.046.

Foster, D.H. and Amano, K. 2019. Hyperspectral imaging in color vision research: tutorial. J. Opt. Soc. Am. A 36: 606. https://doi.org/10.1364/JOSAA.36.000606.

Gałuszka, A., Migaszewski, Z. and Namieśnik, J. 2013. The 12 principles of green analytical chemistry and the significance mnemonic of green analytical practices. TrAC - Trends Anal. Chem. 50: 78–84. https://doi.org/10.1016/j.trac.2013.04.010.

Gao, J., Li, X., Zhu, F. and He, Y. 2013. Application of hyperspectral imaging technology to discriminate different geographical origins of *Jatropha curcas* L. seeds. Comput. Electron. Agric 99: 186–193. https://doi.org/10.1016/j.compag.2013.09.011.

Garrido-Novell, C., Garrido-Varo, A., Pérez-Marin, D. et al. 2012. Grading and color evolution of apples using RGB and hyperspectral imaging vision cameras. J. Food Eng. 113: 281–288. https://doi.org/10.1016/j.jfoodeng.2012.05.038.

Giraudo, A., Calvini, R., Orlandi, G. et al. 2018. Development of an automated method for the identification of defective hazelnuts based on RGB image analysis and colourgrams. Food Control 94: 233–240. https://doi.org/10.1016/j.foodcont.2018.07.018.

Godinho, M.S., Oliveira, A.E. and Sena, M.M. 2010. Determination of interfacial tension of insulating oils by using image analysis and multi-way calibration. Microchem. J. 96: 42–45.

Gou, J., Ma, H., Ou, W. et al. 2019. A generalized mean distance-based k-nearest neighbor classifier. Expert Syst. Appl. 115: 356–372. https://doi.org/10.1016/j.eswa.2018.08.021.

Granato, D., Santos, J.S., Escher, G.B. et al. 2018. Use of principal component analysis (PCA) and hierarchical cluster analysis (HCA) for multivariate association between bioactive compounds and functional properties in foods: A critical perspective. Trends Food Sci. Technol. 72: 83–90. https://doi.org/10.1016/j.tifs.2017.12.006.

Grassi, S., Amigo, J.M., Lyndgaard, C.B. et al. 2014. Beer fermentation: Monitoring of process parameters by FT-NIR and multivariate data analysis. Food Chem. 155: 279–286. https://doi.org/10.1016/j.foodchem.2014.01.060.

Grassi, S. and Alamprese, C. 2018. Advances in NIR spectroscopy applied to process analytical technology in food industries. Curr. Opin. Food Sci. 22: 17–21. https://doi.org/10.1016/j.cofs.2017.12.008.

Gukowsky, J.C., Xie, T., Gao, S. et al. 2018. Rapid identification of artificial and natural food colorants with surface enhanced Raman spectroscopy. Food Control 92: 267–275. https://doi.org/10.1016/j.foodcont.2018.04.058.

Haq, N., Iqbal, M., Alanazi, F.K. et al. 2017. Applying green analytical chemistry for rapid analysis of drugs: Adding health to pharmaceutical industry. Arab J. Chem. 10: S777–S785. https://doi.org/10.1016/j.arabjc.2012.12.004.

Harting, J. and Kleinebudde, P. 2019. Optimisation of an in-line Raman spectroscopic method for continuous API quantification during twin-screw wet granulation and its application for process characterisation. Eur. J. Pharm. Biopharm. 137: 77–85. https://doi.org/10.1016/j.ejpb.2019.02.015.

Hein, P.R.G., Chaix, G., Clair, B. et al. 2016. Spatial variation of wood density, stiffness and microfibril angle along Eucalyptus trunks grown under contrasting growth conditions. Trees - Struct. Funct. 30: 871–882. https://doi.org/10.1007/s00468-015-1327-8.

Hell, J., Prückler, M., Danner, L. et al. 2016. A comparison between near-infrared (NIR) and mid-infrared (ATR-FTIR) spectroscopy for the multivariate determination of compositional properties in wheat bran samples. Food Control 60: 365–369. https://doi.org/10.1016/j.foodcont.2015.08.003.

Hu, Y., Pan, Z.J., Liao, W. et al. 2016. Determination of antioxidant capacity and phenolic content of chocolate by attenuated total reflectance-Fourier transformed-infrared spectroscopy. Food Chem. 202: 254–261. https://doi.org/10.1016/j.foodchem.2016.01.130.

Huang, J., Wium, H., Qvist, K.B. and Esbensen, K.H. 2003. Multi-way methods in image analysis - relationships and applications. Chemom Intell. Lab Syst. 66: 141–158.

Hwang, S.W., Horikawa, Y., Lee, W.H. and Sugiyama, J. 2016. Identification of Pinus species related to historic architecture in Korea using NIR chemometric approaches. J. Wood Sci. 62: 156–167. https://doi.org/10.1007/s10086-016-1540-0.

Iqbal, Z. and Bjorklund, R.B. 2011. Assessment of a mobile phone for use as a spectroscopic analytical tool for foods and beverages. Int. J. Food Sci. Technol. 46: 2428–2436.

Kucheryavskiy, S., Melenteva, A. and Bogomolov, A. 2014. Determination of fat and total protein content in milk using conventional digital imaging. Talanta 121: 144–152. https://doi.org/10.1016/j.talanta.2013.12.055.

Kumar, R. and Sharma, V. 2018. Chemometrics in forensic science. TrAC Trends Anal. Chem. 105: 191–201. https://doi.org/10.1016/J.TRAC.2018.05.010.

Larsen, F.H., Blennow, A. and Engelsen, S.B. 2008. Starch granule hydration-A MAS NMR investigation. Food Biophys 3: 25–32. https://doi.org/10.1007/s11483-007-9045-4.

Lee, K., Baek, S., Kim, D. and Seo, J. 2019. A freshness indicator for monitoring chicken-breast spoilage using a Tyvek® sheet and RGB color analysis. Food Packag Shelf Life 19: 40–46. https://doi.org/10.1016/j.fpsl.2018.11.016.

Lee, K.A. 2004. Review of applications of near infrared spectroscopy to food analysis. NIR Spectr. 2: 11–16.

Li, H., He, J., Li, F. et al. 2016. Application of NIR and MIR spectroscopy for rapid determination of antioxidant activity of Radix Scutellariae from different geographical regions. Phytochem Anal. 27: 73–80. https://doi.org/10.1002/pca.2602.

Lima, M. dos, S., da Conceição Prudêncio Dutra, M., Toaldo, I.M. et al. 2015. Phenolic compounds, organic acids and antioxidant activity of grape juices produced in industrial scale by different processes of maceration. Food Chem. 188: 384–392. https://doi.org/10.1016/j.foodchem.2015.04.014.

Liu, Y., Pu, H. and Sun, D.-W. 2017. Hyperspectral imaging technique for evaluating food quality and safety during various processes: A review of recent applications. Trends Food Sci. Technol. 69: 25–35. https://doi.org/10.1016/j.tifs.2017.08.013.

Lobato, K.B. de S., Alamar, P.D., Caramês, E.T. dos S. and Pallone, J.A.L. 2018. Authenticity of freeze-dried açaí pulp by near-infrared spectroscopy. J. Food Eng. 224: 105–111. https://doi.org/10.1016/J.JFOODENG.2017.12.019.

Lohumi, S., Lee, S., Lee, H. and Cho, B.-K. 2015a. A review of vibrational spectroscopic techniques for the detection of food authenticity and adulteration. Trends Food Sci. Technol. 46: 85–98. https://doi.org/10.1016/J.TIFS.2015.08.003.

Lohumi, S., Lee, S., Lee, H. and Cho, B.-K. 2015b. A review of vibrational spectroscopic techniques for the detection of food authenticity and adulteration. Trends Food Sci. Technol. 46: 85–98. https://doi.org/10.1016/j.tifs.2015.08.003.

Long, Y., Wang, Y., Zhai, Z. et al. 2018. Potato volume measurement based on RGB-D camera. IFAC-PapersOnLine 51: 515–520. https://doi.org/10.1016/j.ifacol.2018.08.157.

Ma, J., Sun, D.W., Qu, J.H. and Pu, H. 2017. Prediction of textural changes in grass carp fillets as affected by vacuum freeze drying using hyperspectral imaging based on integrated group wavelengths. LWT - Food Sci. Technol. 82: 377–385. https://doi.org/10.1016/j.lwt.2017.04.040.

Ma, J., Sun, D.W., Pu, H. et al. 2019. Protein content evaluation of processed pork meats based on a novel single shot (snapshot) hyperspectral imaging sensor. J. Food Eng. 240: 207–213. https://doi.org/10.1016/j.jfoodeng.2018.07.032.

Mabood, F., Jabeen, F., Ahmed, M. et al. 2017a. Development of new NIR-spectroscopy method combined with multivariate analysis for detection of adulteration in camel milk with goat milk. Food Chem. 221: 746–750. https://doi.org/10.1016/j.foodchem.2016.11.109.

Mabood, F., Jabeen, F., Ahmed, M. et al. 2017b. Development of new NIR-spectroscopy method combined with multivariate analysis for detection of adulteration in camel milk with goat milk. Food Chem. 221: 746–750. https://doi.org/10.1016/j.foodchem.2016.11.109.

Magdas, D.A., Pirnau, A., Feher, I. et al. 2019. Alternative approach of applying 1 H NMR in conjunction with chemometrics for wine classification. Lwt 109: 422–428. https://doi.org/10.1016/j.lwt.2019.04.054.

Mandrile, L., Zeppa, G., Giovannozzi, A.M. and Rossi, A.M. 2016. Controlling protected designation of origin of wine by Raman spectroscopy. Food Chem. 211: 260–267. https://doi.org/10.1016/j.foodchem.2016.05.011.

Medina, S., Perestrelo, R., Silva, P. et al. 2019. Current trends and recent advances on food authenticity technologies and chemometric approaches. Trends Food Sci. Technol. 85: 163–176. https://doi.org/10.1016/j.tifs.2019.01.017.

Méndez Perez, R., Cheein, F.A. and Rosell-Polo, J.R. 2017. Flexible system of multiple RGB-D sensors for measuring and classifying fruits in agri-food Industry. Comput. Electron. Agric 139: 231–242. https://doi.org/10.1016/j.compag.2017.05.014.

Merkx, D.W.H., Westphal, Y., van Velzen, E.J.J. et al. 2018. Quantification of food polysaccharide mixtures by 1H NMR. Carbohydr. Polym. 179: 379–385. https://doi.org/10.1016/j.carbpol.2017.09.074.

Miaw, C.S.W., Sena, M.M., Souza, S.V.C. de et al. 2018. Detection of adulterants in grape nectars by attenuated total reflectance Fourier-transform mid-infrared spectroscopy and multivariate classification strategies. Food Chem. 266: 254–261. https://doi.org/10.1016/j.foodchem.2018.06.006.

Mir-Marqués, A., Elvira-Sáez, C., Cervera, M.L. et al. 2016. Authentication of protected designation of origin artichokes by spectroscopy methods. Food Control 59: 74–81. https://doi.org/10.1016/j.foodcont.2015.05.004.

Monakhova, Y.B., Kuballa, T., Tschiersch, C. and Diehl, B.W.K. 2017. Rapid NMR determination of inorganic cations in food matrices: Application to mineral water. Food Chem. 221: 1828–1833. https://doi.org/10.1016/j.foodchem.2016.10.095.

Nolasco-Perez, I.M., Rocco, L.A.C.M., Cruz-Tirado, J.P. et al. 2019a. Comparison of rapid techniques for classification of ground meat. Biosyst. Eng. https://doi.org/10.1016/j.biosystemseng.2019.04.013.

Nolasco-Perez, I.M., Rocco, L.A.C.M., Cruz-Tirado, J.P. et al. 2019b. Comparison of rapid techniques for classification of ground meat. Biosyst. Eng. 183: 151–159. https://doi.org/10.1016/j.biosystemseng.2019.04.013.

Noviyanto, A. and Abdulla, W.H. 2019. Segmentation and calibration of hyperspectral imaging for honey analysis. Comput. Electron. Agric 159: 129–139. https://doi.org/10.1016/j.compag.2019.02.006.

Oliveira, L.F., Canevari, N.T., Guerra, M.B.B. et al. 2013. Proposition of a simple method for chromium (VI) determination in soils from remote places applying digital images: A case study from Brazilian Antarctic Station. Microchem J. 109: 165–169.

Oliveri, P. and Simonetti, R. 2016. Chemometrics for food authenticity applications. Adv. Food Authent. Test 701–728. https://doi.org/10.1016/B978-0-08-100220-9.00025-4.

Orrillo, I., Cruz-Tirado, J.P., Cardenas, A. et al. 2019. Hyperspectral imaging as a powerful tool for identification of papaya seeds in black pepper. Food Control 101: 45–52. https://doi.org/10.1016/j.foodcont.2019.02.036.

Osborne, B.G. 1999. Near-Infrared Spectroscopy in Food Analysis, Encyclopedia of Analytical Chemistry. Wiley, Chichester.

Pallone, J.A.L., Caramês, E.T. dos S. and Alamar, P.D. 2018. Green analytical chemistry applied in food analysis: alternative techniques. Curr. Opin. Food Sci. 22: 115–121. https://doi.org/10.1016/J.COFS.2018.01.009.

Pavia, D., Lampman, G., Kriz, G. and Vyvyan, J. 2008. Introduction to Spectroscopy, 4th edn. Cengage Learning, Stamford.

Pereira, C.G., Leite, A.I.N., Andrade, J. et al. 2019. Evaluation of butter oil adulteration with soybean oil by FT-MIR and FT-NIR spectroscopies and multivariate analyses. Lwt 107: 1–8. https://doi.org/10.1016/j.lwt.2019.02.072.

Pissard, A., Fernández Pierna, J.A., Baeten, V. et al. 2013. Non-destructive measurement of vitamin C, total polyphenol and sugar content in apples using near-infrared spectroscopy. J. Sci. Food Agric 93: 238–244. https://doi.org/10.1002/jsfa.5779.

Płotka-Wasylka, J., Kurowska-Susdorf, A., Sajid, M. et al. 2018. Green chemistry in higher education: state of the art, challenges, and future trends. ChemSusChem. 2845–2858. https://doi.org/10.1002/cssc.201801109.

Polder, G. and Heijden, G. van der. 2010. Measuring Ripening of Tomatoes using Imaging Spectrometry, First Edit. Elsevier Inc.

Porep, J.U., Kammerer, D.R. and Carle, R. 2015. On-line application of near infrared (NIR) spectroscopy in food production. Trends Food Sci. Technol. 46: 211–230. https://doi.org/10.1016/j.tifs.2015.10.002.

Qu, J.H., Sun, D.W., Cheng, J.H. and Pu, H. 2017. Mapping moisture contents in grass carp (Ctenopharyngodon idella) slices under different freeze drying periods by Vis-NIR hyperspectral imaging. LWT - Food Sci. Technol. 75: 529–536. https://doi.org/10.1016/j.lwt.2016.09.024.

Richter, B., Rurik, M., Gurk, S. et al. 2019. Food monitoring: Screening of the geographical origin of white asparagus using FT-NIR and machine learning. Food Control 104: 318–325. https://doi.org/10.1016/j.foodcont.2019.04.032.

Ríos-Reina, R., García-González, D.L., Callejón, R.M. and Amigo, J.M. 2018. NIR spectroscopy and chemometrics for the typification of Spanish wine vinegars with a protected designation of origin. Food Control 89: 108–116. https://doi.org/10.1016/j.foodcont.2018.01.031.

Roberts, J., Power, A., Chapman, J. et al. 2018. A short update on the advantages, applications and limitations of hyperspectral and chemical imaging in food authentication. Appl. Sci. 8: 505. https://doi.org/10.3390/app8040505.

Rodrigues Júnior, P.H., De Sá Oliveira, K., Almeida, C.E.R. De et al. 2016. FT-Raman and chemometric tools for rapid determination of quality parameters in milk powder: Classification of samples for the presence of lactose and fraud detection by addition of maltodextrin. Food Chem. 196: 584–588. https://doi.org/10.1016/j.foodchem.2015.09.055.

Rodríguez, S.D., Rolandelli, G. and Buera, M.P. 2019. Detection of quinoa flour adulteration by means of FT-MIR spectroscopy combined with chemometric methods. Food Chem. 274: 392–401. https://doi.org/10.1016/j.foodchem.2018.08.140.

Santiago-Ramos, D., Figueroa-Cárdenas, J. de D., Véles-Medina, J.J. and Salazar, R. 2018. Physicochemical properties of nixtamalized black bean (*Phaseolus vulgaris* L.) flours. Food Chem. 240: 456–462. https://doi.org/10.1016/j.foodchem.2017.07.156.

Santos, A.D.C., Fonseca, F.A., Lião, L.M. et al. 2015. High-resolution magic angle spinning nuclear magnetic resonance in foodstuff analysis. TrAC - Trends Anal. Chem. 73: 10–18. https://doi.org/10.1016/j.trac.2015.05.003.

Santos, D.M. and Royer, M.R. 2004. Análise da percepção dos alunos sobre a química verde e a educação ambiental no ensino de química. 142–164.

Santos, P.M., Wentzell, P.D. and Pereira-Filho, E.R. 2012. Scanner digital images combined with color parameters: A case study to detect adulterations in liquid cow's milk. Food Anal. Meth. 5: 89–95.

Schrader, B. and Bougeard, D. 1995. Infrared and Raman Spectroscopy: Methods and Applications. VCH, Weinheim.

Shieber, A. 2008. Modern Techniques for Food Authentication, First. Da-Wen Sun.

Silva, C.R. 2009. Ressonância magnética nuclear de baixo campo em estudos de petróleos. Universidade Federal do Espírito Santo.

Silveira, L., Motta, E.D.C.M., Zângaro, R.A. et al. 2016. Characterization of nutritional parameters in bovine milk by Raman spectroscopy with least squares modeling. Instrum. Sci. Technol. 44: 85–97. https://doi.org/10.1080/10739149.2015.1055578.

Simpson, J.H. 2008. Organic Structure Determination using 2D NMR Spectroscopy: A Problem-based Approach. Elsevier, Amsterdam.

Skoog, D.A., Holler, F.J. and Crouch, S.R. 2009. Princípios de Análise Instrumental, 6th edn. Bookman, Porto Alegre.

Soares, L.V. 2006. Curso basico de instrumentação para analistas de alimentos e fármacos. Manole, Barueri.

Solomons, T.W., Graham, F. and Craig, B. 1996. Química Orgânica, 6th edn. LTC.

Su, W.H., Bakalis, S. and Sun, D.W. 2019. Chemometrics in tandem with near infrared (NIR) hyperspectral imaging and Fourier transform mid infrared (FT-MIR) microspectroscopy for variety identification and cooking loss determination of sweet potato. Biosyst. Eng. 180: 70–86. https://doi.org/10.1016/j.biosystemseng.2019.01.005.

Sun, D.W. 2008. Modern Techniques for Food Authentication, 2nd edn. Academic Press.

Sun, D.W. 2009. Infrared Spectroscopy for Food Quality Analysis and Control, 1st edn. Academic Press, New York.

Teye, E., Amuah, C.L.Y., McGrath, T. and Elliott, C. 2019. Innovative and rapid analysis for rice authenticity using hand-held NIR spectrometry and chemometrics. Spectrochim Acta - Part A Mol. Biomol. Spectrosc. 217: 147–154. https://doi.org/10.1016/j.saa.2019.03.085.

Tobiszewski, M. and Namieśnik, J. 2017. Greener organic solvents in analytical chemistry. Curr. Opin. Green Sustain. Chem. 5: 1–4. https://doi.org/10.1016/j.cogsc.2017.03.002.

Tsuchikawa, S. and Kobori, H. 2015. A review of recent application of near infrared spectroscopy to wood science and technology. J. Wood Sci. 61: 213–220. https://doi.org/10.1007/s10086-015-1467-x.

Valderrama, L., Gonçalves, R.P., Março, P.H. and Valderrama, P. 2015. UV-Vis spectrum fingerprinting and chemometric method in the evaluation of extra virgin olive oil adulteration and fraud Espectroscopia Uv-Vis e Método Quimiométrico na Avaliação de Adulterações e Fraudes em Azeite de Oliva Extra Virgem. https://doi.org/10.14685/rebrapa.v5i2.171.

Verdú, S., Barat, J.M. and Grau, R. 2019. Fresh-sliced tissue inspection: Characterization of pork and salmon composition based on fractal analytics. Food Bioprod. Process. https://doi.org/10.1016/j.fbp.2019.04.008.

Wang, T., Wu, H.L., Long, W.J. et al. 2019a. Rapid identification and quantification of cheaper vegetable oil adulteration in camellia oil by using excitation-emission matrix fluorescence spectroscopy combined with chemometrics. Food Chem. 293: 348–357. https://doi.org/10.1016/j.foodchem.2019.04.109.

Wang, T., Wu, H.L., Long, W.J. et al. 2019b. Rapid identification and quantification of cheaper vegetable oil adulteration in camellia oil by using excitation-emission matrix fluorescence spectroscopy combined with chemometrics. Food Chem. https://doi.org/10.1016/j.foodchem.2019.04.109.

Weng, H., Lv, J., Cen, H. et al. 2018. Hyperspectral reflectance imaging combined with carbohydrate metabolism analysis for diagnosis of citrus Huanglongbing in different seasons and cultivars. Sensors Actuators, B Chem. 275: 50–60. https://doi.org/10.1016/j.snb.2018.08.020.

Williamson, K. and Hatzakis, E. 2019. NMR analysis of roasted coffee lipids and development of a spent ground coffee application for the production of bioplastic precursors. Food Res. Int. 119: 683–692. https://doi.org/10.1016/j.foodres.2018.10.046.

Wu, D. and Sun, D.W. 2013. Colour measurements by computer vision for food quality. Trends Food Sci. Technol. 29: 5–20.

Wu, Z., Xu, E., Long, J. et al. 2015. Rapid measurement of antioxidant activity and γ-aminobutyric acid content of chinese rice wine by fourier-transform near infrared spectroscopy. Food Anal. Methods 8: 2541–2553. https://doi.org/10.1007/s12161-015-0144-4.

Xie, A., Sun, D.W., Zhu, Z. and Pu, H. 2016. Nondestructive measurements of freezing parameters of frozen porcine meat by NIR hyperspectral imaging. Food Bioprocess Technol. 9: 1444–1454. https://doi.org/10.1007/s11947-016-1766-2.

Xu, Z.M. and Lam, E.Y. 2010. Image reconstruction using spectroscopic and hyperspectral information for compressive terahertz imaging. J. Opt. Soc. Am. A 27: 1638–1646.

Zhang, Y. 2018. On study of application of micro-reactor in chemistry and chemical field. IOP Conf. Ser. Earth Environ. Sci. 113. https://doi.org/10.1088/1755-1315/113/1/012003.

CHAPTER 12

Green Methods for Agrochemical Residues Analysis in Agriculture

Elem Tamirys dos Santos Caramês, Ana Paula Rebellato,
Joyce Grazielle Siqueira Silva, José Luan da Paixão Teixeira and
*Juliana Azevedo Lima Pallone**

1. Introduction

Agrochemical residues are found in the environment for many reasons. Some of them result from the use of agrochemicals in agriculture to improve crop yield and prevent, destroy or mitigate pests (Sabarwal et al. 2018, Elahi et al. 2019). Other agrochemical residues result from chemical analyzes used to evaluate toxic and non-toxic compounds in different matrices, which could be more toxic than the residues analyzed (Armenta et al. 2008).

Agrochemical residues include fungicides, insecticides, herbicides, pesticides (organochloride, organophosphorus, carbamates, and triazines), synthetic pyrethroids and plant growth regulators (Szpyrka et al. 2015, Nicolopoulou-Stamati et al. 2016, Carabajal et al. 2016).

The evaluation and control of agrochemical residues in different matrices are important since they could have harmful effects, such as causing disorders in humans and wildlife and contamination of soil, water, air, and food (Szpyrka et al. 2015, Loughlin et al. 2018, Rajput et al. 2018, Sabarwal et al. 2018, Elahi et al. 2019).

Pesticides, for example, are related to harmful effects and the appearance of human disorders. They can cause coughing, nausea, dizziness, diarrhoea and irritation in the eyes and skin (Elahi et al. 2019). They are involved in the pathogenesis of

Department of Food Science and Nutrition, School of Food Engineering, University of Campinas, Monteiro Lobato Street, 80, Zip Code: 13083-862, Campinas, São Paulo, Brazil.
* Corresponding author: jpallone@unicamp.br

neurodegenerative diseases, such as Parkinson's and Alzheimer's and reproductive and endocrine system disorders (Sabarval et al. 2018). Besides this, pesticides have also been correlated with a variety of cancers, including colon, breast and blood cancers (Attaullah et al. 2018, Martin et al. 2018, Ventura et al. 2019).

Government regulatory agencies and research groups have been evaluating the number of agrochemical residues in different industrial, agricultural and research practices. The residues are normally determined using gas chromatography (GC), High Performance Liquid Chromatography (HPLC) and spectrophotometry (Szpyrka et al. 2015, Loughlin et al. 2018, Rajput et al. 2018). However, these techniques, in most cases, use toxic solvents like dichloromethane, acetone and others (Szpyrka et al. 2015). In order to evaluate the risks, monitor chemical residues in agriculture and reduce the negative impact of chemical analyzes on the environment, some green-analytical chemistry (GAC) methods have been developed (Gałuszka et al. 2013, Carabajal et al. 2016).

GAC may make laboratory practices more environmentally friendly by reducing reagents and waste during analyzes and substituting toxic reagents with less harmful alternatives (Armenta et al. 2008). The concepts of GAC, however, are still in development and in this context, other parameters such as operator safety, increased efficiency, reagents from renewable sources and multi-analyte or multi-parameter methods are considered (Galuszka et al. 2013).

One of the ways to evaluate chemical compounds, according to GAC, is to replace polluting methodologies with clean ones (Armenta et al. 2008). Other options are related to the reduction of solvents used in the sample pre-treatment step or the reduction in the amount and toxicity of solvents and reagents during measurement, mainly through miniaturization and automatization of the process. Research and development for analytical methodologies that do not require solvents or reagents may also be included (Armenta et al. 2008).

Chromatography is the technique most often chosen to detect and quantify multiple agrochemical residues in food and food products. However, foods are complex matrices to analyze and the development of a method that assures full selectivity for the analytes in the detector is difficult, especially since the most commonly used detectors are the diode array (DAD) and the fluorescence detector (FLU), which are simple and non-selective (Sousa et al. 2017).

Global environmental problems have changed the perspective of analytical chemistry. Recently, the principles of green chemistry were determined and several efforts to apply them have been made, predominantly as a reduction and/or substitution for organic solvents. In this context, chemometric tools, such as multivariate curve resolution, have been used to resolve multiple component responses and unknown mixtures instead of the application of derivatizations and multiple extraction methods, avoiding the need for an extensive protocol, saving money and time (Jaumot et al. 2005, Gałuszka et al. 2013).

In this context, a few modifications to the instruments used can contribute to and conform to the GAC analyzes. Among the chromatographic analyzes, the reduction of the size of the chromatographic columns and the particle size of the stationary phase are some options (Armenta et al. 2008).

Despite that, the use of chemometric treatments, such as the multivariate curve resolution, is associated with some spectrometric techniques (near-infrared (NIR), mid-infrared (MIR) or Raman, nuclear magnetic resonance (NMR) and fluorescence and UV-spectroscopy), which could permit direct measurement, avoid the pre-treatment step of the sample and contribute to the reduction of solvents, reagents and time (Armenta et al. 2008, Galuszka et al. 2013).

2. Agrochemical Residues in Agriculture

According to the Guidelines for predicting dietary intake of pesticide residues, in collaboration with the Codex Committee on Pesticide Residues (WHO 1997), a pesticide is "any substance or mixture of substances intended to prevent, destroy, attract, repel or control any pest, including undesirable species of plants or animals during the production, storage, transport, distribution and processing of food, agricultural products, animal feed or which can be administered to animals for the control of ectoparasites". The term includes substances intended to be used as a plant growth regulator, defoliant, desiccant, fruit weakening agent or germination inhibitor and substances applied to cultures before or after transport. The term normally excludes fertilizers, plant and animal nutrients, food additives and animal medicines. Still, according to the Guidelines, pesticide residue is "any specific substance in food, agricultural products or animal feed resulting from the use of a pesticide. The term includes any derivatives of a pesticide such as conversion products, metabolites, reaction products and impurities which are considered to be of toxicological significance" (WHO 1997).

Agrochemicals can be classified according to their effects, such as acaricide, bactericide, insecticide, nematicide, herbicides, insect growth regulators, fungicides and plant growth regulators. They can also be classified as synthetic, organic (carbamates, chlorinated, phosphorous and chlorophosphorated), inorganic and botanical. Hundreds of agrochemicals are available for commercialization and later use in agriculture (Nicolopoulou-Stamati et al. 2016).

The discovery and rapid dissemination of agrochemicals to combat pests, insects, fungi, pathogens, parasites, and weeds in agriculture is increasing and deserves attention. At first, human health has not been considered and possible toxic residues deposited in plants and animals produced for human consumption have been causing human health damage. These include congenital malformation, endocrine, neurological and mental disorders, immunodepression, Parkinson's, infertility and some types of cancers. In addition, there is concern about the damage that the toxic waste can cause to the environment, contaminating soil and water sources (Elahi et al. 2019, Sabarval et al. 2018, Attaullah et al. 2018, Martin et al. 2018, Ventura et al. 2019).

For the monitoring of residues of agrochemicals in food, a preliminary study of the main foods consumed by the population is carried out, and later the samples are collected and then sent to the laboratories accredited to the inspection agencies for further analysis (Nougadère et al. 2012). Monitoring agrochemical residues in food is not an easy task. For each sample evaluated, the participating laboratories search for hundreds of residues, including their metabolites, in addition to prohibited residues in different countries.

As a result, each country has maximum permitted levels of residues in food to protect the health of the consumer. The Maximum Residue Levels (MRLs) are established by the Codex Committee on Pesticide Residues (CCPR) in food or feed on the international market. In addition, human health risk assessments are conducted to ensure the delivery of safe food before MRLs are established (FAO/WHO 2019, USDA 2017). It is still worth mentioning that some countries use the MRLs already proposed by the United States of America (USA) and the European Union (EU) if they do not have their set values for pesticide residues in food.

Since 1963, the United Nations Food and Agriculture Organization (FAO) and the World Health Organization (WHO) have held regular meetings to review waste and analytical methods for the evaluation of pesticides, estimate the MRLs of pesticides in food (including drinking water) and review toxicological data in addition to estimating the acceptable daily intake (ADI) for humans in relation to the pesticide under consideration (FAO/WHO 2019). In the USA, since 1991, the Department of Agriculture (USDA) has administered the Pesticide Data Program (PDP). This program provides support for the formation of a database of pesticide residues in food in the USA (USDA 2017). In the European Union, the European Food Safety Authority (EFSA), through EU-coordinated programs (EUCP), monitors the levels of pesticide residues in food. The harmonized MRLs by the EU are set for more than 500 pesticides in more than 370 food samples. The standard MRL of 0.01 mg/kg (limit of quantification) applies to pesticides that do not have values defined in the legislation. The EU also has specific MRLs for foods for infants and children (UFSA 2016).

The USDA's Agricultural Marketing Service (AMS), regularly collects data on agrochemical residues, with an emphasis on pesticides. In 2017, the 27th Annual Summary of the PDP was presented (USDA 2017). In this work, the PDP evaluated a variety of foods, including fresh and processed fruits and vegetables, honey, milk and mineral water. A total of 10,541 samples were analyzed of which 8,759 correspond to fresh and processed fruits and vegetables, 315 samples of honey, 711 samples of milk and 756 samples of bottled water. Of the samples evaluated, 99% of the monitored samples showed pesticide residues below the tolerance levels established by the Environmental Protection Agency (EPA). Of the honey samples evaluated, the PDP detected six different residues, including metabolites from six pesticides (2,4-dimethylphenyl formamide (2,4-DMPF), 2,6-dichlorobenzamide, Alachlor, Carbendazim and Coumaphos oxygen analog). All residue detections were lower than the tolerances established for that compound. For the milk samples, only 1 pesticide was detected (insecticide Flubendiamide) in 12 samples and concentrations ranged from 0.003 to 0.005 ppm with the tolerance set at 0.15 ppm. Of the 756 samples of bottled water, 10 different residues including metabolites were detected from 7 pesticides (Acetochlor ethanesulfonic acid, Alachlor ethanesulfonic acid, Atrazine, Desethyl atrazine, Desisopropyl atrazine, Imidacloprid, Metolachlor, Metolachlor ethanesulfonic acid, Simazine and Tebuthiuron). All the detections were below the levels set by the FDA.

In 2016, the EFSA released data on agrochemical residues with an emphasis on pesticides in foods. In total, 12 food kinds were evaluated (beans with pods, carrot, cucumber, orange, pear, potato, spinach, rice, wheat flour, ruminant liver, pork and poultry) and the number of samples per type varied from 15 to 93, depending on the population of the country inserted in the survey and totaling 12,850 samples. A total of 213 pesticides were analyzed, 191 of which were present in foods of plant origin and 58 in foods of animal origin and 40 new pesticides were also included (2-phenylphenol, biphenyl, chlorantraniliprole, cymoxanil, cyromazine, dichlorprop, diethofencarb, diflubenzuron, diniconazole-M, dithianon, dodine, ethoprophos, famoxadone, fenamidone, fenpropidin, fenpyroximate, flonicamid, flubendiamide, fluopyram, formothion, glufosinate, ioxynil, isocarbophos, isofenphos-methyl, isoprocarb, maleic hydrazide, mandipropamid, meptyldinocap, metaflumizone, metconazole, metobromuron, propoxur, pymetrozine, pyraclostrobin, rotenone, spirodiclofen, spiromesifen, terbuthylazine, tetramethrin and topramezone).

Of the 191 pesticides analyzed in foods of plant origin, 37 were not detected in any of the samples. The agrochemical residues most frequently detected in more than 4% of the analyzed samples were imazalil, boscalid, dithiocarbamates, chlorpyrifos, chlormequat, propamocarb, bromide ion, thiabendazole, pyrimethanil and cyprodinil.

For foods of animal origin, 46 of 58 pesticides investigated by the EUCP were not detected in any of the samples evaluated. The remainder of the pesticide residues were detected sporadically, and the DDT compound was most frequently detected in 1.3% of the samples analyzed. Other pesticide residues such as hexachlorobenzene, chlordane, lindane, alpha- and beta-HCH and endosulfan (unapproved pesticides present in the food chain due to their persistence) were detected in less than 0.75% of the samples.

Of the samples analyzed (82,649), 97.1% were within the legal limits; 53.6% of these samples (44,333) contained no detectable residues (results for all pesticides analyzed were below the limit of quantification (LOQ); 43.4% of the samples (35,895) contained measurable residues but did not exceed the legal limits).

The MRLs were exceeded in 2.9% of the analyzed samples (2,421 samples); these samples were considered not to comply with the legal limits. The surplus MRLs were most frequently detected in samples of tea, pepper (sweet or chilli), beans with pods (including beans) and celery leaves (including coriander leaves or other products which according to the classification of foods for pesticide residues are in the same category). Among these cases, 185 of them exceeded MRLs were caused by non-approved substances, most often carbendazim, followed by procymidone, dieldrin and anthraquinone. Among the approved pesticides, chlorpyrifos was the most frequently found substance in concentrations above legal limits, especially in carrot, potato, apple, parsley and cucumber samples; followed by the dimethoate compound present in samples of cherries, apples, radishes and cucumbers.

In September 2018, the Joint Meeting of the FAO Panel of Experts on Pesticide Residues (JMPR) in Food and the Environment and the WHO Core Assessment Group on Pesticide Residues was held in Berlin. Here, they established the acceptable daily intake, short-term dietary intake, acute reference doses, recommended maximum residue limits and average residual values for the compounds Abamectin (177),

Bentazone (172), Benzovindiflupyr (261), Chlorfenapyr (254), Cyantraniliprole (263), Cyazofamid (281), Cyprodinil (207), Diquat (031), Ethiprole (304), Fenpicoxamid (305), Fenpyroximate (193), Fluazinam (306), Fludioxonil (211), Fluopyram (243), Fluxapyroxad (256), Imazalil (1108), Isofetamid (290), Kresoxim-methyl (199), Lambda-cyhalothrin (146), Lufenuron (286), Mandestrobin (307), Mandipropamid (231), Norflurazon (308), Oxathiapiprolin (291), Profenofos (171), Propamocarb (148), Propiconazole (160), Pydiflumetofen (309), Pyraclostrobin (210), Pyriofenone (310), Pyriproxyfen (200), Sulfoxaflor (252) and Tioxazafen (311) (FAO/WHO 2019).

Studies and evaluations of the use of agrochemicals in products of plant and animal origin and consequently the analysis of residues of pesticides in foods should be constantly updated. Experts from the United Nations Organization reported in 2017 that about 200,000 people die each year from acute poisoning by pesticides (ONU 2017).

3. Separation Techniques, Extraction and GAC Principles

Monitoring agrochemical residues requires separation, identification and quantification techniques that must be reliable and previously validated. The time and money expended are also relevant (Szpyrka et al. 2015).

The GAC principles are important for the increasing emergence of new and environmentally friendly techniques and methodologies. During agrochemical residue analysis, there are five basic steps, which are extraction of the analytes from the sample into an aqueous or organic medium, the exclusion of co-extractives (cleaning), separation, identification and quantification of the compounds of interest (Omeroglu et al. 2012). In this context, methodologies that involve fewer toxic solvents are extremely relevant to human health and the environment.

The extraction of agrochemical residues from a food matrix can be achieved with several techniques and methodologies. Regarding solid samples, liquid-solid extraction (SLE) demands a large number of organic solvents and higher costs. In this context, new techniques that can achieve the same performance and efficiency using less solvent have been developed and applied (Nantia et al. 2017). These include QuEChERS (Quick, Easy, Cheap, Effective, Rugged and Safe) and Mini-luke methods. These provide good extractions using reduced amounts of reagents and samples. The QuEChERS method has been widely studied and has shown great results. It was initially proposed by Anastassiades et al. (2003) to extract pesticides from fruits and vegetables. The QuEChERS method is simple and easily adapted and consequently has been optimized and adjusted since 2003 to extract residual compounds from complex matrices, such as egg, avocado, chocolate, coffee, cereals, baby food, linseed and peanut seeds, olives and olive oil, animal tissue, milk and others.

Other methods of detection and quantification are GC and HPLC coupled with a mass detector, which has an excellent capacity to simultaneously detect a great number of distinct compounds. These techniques are the most utilized to analyze agrochemical residues in food. Besides this, these techniques are more sensitive with lower limits of detection due to the mass detector. Normally, these techniques are applied to monitor multi-residues (MRM) in food, which consists of simultaneously analyzing different active agrochemical residues in the same sample. It also detects several metabolites, therefore contributing to a quick, efficient, and low-cost monitoring method that has been applied in Germany, Australia, Canada, the USA, Netherland and other countries (Nougadère et al. 2012, Pang et al. 2015, Carabajal et al. 2016).

Several more efforts have been applied to develop an analytical method using fewer toxic reagents with lower required volumes or reagents considered green (good biodegradability, low risk to operator and other parameters). In addition, chemometric techniques like Multivariate Curve Resolution (MCR) have been applied to amplify the detection capacity of chromatography techniques without the additional steps of cleaning and extraction which would increase the volume of toxic waste and environmental damage (Carabajal et al. 2016, Gałuszka et al. 2013).

3.1 Green Solvents

Techniques, like liquid chromatography (LC), GC and mass spectrometry (MS) need judicious and effective methods for the extraction of compounds of interest, including agrochemical residues. A high amount of organic solvents that have high toxicity and high cost are required for this. In this context, several researchers are directing their attention to the development of new techniques of extraction that use less amount of toxic solvent and/or a greener solvent. Besides this, for LC the relevance of using green solvents is reinforced regarding the amount that is used during the separation technique as the acetonitrile is mainly used in the mobile phase (Pacheco-Fernández and Pino 2019).

Green solvents can be defined as functional solvents with minimal toxicity, high biodegradability under environmental conditions and are produced from renewable sources; they also exhibit low vapour pressure and a high boiling point. Another important point for suitable solvent choice is the ease of solvent recovery after extraction or any solvent-based separation process, which significantly reduces energy waste in this process (Lomba et al. 2019, Schuur et al. 2019, Vian et al. 2017).

The principal types of green solvents are amphiphilic solvents, ionic liquids (ILs), derivatives and deep eutectic solvents (DES). These have been used in separation techniques as the mobile phase and sample preparation as an extraction solvent (Pacheco-Fernández and Pino 2019, Schuur et al. 2019).

The amphiphilic solvents are mainly composed of ionic and non-ionic surfactants, alcohols and carboxylic acids. Among them, surfactants are the most successfully used in analytical chemistry. When these substances are added to an aqueous medium, micelles are formed and this promotes a better interaction between compounds of different polarity and gives the solvent excellent solvation properties (Pacheco-Fernández and Pino 2019).

ILs are salts that are liquid in temperatures below 100°C and whose ionic properties can be customized by changing the structure of the organic cation and/or the organic or inorganic anion. The ILs also have interesting properties, such as low vapor pressure at room temperature, high stability as a chemical and thermal and expressive versatility. Thus, ILs have been used in the past decade in separation procedures, aromatic-aliphatic separation, acid extraction and dissolution and processing of biopolymers. They have become the most popular choice to replace conventional organic solvents in analytical methods (Prasad and Sharma 2019, Schuur et al. 2019).

The DESs are eutectic mixtures of a hydrogen bond acceptor (HBA) and a hydrogen bond donor (HBD) at various rates. Ammonium and phosphonium salts are the main substances used as the HBA, while alcohols, carboxylic acids and amines are commonly applied as the HBD. When DESs are composed using derivatives of sugar, organic acids or bases and amino acids they can be classified as Natural Deep Eutectic Solvent (NADES). These NADES are obtained through biomass fermentation and enzymatic or esterification processes, and this type of solvent have wide applicability. These include additives for gasoline, flavoring, fragrance, extractor solution and reaction media. The majority is composed of fatty acid esters, furfural, terpenes, glycerol derivatives, glycols and others (Lomba et al. 2019). The DES substances show equivalent advantages that ILs present, but they have cheaper preparation when compared. While NADES are a solvent with low toxicity, biodegradability and sustainability, its properties enable application in chemical speciation (Prasad and Sharma 2019, Santana et al. 2019, Schuur et al. 2019).

In the general context, green solvents are an adequate choice for environmentally friendly analytical chemistry due to their technical features like biodegradability, sustainability, the ability for recovery and reuse and versatility in the application (mobile phase, sole extraction solution or coupled with other techniques) (Pacheco-Fernández and Pino 2019, Prasad and Sharma 2019).

3.2 Extraction Techniques and Chromatography Methods to Analyze Chemical Residues

Pesticides are substances used in agriculture to kill pests, remove weeds and reduce diseases in crops; however, these chemicals bring risks to human health and increase the chances of cancer, Parkinson's and Alzheimer's diseases and immune and neuropsychological disorders (Huang et al. 2019). Generally, these compounds are found at low concentration levels but can still be harmful to humans. In this context, efforts to develop new, green and reliable methods to quantify these chemicals in the food industry are extremely relevant (Farajzadeh et al. 2018).

Considering that, the target analytes (agrochemical residues) analyzed by separation techniques (GC and LC) are commonly present in low concentrations. The sample preparation often involves an extraction cleanup and/or concentration step followed by a derivatization process. Illustrating this, Braun et al. (2019) developed a protocol in a study about pesticides and antibiotics in permanent rice production. To quantify these pesticide compounds, an extraction step using hexane and acetone in an accelerated extraction process was necessary; next, was a clean-up process with aluminum oxide. The pesticides were analyzed using GC-MS, while the antibiotics were determined using LC-MS/MS after an extensive extraction process using methanol, acetonitrile and phosphoric acid.

In chromatography analysis, however, the sample preparation step is the most time and toxic solvent consuming stage. These techniques usually require organic reagents are tedious, time-consuming and sometimes lead to some loss of the components of interest. Liquid-Liquid Extraction (LLE) is a classic extraction method based on the principle of partitioning one or more analytes from a feed solution (aqueous part) into an immiscible solvent (organic part). This technique relies on such parameters as solubility and polarity of the chosen solvent; multiple and sequential extractions are the most efficient form of increasing the recovery rates. To guarantee an effective extraction, a high amount of an organic solvent is necessary (Chormey and Bakirdere 2018). Solid-Phase Extraction (SPE) consists of holding the analytes present in a liquid sample within a solid phase or in a liquid phase loaded into solid support; unlike LLE, this kind of extraction is not based on equilibrium but sorption principles (Chormey and Bakirdere 2018, Samanidou, 2017).

To apply the principles of GAC in this context, LC and GC are being processed using microextraction techniques like Solid Phase MicroExtraction (SPME) and Liquid Phase MicroExtraction (LPME) that are sometimes coupled to a non-toxic ionic liquid and using ultrasound as a dispersing agent. The QuEChERS protocol is widely used as a cleanup step (Huang et al. 2019, Nowak et al. 2018, Werner et al. 2018).

The QuEChERS (acronym of Quick, Easy, Cheap, Effective, Rugged and Safe) method is characterized as a cleanup step. Usually, there is an additional cleanup step with a dispersive solid-phase extraction (d-SPE) that has the functionality to remove residual water and matrix co-extracts. It was originally used to extract pesticide residue in vegetables and fruits but is now widely used as a multi-residue extraction technique (Anastassiades et al. 2003, Dong et al. 2018). For example, QuEChERS in conjunction with d-SPE was applied to recover 44 pesticides out of 60 present in samples of cinnamon bark and to simultaneously extract 102 pesticides found in green tea (Dong et al. 2018, Huang et al. 2019).

In order to follow the demand for GAC, micro-extraction has emerged as the solution to the problems presented by traditional LLE and SPE. Both these methods are time-consuming, generate toxic waste and use a huge amount of organic solvent to reach an efficient yield of recovery. Microextractions, however, and their use of smaller volumes, efficient and simple protocols, ability to adapt to automated systems, relative portability and inexpensiveness are presently justifying their classification as environment-friendly. The different and predominant modifications and innovations developed in this area within the past decade are Dispersive liquid-liquid microextraction (DLLME), Single-Drop microextraction (SDME), fiber liquid-phase microextraction (LPME) and Multistate stir bar sorptive extraction (MSBSE). Typically, these micro-extractions are assisted by an auxiliary technique, such as vortex (VA), ultrasound (UA), up-and-down shaker (UDSA) and magnetic string (MSA) (Chormey and Bakirdere 2018, Farajzadeh et al. 2018, Samanidou 2017).

Piri-Moghadam et al. (2018) quantified enrofloxacin and ciprofloxacin residues using DLLME in five different kinds of chicken tissues. Studies were developed to optimize the use of graphene oxide/polyaniline as a fiber membrane in the LPME method to quantify and preconcentrate Ivermectin, and a multi-class pesticide analysis in mango using the method SDME was also undertaken. MSBSE was applied to the extraction of odor compounds in aqueous samples with fifteen compounds were extracted in total (Ochiai et al. 2013, Pano-Farias et al. 2017, Rezazadeh et al. 2018).

As discussed, the micro-extractions and multi-compound extractions, like QuEChERS, are consistent with the proposed GAC. They are quick with reduced organic solvent volume, multi-extraction methods and cheap and reliable. Therefore, they should be the standard extraction and cleanup steps for the quantification of low concentration analytes like agrochemical residues (Chormey and Bakirdere 2018).

Regarding the chromatography method, in the last decade use of LC techniques, such as UPLC, the HPLC has instead become more common, allowing for reduced mobile phase flow rates during the separation process. Consequently, there was a reduction in the amount of toxic waste and the use of organic solvents without losses in resolution quality. Besides this, these eco-friendly methods were also developed in order to decrease the run time and solvent usage. For example, in a study that aimed to determine a protocol for carcinogenic arylamine, the authors achieved a reduction in the gradient analysis time from 40 to 6 minutes (Suresh et al. 2016). The tendency to replace organic solvents used in the mobile phase for those considered green, for example, has also been implemented in the simultaneous determination of three benzodiazepines using a green micellar mobile phase (non-ionic surfactant) (Elmansi and Belal 2019).

The agrochemical residues are a miscellaneous class of compounds that can be detected by LC techniques. Using different types of detectors, some are visible in UV-Vis with a DAD and some are detected with an FLD-MS. The analysis times and the equipment required to determine the presence and/or concentration of these compounds in food matrices are doubled with these techniques, making the efforts for GAC principles in sample preparation useless. So, in order to change this scenario, other strategies besides the reduction of flow rates and the use of green solvents are also being investigated to reduce the delay time between the two modes of detection, these include chemometric tools (Carabajal et al. 2016).

In summary, the agrochemical residues are compounds with distinct characteristics and are frequently found in low concentrations; because of this, separation techniques, such as LC, are commonly used. In this context, sample preparation is a critical step requiring organic solvent use and consequently generates increased toxic waste. To change, several extraction techniques have been developed to reduce solvent use and toxic waste, while chemometric tools help to improve detection.

4. Multivariate Curve Resolution

Multivariate curve resolution and alternative least square (MCR-ALS) have become popular chemometric tools in the last few years. This method can be summarized as the bilinear decomposition that assumes additivity of the recorded signal as expressed analogically Lambert-Beer's law, presented in Equation 1:

$$D = CS^T + E \tag{1}$$

Considering chromatographic data obtained in agrochemical residues analyses, D is a matrix of size $j \times k$, j is the elution times, k is the wavelengths, C is a matrix with the concentration values, S^T is a matrix of size $n \times k$ where n is the number of factors determined for the MCR-ALS model and E is the matrix that contains the data, which is not modeled by MCR-ALS. The values of C or S^T can be evaluated by SIMPLISMA (Self Modeling Mixture Analysis) method or Evolving Factor Analysis (EFA) (Jaumot et al. 2005, Pinto et al. 2016).

On the other hand, if distinct data sets have been analyzed by the same method, the possible data arrangement and bilinear model extension can be described in Equation 2 or Equation 3:

$$D_{aug} = C_{aug} S^T + E_{aug} \qquad (2)$$

$$[D_{cal1} \ D_{cal2} \ D_{cal3} \ldots D_{test}] = [C_{cal1} \ C_{cal2} \ C_{cal3} \ldots C_{test}] S^T + [E_{cal1} \ E_{cal2} \ E_{cal3} \ldots E_{test}] \quad (3)$$

where D_{aug} is an augment matrix produced from I individual matrices corresponding to calibration set and test set. C_{aug} is the matrix augmented by the concentration profiles, the S^T is the matrix that corresponds to loadings in the vector space and E_{aug} is the collected residuals (Boeris et al. 2014).

MCR-ALS can be also coupled to other chemical analyses such as electrochemical and vibrational, such as Raman, used to develop a method of detection for the pesticide malathion on fruit peels. MCR helped to recover the spectral data of the pesticide spectrum, allowing for the detection of malathion in concentrations below the maximum residue limit (Albuquerque and Poppi 2015).

Rapid quantitative analysis of three pesticides in cherry tomatoes and red grape samples was developed using HPLC-DAD coupled with the MCR technique; the application of the MCR technique provided robust and reliable multi-resolution models (Lu et al. 2019).

The MCR-ALS, in this case, a greener chemometric cleanup was used as a replacement method to QuEChERS coupled with HPLC-DAD for quantification of pesticide residues in vegetables. Quantification of seven pesticide residues was achieved (Carbendazin, Thiabendazole, Fuberidazole, Carbofuran, Carbaryl, 1-naphthol and Flutriafol) with good performance, reliable quantification of co-eluted compounds and avoided analyte loss during the cleanup step (Sousa et al. 2017).

Five pesticides (propoxur, carbaryl, carbendazim, thiabendazole and fuberidazole) were determined in tomato, Orange juice, grapefruit juice, lemon and tangerine using HPLC-DAD coupled to MCR-ALS; the use of this chemometric tool allowed reduction of the chromatography run to less than 10 minutes, saving time, money and making the method greener (Boeris et al. 2014).

HPLC-excitation-emission fluorescence was used to simultaneously determine the quantity benzo[*a*]anthracene, chrysene, bezo[*b*]fluoranthene and benzo[*a*]pyrene in leaves from different types of tea. The samples were extracted with ethyl acetate accessorized by an ultrasonic bath. In order to establish reliable results, elution time and excitation wavelength values were processed using MCR-ALS, which determined the limit of detection values in the range 1.0–1.4 ng mL^{-1} (Carabajal et al. 2018).

Therefore, the MCR-ALS chemometric tool can be applied, coupled with various methods of analysis (electrochemical, chromatography and spectroscopic); when considering agrochemical residues, the most frequently chosen is LC. However, most common detectors DAD and FLU are not selective enough for complex matrices, such as food and food products, which require long chromatography runs, high organic solvent usage and toxic waste production. In this context, the mass detector is extremely efficient but also comes with an elevated purchase price. Thus, the MCR-ALS can be applied to possibly solve such problems as co-eluted compounds and the demand for green analytical methods (Achir et al. 2019, Carabajal et al. 2018).

References

Achir, N., Sinela, A., Mertz, C., Fulcrand, H. and Dornier, M. 2019. Monitoring anthocyanin degradation in Hibiscus sabdariffa extracts with multi-curve resolution on spectral measurement during storage. Food Chemistry 271(July 2018): 536–542. https://doi.org/10.1016/j.foodchem.2018.07.209.

Agnieszka Gałuszka, A., Migaszewski, Z. and Namieśnik, J. 2013. The 12 principles of green analytical chemistry and the SIGNIFICANCE mnemonic of green analytical practices. TrAC Trends in Analytical Chemistry 50: 78–84.

Albuquerque, C.D.L. and Poppi, R.J. 2015. Detection of malathion in food peels by surface-enhanced Raman imaging spectroscopy and multivariate curve resolution. Analytica Chimica Acta 879: 24–33. https://doi.org/10.1016/j.aca.2015.04.019.

Alexandre Nougadère, Véronique Sirot, Ali Kadar, Antony Fastier, Eric Truchot, Claude Vergnet, Frédéric Hommet, Joëlle Baylé, Philippe Gros and Jean-Charles Leblanc. 2012. Total diet study on pesticide residues in France: Levels in food as consumed and chronic dietary risk to consumers. Environment International. 45: 135–150. ISSN 0160-4120. https://doi.org/10.1016/j.envint.2012.02.001.

Anastassiades, M., Lehotay, S., Stajnbaher, D. and Schenck, F.J. 2003. Fast and easy multiresidue method employing acetonitrile extraction/partitioning and "dispersive solid-phase extraction" for the determination of pesticide residues in produce. Journal of AOAC International 86(2): 412–431.

Armenta, S., Garrigues, S. and Guardia, M. 2008. Green analytical chemistry. TrAC Trends in Analytical Chemistry 27(6): June 2008, 497–511.

Attaullah, M., Yousuf, M.J., Shaukat, S., Anjum, S.I., Ansari, M.J., Buneri, I.D., Tahir, M., Amin, M., Ahmad, N. and UllahKhan, S. 2018. Serum organochlorine pesticides residues and risk of cancer: A case-control study. Saudi Journal of Biological Sciences 25(7): 1284–1290.

Boeris, V., Arancibia, J.A. and Olivieri, A.C. 2014. Determination of five pesticides in juice, fruit and vegetable samples by means of liquid chromatography combined with multivariate curve resolution. Analytica Chimica Acta 814: 23–30. https://doi.org/10.1016/j.aca.2014.01.034.

Braun, G., Braun, M., Kruse, J., Amelung, W., Renaud, F.G., Khoi, C.M. and Sebesvari, Z. 2019. Pesticides and antibiotics in permanent rice, alternating rice-shrimp and permanent shrimp systems of the coastal Mekong Delta, Vietnam. Environment International 127(February): 442–451. https://doi.org/10.1016/j.envint.2019.03.038.

Carabajal, M.D., Arancibia, J.A. and Escandar, G.M. 2016. A green-analytical chemistry method for agrochemical-residue analysis in vegetables. Microchemical Journal 128: 34–41. https://doi.org/10.1016/j.microc.2016.03.006.

Carabajal, M., Arancibia, J.A. and Escandar, G.M. 2016. A green-analytical chemistry method for agrochemical-residue analysis in vegetables. Microchemical Journal 128: 34–41.

Carabajal, M.D., Arancibia, J.A. and Escandar, G.M. 2018. Multivariate curve resolution strategy for non-quadrilinear type 4 third-order/four way liquid chromatography–excitation-emission fluorescence matrix data. Talanta 189(July): 509–516. https://doi.org/10.1016/j.talanta.2018.07.017.

Chormey, D.S. and Bakirdere, S. 2018. Principles and recent advancements in microextraction techniques. Comprehensive Analytical Chemistry 81: 257–294. https://doi.org/10.1016/bs.coac.2018.03.011.

Dong, M., Song, S., Han, L., Yao, W., Hao, X. and Zhang, Z. 2018. Evaluation of cleanup procedures in pesticide multi-residue analysis with QuEChERS in cinnamon bark. Food Chemistry 276(April 2018): 140–146. https://doi.org/10.1016/j.foodchem.2018.10.019.

EFSA (European Food Safety Authority), 2016. The 2014 European Union report on pesticide residues in food. EFSA Journal 14(10): 4611, 139 pp. doi:10.2903/j.efsa.2016.4611.

Elahi, E., Weijun, C., Zhang, H. and Nazeer, M. 2019. Agricultural intensification and damages to human health in relation to agrochemicals: Application of artificial intelligence. Land Use Policy 83: 61–474.

Elmansi, H. and Belal, F. 2019. Development of an Eco-friendly HPLC method for the simultaneous determination of three benzodiazepines using green mobile phase. Microchemical Journal 145(October 2018): 330–336. https://doi.org/10.1016/j.microc.2018.10.059.

Ewa Szpyrka, Anna Kurdziel, Aneta Matyaszek, Magdalena Podbielska, Julian Rupar and Magdalena Słowik-Borowiec. 2015. Evaluation of pesticide residues in fruits and vegetables from the region of south-eastern Poland. Food Control. 48: 137–142.

FAO/WHO Codex Alimentarius Commission. Committee on Pesticide Residues. 1997. Guidelines for predicting dietary intake of pesticide residues/prepared by the Global Environment Monitoring System - Food Contamination Monitoring and Assessment Programme (GEMS/Food); in collaboration with the Codex Commitee on Pesticide Residues, Rev. ed. World Health Organization.

FAO and WHO. 2019. Pesticide residues in food 2018 - Report 2018 - Joint FAO/WHO Meeting on Pesticide Residues. FAO Plant Production and Protection Paper no. 234. Rome. 668 pp.

Farajzadeh, M.A., Sadeghi Alavian, A. and Sattari Dabbagh, M. 2018. Application of vortex-assisted liquid-liquid microextraction based on solidification of floating organic droplets for determination of some pesticides in fruit juice samples. Analytical Methods 10(48): 5842–5850. https://doi.org/10.1039/c8ay01766b.

Gałuszka, A., Migaszewski, Z. and Namieśnik, J. 2013. The 12 principles of green analytical chemistry and the SIGNIFICANCE mnemonic of green analytical practices. TrAC - Trends in Analytical Chemistry 50: 78–84. https://doi.org/10.1016/j.trac.2013.04.010.

Huang, Y., Shi, T., Luo, X., Xiong, H., Min, F., Chen, Y. and Xie, M. 2019. Determination of multi-pesticide residues in green tea with a modified QuEChERS protocol coupled to HPLC-MS/MS. Food Chemistry 275(August 2018): 255–264. https://doi.org/10.1016/j.foodchem.2018.09.094.

Jaumot, J., Gargallo, R., De Juan, A. and Tauler, R. 2005. A graphical user-friendly interface for MCR-ALS: A new tool for multivariate curve resolution in MATLAB. Chemometrics and Intelligent Laboratory Systems 76(1): 101–110. https://doi.org/10.1016/j.chemolab.2004.12.007.

Lomba, L., Zuriaga, E. and Giner, B. 2019. Solvents derived from biomass and their potential as green solvents. Current Opinion in Green and Sustainable Chemistry 18: 51–56. https://doi.org/10.1016/j.cogsc.2018.12.008.

Lu, S.H., Li, S.S., Yin, B., Mi, J.Y. and Zhai, H.L. 2019. The rapid quantitative analysis of three pesticides in cherry tomatoes and red grape samples with Tchebichef image moments. Food Chemistry 290(March): 72–78. https://doi.org/10.1016/j.foodchem.2019.03.118.

Martin, F.L., Martinez, E.Z., Stopper, H., Garcia, S.B., Uyemura, S.A. and Kannen, V. 2018. Increased exposure to pesticides and colon cancer: Early evidence in Brazil. Chemosphere 209: 623–631.

Nantia, E.A., Moreno-González, D., Manfo, F.P., Gámiz-Gracia, L. and García-Campaña, A.M. 2017. QuEChERS-based method for the determination of carbamate residues in aromatic herbs by UHPLC-MS/MS. Food Chemistry 216(2017): 334–341. doi: 10.1016/j.foodchem.2016.08.038.

Nicolopoulou-Stamati, P., Maipas, S., Kotampasi, C., Stamatis, P. and Hens, L. 2016. Chemical pesticides and human health: the urgent need for a new concept in agriculture. Frontiers in Public Health. 2016.

Nougadère, A., Sirot, V., Kadar, A., Fastier, A., Truchot, E., Vergnet, C., Hommet, F., Baylé, J., Gros, P. and Leblanc, J.C. 2012. Total diet study on pesticide residues in France: levels in food as consumed and chronic dietary risk to consumers. Environ Int. 2012 Sep 15(45): 135–50. doi: 10.1016/j.envint.2012.02.001. Epub 2012 May 15. PMID: 22595191.

Nowak, P.M., Kłodzińska, E., Namieśnik, J., Woźniakiewicz, M., Płotka-Wasylka, J. and Woźniakiewicz, A. 2018. CE-MS and GC-MS as "green" and complementary methods for the analysis of biogenic amines in wine. Food Analytical Methods 11(9): 2614–2627. https://doi.org/10.1007/s12161-018-1219-9.

Ochiai, N., Sasamoto, K., Ieda, T., David, F. and Sandra, P. 2013. Multi-stir bar sorptive extraction for analysis of odor compounds in aqueous samples. Journal of Chromatography A 1315: 70–79. https://doi.org/10.1016/j.chroma.2013.09.070.

Omeroglu, P., Boyacioglu, D., Ambrus, A., Karaali, A. and Saner, S. 2012. An overview on steps of pesticide residue analysis and contribution of the individual steps to the measurement uncertainty. Food Analytical Methods 5(6): 1469–1480.

ONU: Consejo de Derechos Humanos, informe de la Relatora Especial sobre el derecho a la alimentación, acerca de su misión al Paraguay del 4 al 10 de noviembre de 2016, 27 Enero 2017, A/HRC/34/48/Add.2, disponible en esta dirección: https://www.refworld.org.es/docid/58ad9d284.html [Accesado el 27 Mayo 2019].

Pacheco-Fernández, I. and Pino, V. 2019. Green solvents in analytical chemistry. Current Opinion in Green and Sustainable Chemistry 18: 42–50. https://doi.org/10.1016/j.cogsc.2018.12.010.

Pang, G.F., Fan, C.L., Cao, Y.Z., Yan, F., Li, Y., Kang, J., Chen, H. and Chang, Q.Y. 2015. High throughput analytical techniques for the determination and confirmation of residues of 653 multiclass pesticides and chemical pollutants in tea by GC/MS, GC/MS/MS, and LC/MS/MS: Collaborative study, first action 2014.09. Journal of AOAC International 98(5): 1428–1454.

Pano-Farias, N.S., Ceballos-Magaña, S.G., Muñiz-Valencia, R., Jurado, J.M., Alcázar, Á. and Aguayo-Villarreal, I.A. 2017. Direct immersion single drop micro-extraction method for multi-class pesticides analysis in mango using GC–MS. Food Chemistry 237: 30–38. https://doi.org/10.1016/j.foodchem.2017.05.030.

Pinto, L., Díaz Nieto, C.H., Zón, M.A., Fernández, H. and de Araujo, M.C.U. 2016. Handling time misalignment and rank deficiency in liquid chromatography by multivariate curve resolution: Quantitation of five biogenic amines in fish. Analytica Chimica Acta 902: 59–69. https://doi.org/10.1016/j.aca.2015.10.043.

Piri-Moghadam, H., Gionfriddo, E., Grandy, J.J., Alam, M.N. and Pawliszyn, J. 2018. Development and validation of eco-friendly strategies based on thin film microextraction for water analysis. Journal of Chromatography A 1579: 20–30. https://doi.org/10.1016/j.chroma.2018.10.026.

Prasad, K. and Sharma, M. 2019. Green solvents for the dissolution and processing of biopolymers. Current Opinion in Green and Sustainable Chemistry 18: 72–78. https://doi.org/10.1016/j.cogsc.2019.02.005.

Rajput, S., Kumari, A., Arora, S. and Kaur, R. 2018. Multi-residue pesticides analysis in water samples using reverse phase high performance liquid chromatography (RP-HPLC). MethodsX 5: 744–751.

Rezazadeh, T., Dalali, N. and Sehati, N. 2018. Investigation of adsorption performance of graphene oxide/polyaniline reinforced hollow fiber membrane for preconcentration of Ivermectin in some environmental samples. Spectrochimica Acta - Part A: Molecular and Biomolecular Spectroscopy 204: 409–415. https://doi.org/10.1016/j.saa.2018.06.040.

Sabarwal, Akash, Kumar, Kunal and Singh, Rana P. 2018. Hazardous effects of chemical pesticides on human health–Cancer and other associated disorders. Environmental Toxicology and Pharmacology 63: October 2018, 103–114.

Samanidou, V. 2017. Trends in microextraction techniques for sample preparation. Separations 5(1): 1. https://doi.org/10.3390/separations5010001.

Santana, A.P.R., Andrade, D.F., Mora-Vargas, J.A., Amaral, C.D.B., Oliveira, A. and Gonzalez, M.H. 2019. Natural deep eutectic solvents for sample preparation prior to elemental analysis by plasma-based techniques. Talanta 199(January): 361–369. https://doi.org/10.1016/j.talanta.2019.02.083.

Schuur, B., Brouwer, T., Smink, D. and Sprakel, L.M.J. 2019. Green solvents for sustainable separation processes. Current Opinion in Green and Sustainable Chemistry 18: 57–65. https://doi.org/10.1016/j.cogsc.2018.12.009.

Sousa, E.S., Pinto, L. and de Araujo, M.C.U. 2017. A chemometric cleanup using multivariate curve resolution in liquid chromatography: Quantification of pesticide residues in vegetables. Microchemical Journal 134: 131–139. https://doi.org/10.1016/j.microc.2017.05.017.

Suresh, S., Ganeshjeevan, R., Priya, N. and Muralidharan, C. 2016. Eco-friendly, rapid and efficient analytical procedure for carcinogenic aryl amines in dyes and consumer products. Journal of the American Leather Chemists Association 111(1): 17–23.

Szpyrka, E., Kurdziel, A., Rupar, J. and Słowik-Borowiec, M. 2015. Pesticide residues in fruit and vegetable crops from the central and eastern region of Poland. Rocz Panstw Zakl Hig 66(2): 107–113.

Tomás M. Mac Loughlin, Ma. LeticiaPeluso, Ma. AgustinaEtchegoyen, Lucas L. Alonso, Ma. Cecilia de Castro, Ma. CeciliaPercudani and Damián J.G. Marino. 2018. Pesticide residues in fruits and vegetables of the argentine domestic market: Occurrence and quality. Food Control 93: 129–138, November 2018.

United States Department of Agriculture (USDA). 2017. Pesticide Data Program, Annual Summary, Calendar Year 2017. p. 203. www.ams.usda.gov/pdp.

Ventura, C., Zappia, C.D., Lasagna, M., Pavicic, W., Richard, S., Bolzan, A.D., Monczor, F., Núñe, M. and Cocca, C. 2019. Effects of the pesticide chlorpyrifos on breast cancer disease. Implication of epigenetic mechanisms. The Journal of Steroid Biochemistry and Molecular Biology 186: 96–104, February 2019.

Vian, M., Breil, C., Vernes, L., Chaabani, E. and Chemat, F. 2017. Green solvents for sample preparation in analytical chemistry. Current Opinion in Green and Sustainable Chemistry 5: 44–48. https://doi.org/10.1016/j.cogsc.2017.03.010.

Werner, J., Grześkowiak, T., Zgoła-Grześkowiak, A. and Stanisz, E. 2018. Recent trends in microextraction techniques used in determination of arsenic species. TrAC - Trends in Analytical Chemistry 105: 121–136. https://doi.org/10.1016/j.trac.2018.05.006.

World Health Organization. Programme of Food Safety and Food Aid, Global Environment Monitoring System. Food Contamination and Monitoring Programme & Joint FAO/WHO Codex Alimentarius Commission. Committee on Pesticide Residues. (1997). Guidelines for predicting dietary intake of pesticide residues/prepared by the Global Environment Monitoring System - Food Contamination Monitoring and Assessment Programme (GEMS/Food); in collaboration with the Codex Commitee on Pesticide Residues, Rev. ed. World Health Organization. https://apps.who.int/iris/handle/10665/63787.

CHAPTER 13

Functional Food Ingredients Production Using Green Technologies

Kleopatra Tsatsaragkou,[1,]* *Paraskevi Paximada,*[2]
Styliani Protonotariou[3] and *Olga Kaltsa*[4]

1. Introduction

Consumer awareness and promotion of healthy eating and lifestyle led to the growth of the functional ingredients and foods market in recent years. Consumers demand not only healthy nutritious foods but also additional health-promoting functions, i.e., functional foods (Day et al. 2009). There is not a formal definition of functional foods; however, researchers define 'functional food' as foods or ingredients of foods that together with their basic nutritional impact have beneficial effects on one or more functions of the human organism, thus either improving the general and physical conditions or/and decreasing the risk of the evolution of diseases (Martirosyan and Singh 2015). The main categories of functional food ingredients are bioactive compounds, such as phenolics, antioxidants, vitamins, prebiotic and probiotic compounds and essential oils. The challenges for food scientists and researchers are associated with the processes to obtain functional ingredients so that they maintain not only their biological functionality but also their quality and sensory attributes. Traditionally, functional ingredients were recovered using conventional extraction techniques, such as hydrodistillation, steam distillation or

[1] Independent Researcher, 7 Polymnias street, Melissia, 15127, Athens Greece.
[2] The University of Leeds, Woodhouse, Leeds LS2 9JT, UK.
[3] Agricultural University of Athens, Greece.
[4] Technological Institute of Thessaly, Greece.
Emails: paximadapar@gmail.com; s.protonotariou@aua.gr; olgakalt@teilar.gr
* Corresponding author: kleotsat@gmail.com

Soxhlet extraction. The main drawbacks of the traditional extraction techniques are the chemical modifications of the functional ingredients due to the high temperatures involved, and the formation of unwanted by-products due to thermal degradation and hydrolysis as well as the presence of organic solvents which are often toxic (Rezzoug et al. 2005, Mejri et al. 2018). Emerging or novel extraction methods often called 'Green Technologies' that have been recently developed are Microwave-Assisted Extraction (MAE), Ultrasounds-Assisted Extraction (USAE), Pulsed Electric Fields (PEF), High voltage electrical discharges, Pulsed Ohmic Heating, High-Pressure Extraction, Pressurized Liquid Extraction, Sub- and Supercritical Fluid Extraction, Accelerated Solvent Extraction, extraction assisted by hydrotropic solvents and Deep Eutectic Solvent Extraction (DES). The main advantages of green technologies are the relatively low temperatures involved, the reduced equipment size, ease of use, speed, and low solvent consumption, all of which contribute to reducing environmental impact and cost (Figure 1) (Périno-Issartier et al. 2013). This chapter discusses the production of functional food ingredients using green technologies. The conventional and green production technologies for each functional ingredient category will be described and the characteristics of the ingredients produced using green technologies will be compared to those produced traditionally.

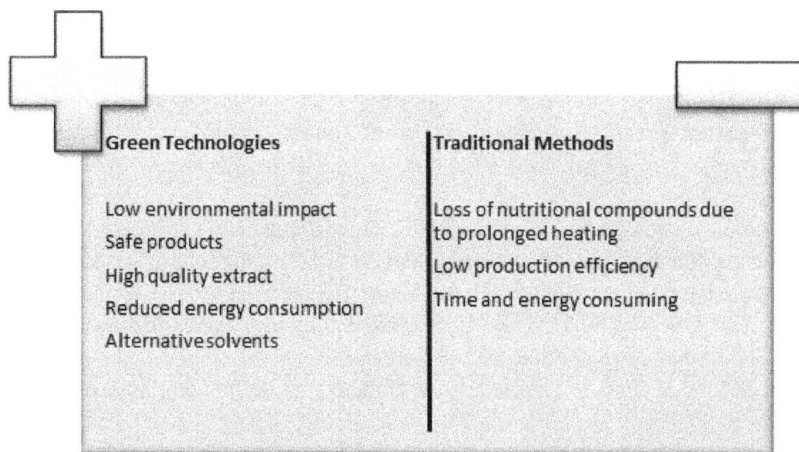

Green Technologies	Traditional Methods
Low environmental impact	Loss of nutritional compounds due to prolonged heating
Safe products	
High quality extract	Low production efficiency
Reduced energy consumption	Time and energy consuming
Alternative solvents	

Figure 1. The advantages of green technologies over the traditional extraction methods.

2. Production of Bioactive Compounds

Bioactive compounds are mainly plant-derived ingredients exerting significant metabolic modulations throughout the human body tissues, despite occurring in low amounts in foods. The whole health-promoting effect is associated with a number of distinct activities, such as gene expression, enzyme regulating, anti-inflammatory and antioxidant capacity (Kris-Etherton et al. 2002); the latter being the most extensively studied. Apart from known antioxidants like vitamins (E and C), b-carotene or selenium, phenols along with their subcategories of phenolic acids (e.g., gallic acid), flavonoids (i.e., quercetin), tannins and stilbenes (i.e., resveratrol) are also of great importance as these secondary metabolites represent up to 0.5–5 g/100 g dry weight of plant tissue. Phenolic compounds are considered a significant class of dietary antioxidants with preventive effects against diseases, such as cardiovascular diseases, cancers, diabetes mellitus, antiageing, etc. (Scalbert et al. 2005, Pandley and Rizvi 2009).

2.1 Microwave-Assisted Extraction (MAE)

Microwave-assisted extraction (MAE) is considered a popular method for the recovery of various active compounds from food materials since it offers reduced extraction time, environmental friendliness, low cost and enables automation or online coupling to other analytical procedures.

Compared to conventional extraction methods, MAE requires less solvent amounts (Proestos and Komaitis 2008, Dahmoune 2015), while the total treatment time may be shortened by a factor of 60 (Kaderides et al. 2019). However, the temperature rises above 100°C during processing may favour the degradation of a series of phenolic compounds like epicatechin, resveratrol and myricetin (Laizid et al. 2007). MAE has been reported to exhibit higher polyphenol extraction yields in a number of comparative studies involving conventional and green extraction technologies. Dahmoune et al. (2014, 2015) and Nayak (2015), who compared the optimized MAE with Ultrasound-Assisted Extraction (USAE) of similar energy consumption as well as conventional solvent extraction, showed that MAE demonstrates higher total phenolic capacity, total flavonoids and tannins from various substrates, such as orange peels, Myrtus and pistachio leaves. Quercetin and rutin recovery from Euonymous allatus stalks was also enhanced with MAE (Yang and Zhang 2008).

Synergies among different technologies can also positively affect the recovery of antioxidants. For example, simultaneous MAE-USAE methodology was successfully implemented for the extraction of resveratrol (Chen et al. 2016), whereas USAE pretreatment followed by MAE led to increased contents of phenolics from cabbage leaves due to extensive cellular destruction caused by the combined cavitation and thermal effect that allow the better release of the compounds (Pongmalai et al. 2015).

2.2 Ultrasound-Assisted Extraction (USAE)

Ultrasounds have been used broadly in the food industry as homogenizing devices (Paximada et al. 2016). Ultrasound-Assisted Extraction (USAE) has been investigated as a potential alternative to traditional extraction techniques. High-intensity sonication is normally selected for extraction and processing applications (frequencies higher than 200 Hz). Acoustic cavitation is the primary cause that drives the extraction outcome of sonication (Tiwari 2015). When ultrasound passes through any medium, it causes the molecules of the medium to compress, leading to the amendment of the pressure changes. In conclusion, the bubbles of the liquid are being collapsed.

USAE is an emerging 'green' extraction technique, which has been tested in many applications due to the fact that it does not exhibit the constraints of the traditional techniques, such as high energy consumption and high running costs. Mnayer et al. (2017) compared USAE with traditional technologies to extract green absolute from thyme. Sunflower oil with ethanol was used in a ratio of 1/10 (w/w) as solvents. The conventional method of extraction with hexane had been compared with USAE operating at 20 kHz frequency and 130 W power. In this study, it was observed that USAE increased in yield by 47% compared to the conventional one; 4.03 g of absolute per 100 g dry weight for the conventional method and 6 g per 100 g dry weight for the USAE. This result is in accordance with other studies (Samaram et al. 2014). The USAE using sunflower oil as solvent resulted in higher recovery of phenols and strongest antioxidant activity compared to the conventional method.

Grape marc and olive pomace are major waste from wine-making and olive oil production, respectively (Paini et al. 2016). They both possess a high concentration of polyphenols, whose extraction can be of utmost importance to the sustainability of the food industry and the reduction of environmental impact. In the study, they are targeting to compare ultrasounds and high pressure as emerging green methods of extracting these bioactive. As far as the grape marc is concerned, high hydrostatic pressure extraction (HHPE) resulted in higher extraction yield but for high ethanol contents (75–100% v/v). On the other hand, USAE was capable to improve the yield for low ethanol contents (25–50% v/v). Taking these into account, the authors proposed USAE as the best extraction technique since high ethanol contents might be unsafe for industrial use. As far as the olive pomace is concerned, it is suggested to use USAE as an extraction technique since high hydrostatic pressure extraction (HHPE) was found to have a negative effect on the stability of oleuropein and caffeic acid.

2.3 High Hydrostatic Pressure Extraction

High hydrostatic pressure extraction (HHPE) is known to be a 'green' method for extracting bioactive compounds from plants and foods. During HHPE foods or plants are exposed to hydrostatic pressure in the range of 100–1,000 Mpa. With the HHPE method, the disadvantages of the traditional extraction methods, such as the use of toxic chemicals, low bioaccessibility and extraction yields, and longer processes can be overcome (Briones-Labarca et al. 2019).

A large number of studies have been examined to test the use of this technology in order to extract various bioactive compounds from food, food waste, herbs and plants. In particular, Briones-Labarca et al. (2019) evaluated the effect of high pressure (250–450 MPa) and solvent mixture (40–60% hexane) on the properties of bioactive extracted from tomato pulp. The optimized conditions were found to be at 450 MPa pressure and a 60% solvent mixture. Under these conditions, the extraction yield was statistically higher (8.78%) compared to the conventional method that had been used as a reference. What is more, higher values of total polyphenols and flavonoids were achieved when using HHPE. Hence, HHPE has proven to be a useful green technology to extract health-promoting ingredients with enhanced bioaccessibility.

Microalgae *Parachlorella kessleri* is known for its high concentration not only in bioactive compounds, such as pigments and polyphenols, but also in triglycerides (TAG) and starch. These are energy-rich ingredients that are currently being extracted with conventional methods, such as Soxhlet. A recent study, however, is focusing on HHPE, USAE and the combination of these techniques (Zhang et al. 2019). After testing various processing conditions, it was concluded that the most efficient way to extract bioactive molecules from microalgae is first an ultrasonic extraction of concentrated solutions (10%) followed by HHPE of less concentrated solutions (1%). By applying that, the extraction yield of proteins and pigments was higher.

2.4 *Pressurized Liquid Extraction*

Pressurized liquid extraction (PLE) is an innovative green solution used to increase the extraction efficiency of bioactive from food and plants (Mustafa et al. 2011). The mechanism that PLE works include extraction using liquid solvents at high pressure and temperature. During PLE, the solvents are being handled at temperatures higher than their boiling point, which leads to improved solubility. The elevated temperature and pressure also lead to the reduction of surface tension and viscosity of extracting agents, causing an enhanced extraction yield.

Gilbert-Lopez et al. (2017) examined the effect of processing conditions (pressure, temperate, and type of solvent) on the extraction properties of bioactive from Phaeodactylum tricornutum. They concluded that by increasing the temperature up to 110°C or the % ethanol, the extraction yield increases as well. At the optimized processing conditions of 170°C, for 20 minutes, using 97% EtOH, the yield was 40%, carotenoids were 31 mg/g extract and the total phenols content was 32 gm gallic acid equivalents/g extract. The same study also compared PLE to microwave-assisted extraction, concluding that PLE achieved higher recoveries of bioactive (fucoxanthin) and lipids and higher extraction yield.

In another study, Santos et al. (2019), explored the extraction of bioactive compounds from feijoa peel extracts using low (Ultrasounds and Soxhlet) and high-pressure techniques (PLE and Supercritical CO_2). After comparing all the methods, they concluded that PLE and Soxhlet both exhibited statistically the same and the highest extraction yield; PLE showed a distinctive asset as the extraction time was much shorter in that case compared to that of Soxhlet. The main phenolics found were ferulic acid, gallic and ellagic acids. PLE was discovered to be the method that gave the highest total phenols content as well as the highest antioxidant activity compared to the other tested techniques and was being proposed as an alternative green extraction method of bioactives from feijoa peel.

2.5 *Pulsed Electric Fields (PEF)*

Pulsed electric field (PEF) is a non-thermal technology that involves the application of moderate-intensity (0.5–10 kV), low energy (1–10 kJ/kg) and short duration electric pulses (typically up to 1 ms) (Bobinaite et al. 2014). The exposure of cell membranes to electric fields is known to cause permeability modifications and/or membrane rupture, which is known as electroporation. Due to this phenomenon, the main applications of PEF in the food sector include product preservation (Qin et al. 1994), enzyme inactivation (Giner et al. 2005) and extraction of numerous constituents from plant cells, such as sugars, proteins, lipids and phenolic compounds (López et al. 2009, Zbinden et al. 2013, Roselló-Sotto et al. 2015).

El Darra et al. (2013), who compared USAE and PEF treatments for extraction of the phenolic compound from Cabernet wine, showed that moderate to high voltage PEF (0.8–5 kV/cm) improved the extraction yield from 51 to 61%, whereas ultrasounds only reached 7%. Ambiguous results with respect to chemical selectivity and extraction efficiency between the two technologies have also been reported in various studies (Medina-Meza et al. 2015).

Comparative studies on high-voltage electric discharge (HVED), PEF and USAE revealed that HVED was superior in time efficiency and energy consumption when used to extract phenolic compounds from olive kernels (Roselló-Sotto et al. 2015). The same trend has also been reported by the team of Barba et al. (2015a,c) who compared the same technologies for the recovery of phenolic compounds from fermented grape pomace and stevia leaves, although it was noted that PEF was more selective towards anthocyanin extraction compared to HVED. The latter remark was also confirmed in the following research on the recovery of phenolics/anthocyanins from blackberries (Barba et al. 2015b). HVED treatment has also been shown to cause greater cellular disintegration and turbulence generation compared to PEF and/ or USAE resulting in enhanced diffusion coefficients, which could explain the higher extraction rates (Rajha et al. 2014, Parniakov et al. 2014, Rahja et al. 2015).

2.6 Deep Eutectic Solvent (DES) Extraction

The term 'green extraction' not only refers to energy-efficient technologies but also encompasses the production of different constituents through the use of alternative solvents with minimum environmental impact and toxicity in contrast to petrochemical and volatile organic compounds. For a solvent to qualify as a 'green' one, along with high solvent power and low cost, several other mandatory criteria should be fulfilled, such as production from renewable resources, inflammable medium, biodegradability as well as the absence of toxicity. Deep eutectic solvents (DES) represent another promising class of eco-friendly media that reflects the vast majority of green extraction principles. DES is prepared as a mixture of naturally occurring metabolites. They are called NADES (natural deep eutectic solvents) and include a mixture of organic acids (lactic, citric, malic and tartaric), amino acids (choline, proline and glycine), sugars (fructose and glucose), polyalcohols (glycerol and sorbitol), urea, choline chloride, etc. (Faggian 2016, González et al. 2017). Apart from the type of compounds used, their extraction capacity can be influenced by several factors like the molar ratio of substances in the mixture, the chemical affinity between the DES and the target components, the water content and extraction conditions (temperature and time) (Bi 2013, Ruesgas-Ramon 2017). Bi et al. (2013) used response surface methodology (RSM) to optimize extraction conditions of flavonoids (myricetin and amentoflavone) with DES mixtures composed of alcohols and choline chloride. They concluded that the use of pure DES and DES-water can increase the solubility of flavonoids, and this eutectic mixture could be used for the extraction of bioactive compounds from plants.

Within the last few years/decade numerous studies related to the extraction of phenolic compounds from various substrates have been reported. DES has been successfully used for recovering phenolic compounds from oils (García et al. 2016). DES composed of choline chloride at various ratios with xylitol and propanediol exhibited a higher extraction capacity of oleacin and oleocanthal, two of the most abundant olive oil phenolic substances (García et al. 2016).

Hybridization of DES with non-conventional/green technologies can also be used to promote phenolic compound recovery from different plant sources. Bosiljkov et al. (2017) concluded that deep eutectic solvents coupled with ultrasounds (USAE) promote anthocyanin extraction from wine lees better than acidified ethanol. Jeong et al. (2015) have also reported that anthocyanin extraction yield from grape skin was double or higher under optimized DES-USAE extraction compared to conventional methanolic extraction. Similar results were obtained regarding the extraction yield of major flavonoid substances, such as quercetin and vitexin from buckwheat sprouts (Rois Mansur et al. 2019).

As a final remark, it is essential to mention that DES/NADES not only suggest a novel, green media to efficiently recover bioactive phenolic extracts but can also act as a tool to enhance the bioavailability and activity of some poorly water-soluble constituents *in vitro* and *in vivo*. For instance, formulations of resveratrol with propanediol-choline chloride DES lead to at least 10-fold matrix metalloprotease activity inhibition in comparison with dimethyl-sulfoxide (DMSO) (Radošević et al. 2016). Studies in mice also showed that orally administered rutin, quercetin and berberine DES solutions resulted in higher plasma concentration, prolonged permanence and/or enhanced absorption of the bioactive compounds compared to water suspensions (Faggian 2016).

3. Production of Prebiotics and Probiotics Using Green Technologies

The two main categories of prebiotics that this chapter will focus on are proteins and fibres, especially inulin. The properties of inulin and proteins produced using various green technologies will be analysed and the efficiency and feasibility of green technologies over the traditional methods will be discussed. For probiotics, green technologies mainly focus on increasing the viability of probiotic microorganisms after fermentation. Knorr (2003) reported that non-thermal pre-treatments such as high hydrostatic pressure, high-intensity electric field pulses (PEF) and supercritical carbon dioxide are able to improve the fermentation process and the quality of the probiotic product.

3.1 Inulin

Inulin is a non-digestible oligosaccharide consisting of 2–60 d-fructose units connected through $\beta(2\rightarrow1)$ glycosidic link¬ages with a terminal $\alpha(1\rightarrow2)$ bonded d-glucose (Niness 1999, Nair et al. 2010). Conventional inulin production involves liquid extraction, filtration, evaporation and drying (Mendes et al. 2005). It is a high-cost procedure that requires non-eco-friendly solvents, high energy consumption and a long extraction time. Therefore, there is a need of developing environmentally friendly methods of inulin recovery.

Recently, the attention of researchers has focused on ultrasound extraction, which is a green, effective and safe procedure. Compared to another alternative method, this one needs lower extraction temperature and cheaper operating costs (Zhu et al. 2016). According to Lingyun et al. (2007), the extraction rate of inulin from Jerusalem artichoke tubers was about 2 times faster when they used a direct ultrasound-assisted method compared to a conventional one. Same results were observed during inulin extraction from Burdock root (Milani et al. 2011). Similarly, Abbasi and Farzanmehr (2009) extracted inulin from Iranian artichokes using direct and indirect ultrasonic, concluding that inulin recovery increased significantly and proportionally to the power input of ultrasound. Pourfarzad et al. (2015) research indicated that indirect sonication treatment is a suitable method for fructan extraction from Eremurus spectabilis tubers. Li et al. (2018) also used the ultrasonic method to increase extraction yield. However, they argued that under high ultrasonic treatment, a decrease in inulin content could be caused because of tissue structure modification. On the other hand, Rubel et al. (2018) found that inulin rich carbohydrates extraction was optimal without ultrasound assistance from Jerusalem artichoke tubers, at 76°C, indicating that ultrasound is a powerful tool for extraction but has limits and needs optimization.

Lou et al. (2009) combined this technique with a microwave in order to achieve a faster extraction of burdock root. Microwave-assisted extraction (MAE) is a cheap and more effective method compared with other extraction techniques, resulting in high-quality products (Ganzler et al. 1986). Xiao et al. (2013), reported that inulin production was faster, giving a 20% higher yield when they used MAE than hot water extraction. Tewari et al. (2015) reached a high level of inulin extraction (63%), applying the same method to chicory roots.

Pulsed electric fields (PEF) are considered a powerful non-thermal process, which can assist extraction from plants as biological membranes lose their consistency because of electrical effects (Barba et al. 2014). Loginova et al. (2010) used this technique to extract inulin from chicory roots, notifying that at lower temperatures diffusion was higher for PEF-treated than for untreated slices. Moreover, PEF treatment gave extract with higher purity than the conventional method did (Zhu et al. 2012).

Supercritical fluid extraction (SFE) is a clean alternative method where the solvent can be recycled. Mendes et al. (2005) used carbon dioxide (CO_2) as a solvent for inulin extraction giving promising results. However, they propose deeper research on this field and optimization of the process.

Mendes et al. (2005) experimental results indicated that when the pressure is increased, the extracted mass increases at constant temperature and when the temperature increases at constant pressure, the extracted mass decreases due to the decrease of the solvent density.

3.2 Proteins

Characterization of protein as a functional ingredient is not only due to its nutritional value but also due to its structural properties (Foegeding 2015) as it has a notable effect on foam formation, gelation and emulsification.

PEF technology can be used offering many advantages over traditional thermal processes, retaining high food quality. Xiang et al. (2011) noticed that PEF treatment led to structure modifications in whey proteins. Similarly, Perez and Pilosof (2004) and Li et al. (2007) suggested that by using PEF, proteins with improved functionality could be produced.

Recently, the PEF technique has been applied to recover proteins from marine microalgae (Coustets et al. 2013, Barba et al. 2015). Grimi et al. (2014) obtained promising results using a combination of different techniques, i.e., PEF, ultrasounds (USAE), high voltage electrical discharges (HVED), and high-pressure extraction (HHPE) to recover protein from microalgae *Nannochloropsis* sp.; even though HHPE had the best results and had the highest energetic cost. Compared to all the other techniques (USAE and HVED), PEF had a significantly increased contribution. Parniakov et al. (2014) combined PEF with pressure to extract clear protein with high stability from mushrooms (*Agaricus bisporus*). Conventional methods gave a higher yield but extracts were cloudy with low stability.

Barsotti et al. (2001) used high voltage electric pulses at a repeat frequency of 1 Hz with no adverse effects on lactoglobulin and ovalbumin. Thus, proteins could be extracted at low temperatures with the application of green technologies, maintaining their functionality.

3.3 Probiotics

Probiotics are live health-promoting microorganisms. Probiotic food products contain mainly bacteria that belong to *Lactobacillus* and *Bifidobacterium* strains (Lin 2003). To benefit from their functionality, they have to be active, viable and at least at a concentration of 107 cfu/g (Vinderola et al. 2000). Industrial bacterial production requires high energy and time consumption since it is difficult to achieve high viability after fermentation (Ross et al. 2005). Thus, processes should ensure their viability and functionality.

High-pressure homogenization (HPH) is usually applied to increase milk self-life and safety (Lanciotti et al. 2004). Patrignani et al. (2007) evaluated the effect of HPH on viability loss of *Lactobacillus paracasei* BFE 5264, concluding that under optimization this technique could produce unique probiotic fermented milk. An increase in pressure level was positively correlated to viability.

Ultrasound application has been used to increase Lactobacillus and Bifidobacterium in mannitol-soymilk upon fermentation at 37°C for 24 hours (Yeo and Liong 2011). Although probiotics decreased immediately after treatment, this effect was transient as after fermentation, an increase in growth was observed. Similar results were observed by Yang et al. (2010), who promoted the increase of *Brevibacterium* sp. by combining ultrasound with fermentation. Ultrasonication also led to increased *Lactobacillus* and *Bifidobacterium* viability in milk (Wang et al. 1996). More recently, Guimarães et al. (2019) concluded that high-intensity ultrasound could increase probiotic viability in dairy products, optimizing the process parameters, i.e., duration, energy, frequency and pulse mode.

4. Production of Essential Oils Using Green Technologies

Essential oils are aromatic and volatile liquids, mixtures of organic compounds extracted from plant materials and characterized by a strong and generally pleasant flavor. The essential oils have been widely used as safe flavoring agents or preservatives in foods, in cosmetic or pharmaceutical products due to their antibacterial, antiviral, antifungal, antioxidant, antiparasitic, and insecticidal effects. They are concentrated and complex substances that have the form of oily drops present in one or more organs of the aromatic plant, i.e., in flowers (Jasmine), leaves (Sage), fruits (Orange), seeds (Fennel), bark (Cinnamon) and in roots (Angelica) (Mejri et al. 2018).

4.1 Microwave-Assisted Extraction Techniques (MAE)

Microwave-assisted extraction techniques include the application of microwave dielectric heating, compressed air microwave distillation, vacuum microwave hydrodistillation, microwave hydrodistillation, solvent-free microwave extraction, microwave-accelerated steam distillation, microwave steam distillation and microwave hydrodiffusion and gravity (MDG) (Farhat et al. 2010).

Essential oils from dried caraway seeds were extracted with the use of microwave dry-diffusion and gravity (MDG) (Farhat et al. 2010). The cost of the essential oils produced with this method was clearly advantageous in terms of energy and time compared to the conventional processes. The hydrodistillation method required an extraction time of 300 minutes, while the MDG method required only 45 minutes. The energy consumed was 163 $kWhkg^{-1}$ essential oil for hydrodistillation and 102 $kWhkg^{-1}$ essential oil for MDG. Regarding the environmental impact, the calculated quantity of carbon dioxide rejected in the atmosphere was higher in the case of hydrodistillation (130 kg CO_2 kg^{-1} of essential oil) than for MDG (82 $kgCO_2$ kg^{-1} of essential oil).

Cardoso-Ugarte et al. (2013) studied the extraction of essential oils from basil (*Ocimum basilicum* L.) and epazote (*Chenopodium ambrosioides* L.) using the microwave-assisted extraction and compared the physical and chemical properties of the essential oils, obtained by microwave-assisted extraction to those obtained by steam distillation. Furthermore, the authors determined the influence of three main factors (amount of water, power and heating time) in the microwave-assisted extraction process of essential oils. More specifically, the amount of solvent (water) and heating time affected the production yield, while the power did not affect the yield. According to the authors, the conditions to optimize the essential oil yield were found to be 30 minutes of heating time and 400 mL of water. The yield of basil essential oil obtained by microwave-assisted extraction was found not to be significantly different to the yield of the oil obtained by steam distillation. In terms of chemical composition, in the case of basil essential oil, the method (microwave-assisted extraction and steam distillation) did not affect its chemical composition and a greater number of compounds were detected in the oil obtained by microwave-assisted extraction, proving the advantage of using microwave extraction for this specific herb. On the other hand, in the essential oil of epazote, two common compounds were identified from both extraction methods; limonene oxide cis and (+)-4-carene, although they were found in different abundance.

The microwave-assisted extraction process was also applied for the production of cardamom essential oil (*Elleteria cardamomum* L.) by Lucchesi et al. (2007). The impact of the three main factors (the moisture of the matrix, the extraction time, and the power) on the yield and chemical properties of cardamom essential oil was analysed and compared to the oil produced by hydro-distillation. In this study, the extraction time is the major factor affecting the yield of the microwave-assisted extraction process. As time increases, the yield increases almost linearly. The power and humidity affect also the yield in the same way as the extraction time, contrary to the study of Cardoso-Ugarte et al. (2013), that the power did not affect the yield of basil essential oil. The optimal microwave extraction conditions of cardamom seeds were 75 min extraction time, 390 W power and 67% humidity level. In terms of chemical composition, higher amounts of oxygenated compounds and lower amounts of monoterpene hydrocarbons were present in the essential oil of cardamom extracted by microwave-assisted extraction in comparison with hydro-distillation. Oxygenated compounds are highly odoriferous and hence are valuable in terms of their contribution to the fragrance of the essential oil. The microwave-assisted extraction could be considered a viable process compared to the traditional hydrodistillation for producing high-quality cardamom essential oil. Similar remarks were found in the study of Chemat et al. (2006), which investigated the application of microwave-accelerated steam distillation for producing lavender (*Lavandula angustifolia* Mill./ Lamiaceae) essential oil and compared the results with conventional steam distillation for the extraction of lavender essential oil. In terms of extraction time, 10 minutes with microwave-accelerated steam distillation provided a yield (8.86%) comparable to that obtained after 90 minutes (yield = 8.75%) by means of steam distillation, suggesting a substantial saving of time, energy and plant material. The energy required to perform the two extraction methods were, respectively, 1.5 kWh for steam distillation, and 0.13 kWh for microwave-accelerated steam distillation. In addition, the microwave-accelerated steam distillation produced lavender essential oil with a similar composition to that produced by steam distillation. Linalool was found to be the main oxygenated component in the essential oil extracted from lavender with equivalent relative amounts for both extraction methods (Chemat et al. 2006). The successful application of microwave-assisted distillation for the isolation of essential oil from citrus fruits was also proven in the study by Ferhat et al. (2007). The authors compared traditional hydrodistillation, cold pressing and microwave-assisted distillation and concluded about the important advantages the microwave method offers over traditional alternatives, such as shorter extraction times (30 minutes versus 3 hours for hydrodistillation and 1 hour for cold pressing), better yields (0.24% versus 0.21% for hydrodistillation and 0.05% for cold pressing), environmental impact (energy cost is appreciably higher for performing hydrodistillation and cold pressing than that required for microwave-assisted distillation) and a more valuable essential oil with high amounts of oxygenated compounds. More importantly, the microwave procedure yields essential oils that can be used directly without any clean-up, solvent exchange or centrifugation steps. The advantages of microwave-assisted extraction in terms of energy and time saving over traditional hydrodistillation were also reported in the recent study by Moradi et al. (2018) for the production of rosemary essential oil (*Rosmarinus officinalis* L.). Moradi et al. (2018) reported that the time needed for the complete extraction of rosemary essential oil was reduced by about 67% by microwave-assisted extraction, namely 30 minutes compared to the 90 minutes needed. The energy requirements to perform the extraction were 1,500 W for hydro-distillation and 180 W for microwave-assisted extraction and the quantity of carbon dioxide released into the atmosphere is higher in the case of hydrodistillation (1,200 g CO_2) than for microwave-assisted extraction (144 g CO_2). These data indicate a substantial saving in the extraction cost when using microwave-assisted extraction instead of hydro-distillation. Furthermore, the quality of rosemary essential oil was improved in the microwave-assisted extraction method due to an increase of 17% in oxygenated compounds, which was also reported in the above study by Lucchesi et al. (2007) for the production of cardamom essential oil.

In general, microwave-assisted extraction techniques are a good alternative to traditional techniques for the production of essential oils since they preserve the quality of the essential oil and reduce the cost of the process by using less energy, time and no solvents.

4.2 Subcritical and Supercritical Fluid Extraction

Grosso et al. (2008) tested supercritical CO_2 extraction for coriander oil extraction and reported that pressure and temperature control in both the yield and the composition of the coriander oil. As the pressure rises (at the same temperature), the solvent solubilizes non-volatile components that contaminate the volatile oil. However, working at pressures between 90 and 100 bar, the composition of the volatile oil is closed to that obtained with hydrodistillation.

Pavlić et al. (2015) compared the qualitative and quantitative properties of essential oil and lipid extracts from coriander seeds produced using the traditional extraction techniques, supercritical fluid extraction and subcritical water extraction. The highest yield was obtained by Soxhlet extraction with methylene chloride (14.45%), while the lowest yield was obtained using subcritical water extraction at 100°C, 30 bar for 20 minutes.

In terms of aromatic compounds, essential oil from hydrodistillation shows the highest content of linalool (835.2 mg/g essential oil), essential oils obtained by subcritical water extraction at 100°C and 150°C had particularly high contents of linalool (477.8 mg/g essential oil and 450.2 mg/g essential oil, respectively) (Pavlić et al. 2015).

Supercritical fluid extraction (SFE) of essential oil from ground black pepper (Piper nigrum), using CO_2 as a solvent, showed a significant increase in extraction rate with the increase of pressure or decrease of temperature (Perakis et al. 2005). The effect of process parameters, namely pressure (7.5, 10, and 15 MPa), temperature (30, 40, and 50°C) and particle size (0.5 mm, 0.75 mm, and whole berries), on the extraction rate, was examined. The essential oil obtained from supercritical CO_2 extracts contained higher levels of sequiterpene hydrocarbons, leading to higher sesquiterpene to monoterpene ratios as compared to that obtained from hydrodistillation. The results showed an increase in extraction rate with the increase of pressure or temperature.

In contrast, the increase in particle size reduced the extract yield and extraction rate (Kumoro et al. 2010). In this work, the smaller particle size generated a higher yield and grinding was found to liberate more pepper oil by destroying the inner structures of the particles. An increasing yield of peppermint oil versus size implies that cellular structure should be broken to get a complete extraction of substances. Moreover, even though larger particles contain more essential oil, the extraction rate is slower than that of smaller particles, resulting in a longer extraction process. Also, the pepper oil extracted using supercritical carbon dioxide contained more sesquiterpenes (the main components of perfumes) compared to that obtained from hydrodistillation. The efficiency of the extraction of fresh and dried leaves of Piper piscatorum was evaluated by employing supercritical CO_2 and co-solvents (10% ethanol and 10% methanol) at 40°C and 70°C and a pressure of 40 MPa (Kumoro et al. 2010).

Almeida et al. (2012) studied the extraction of mint essential oil using different conventional and novel extraction techniques and specifically sub-/supercritical fluid extraction with and without co-solvents, hydrodistillation and Soxhlet extraction and evaluated the methods by extraction yield, chemical profile, and antioxidant activity of the extracts. In terms of extraction yield, the hydro-distillation extraction yield was the lowest of all techniques and conditions evaluated in this study. The extraction yield obtained by Soxhlet extraction was the highest among all other techniques; however, the extracts obtained by these methods required a purification process that increased the cost of the overall process (extraction + purification). The use of co-solvents also increased the extraction yield of the supercritical fluid extraction compared to the use of only supercritical CO_2. The best co-solvent concentration in terms of process yield was 20% w/w for ethanol. The different essential oils extracted using supercritical fluid extraction presented good antioxidant behaviour. The presence of co-solvents in supercritical fluid extraction increased the concentration of some compounds, such as pulegone and carvone when compared to supercritical fluid extraction with only CO_2.

In the studies of Glisic et al. (2010) and Dauksas et al. (2001), the extraction of sage oil using supercritical fluid extraction was investigated and compared to extraction performed by Soxhlet ethanol-water (70:30) mixture extraction and hydrodistillation. Supercritical fluid extraction allowed isolation of a wide spectrum of phytochemicals, while other applied methods were limited to either volatiles (hydrodistillation) or high molecular compounds isolation (Soxhlet). A balance between CO_2 solvent power and selectivity was required to optimize sage oil composition. The volatile fraction could be isolated at low pressure and low CO_2 consumption, whereby the pressures between 10 and 15 MPa followed by increased CO_2 consumption were favourable for obtaining the desired selectivity of specific terpenes.

Marzouki et al. (2008) studied the isolation of volatile and fixed oils from dried berries of the bay laurel (*Laurus nobilis*) and compared them to those produced using hydrodistillation. The yield of the volatile oil of *L. nobilis* berries was obtained by supercritical extraction at 90 bar and 40°C was found to be 0.9%. The yield of the essential oil extracted by hydrodistillation was 0.8%. Chemical analysis revealed that essential oil extracted under supercritical conditions had a high content of (E)-β-ocimene (20.9%), 1,8-cineole (8.8%), α-pinene (8.0%), β-longipinene (7.1%), δ-cadinene (4.7%), linalyl acetate (4.5%), β-pinene (4.2%), α-terpinyl acetate (3.8%) and α-bulnesene (3.5%). The compounds isolated by hydrodistillation were practically the same as those extracted by supercritical CO_2 extraction. Although the oil from hydrodistillation contained a little bit more monoterpenes and the supercritical extracted oil had a higher content of sesquiterpenes. The main differences observed were the content of linalool, which was greater in the oil from hydrodistillation (4.2%) than in the supercritical extracted oil (2.2%), and the content of linalyl acetate showed an opposite trend, 1.3% in oil from hydrodistillation and 4.5% in the supercritical extracted oil.

Dahn et al. (2013) investigated the effect of different extraction methods, namely hydrodistillation, solvent extraction and supercritical extraction on extraction yields and chemical and biological characteristics of lavender essential oil. Solvent extraction gave the highest yield (7.57%), followed by supercritical extraction (6.68%), while hydrodistillation gave the lowest yield (4.57%). In terms of chemical composition, the supercritical extracted oil was not thermally modified as in the case of the hydrodistilled oils and did not include high concentrations of undesired compounds that were clearly present in the solvent extracts. Moreover, the antimicrobial activity of the supercritical extracted oil was higher or equal to that of the other extracts. Finally, the supercritical extracted oil had the highest antioxidant activity compared to essential oils obtained by the other two methods and was comparable to pure α-tocopherol and 2,6 ditert-butyl-4-methylphenol. In their previous study, Dahn et al. (2012) reported the effect of pressure, temperature and time of supercritical CO_2 extraction on yield, antioxidant activity and chemical composition of lavender oil. Pressure and time had a significant linear effect on extracts yield, while temperature had a lesser impact. Generally, the yield of the extracts increased with pressure and time. The extracts obtained at high pressure, high temperature and long extraction time had high yield and high antioxidant activities; however, the three operative parameters did not have any impact on the chemical composition of the extracts.

In the case of a basil (*Ocimum basilicum* L.) chemotype rich in T-cadinol, which is of great importance owing to the biological activity in cancer therapy and as an antibiotic, Occhipinti et al. (2013) compared the chemical composition of essential oils produced by hydrodistillation and supercritical CO_2 extraction. Supercritical extraction yielded a higher percentage of 1,8-cineole (10%; 4-fold), linalool (23.2%; 5.8-fold), eugenol (13.3%; 1.2-fold) and germacrene D (5.6%; 28-fold) with respect to hydrodistillation. On the other hand, the essential oil produced using hydrodistillation was characterised by higher percentages of T-cadinol (27.5%; 3-fold) and some other sesquiterpenes with respect to supercritical CO_2 extraction.

In this case, the supercritical extraction presented limited recovery of T-cadinol from sweet basil compared to hydrodistillation.

4.3 Ultrasound-Assisted Extraction (USAE)

The suitability of ultrasound-assisted extraction (USAE) compared to different conventional methods, such as Soxhlet extraction and solvent extraction (SE) for the recovery of papaya seed oil from the Malaysian Sekaki papaya variety was evaluated by Samaram et al. (2014). The authors concluded that USAE produced the most stable oil with the lightest colour and the lowest level of unsaponifiable matters compared to the traditional extraction techniques. Furthermore, USAE required shorter extraction times (30 minutes) reflecting its economic advantage. Sereshti et al. (2012) compared USAE and hydrodistillation extraction of essential oil from the plant *Elettaria cardamomum* Maton. USAE was performed at reduced temperatures (32°C) and shorter extraction times (10 minutes) protecting the essential oil components, which are sensitive to heat compared to hydrodistillation (100°C and 4–6 hours). Furthermore, USAE needed low amounts of plant material (0.1 g) compared to hydrodistillation (50 g). In a more recent study, Tekin et al. (2015) used response surface methodology (RSM) to optimise the USAE of essential oil from clove (*Syzygium aromaticum*). The most important parameter was found to be the extraction temperature since the higher extraction yields were observed at high temperatures and long extraction times. The clove extract composition and antibacterial activity were found to be similar to clove extracts produced using the traditional extraction techniques. More specifically, the major ingredient of the USAE clove extract was eugenol also reported in the study of Guan et al. (2007). Ultrasound-assisted extraction (USAE) is a promising technique to obtain essential oils from plants with high efficiency that preserve their functional and compositional characteristics.

References

Abbasi, S. and Farzanmehr, H. 2009. Optimization of extracting conditions of inulin from iranian artichoke with/without ultrasound using response surface methodology. J. Agric. Sci. Technol. 13: 423–436.

Almeida, P.P., Mezzomo, N. and Ferreira, S.R. 2012. Extraction of *Mentha spicata* L. volatile compounds: Evaluation of process parameters and extract composition. Food Bioproc. Technol. 5: 548–559.

Barba, F.J., Grimi, N. and Vorobiev, E. 2014. New approaches for the use of non-conventional cell disruption technologies to extract potential food additives and nutraceuticals from microalgae. Food Eng. Rev. 7: 45–62.

Barba, F.J., Brianceau, S., Turk, M., Boussetta, N. and Vorobiev, E. 2015a. Effect of alternative physical treatments (ultrasounds, pulsed electric fields, and high-voltage electrical discharges) on selective recovery of bio-compounds from fermented grape pomace. Food Bioproc. Technol. 8: 1139–1148.

Barba, F.J., Galanakis, C.M., Esteve, M.J., Frigola, A. and Vorobiev, E. 2015b. Potential use of pulsed electric technologies and ultrasounds to improve the recovery of high-added value compounds from blackberries. J. Food Eng. 167: 38–44.

Barba, F.J., Grimi, N. and Vorobiev, E. 2015c. Evaluating the potential of cell disruption technologies for green selective extraction of antioxidant compounds from Stevia rebaudiana Bertoni leaves. J. Food Eng. 149: 222–228.

Barsotti, L., Dumay, E., Mu, T.H., Fernandez Diaz, M.D. and Cheftel, J.C. 2001. Effects of high voltage electric pulses on protein-based food constituents and structures. Trends Food Sci. Technol. 12: 136–144.

Bi, W., Tian, M. and Row, K.H. 2013. Evaluation of alcohol-based deep eutectic solvent in extraction and determination of flavonoids with response surface methodology optimization. J. Chromatog. A 1285: 22–30.

Bobinaitė, R., Pataro, G., Lamanauskas, N., Šatkauskas, S., Viškelis, P. and Ferrari, G. 2014. Application of pulsed electric field in the production of juice and extraction of bioactive compounds from blueberry fruits and their by-products. J. Food Sci. Tech. 52: 5898–5905.

Bosiljkov, T., Dujmić, F., Cvjetko Bubalo, M., Hribar, J., Vidrih, R., Brnčić, M. et al. 2017. Natural deep eutectic solvents and ultrasound-assisted extraction: Green approaches for extraction of wine lees anthocyanins. Food Bioprod. Process 102: 195–203.

Briones-Labarca, V., Giovagnoli-Vicuña, C. and Cañas-Sarazúa, R. 2019. Optimization of extraction yield, flavonoids and lycopene from tomato pulp by high hydrostatic pressure-assisted extraction. Food Chem. 278: 751–759.

Cardoso-Ugarte, G.A., Juárez-Becerra, G.P., SosaMorales, M.E. and López-Malo, A. 2013. Microwave-assisted Extraction of Essential Oils from Herbs. J. Microw Power Electromagn Energy 47: 63–72.

Chemat, F., Lucchesi, M., Smadja, J., Favretto, L., Colnaghi, G. and Visinoni, F. 2006. Microwave accelerated steam distillation of essential oil from lavender: A rapid, clean and environmentally friendly approach. Anal. Chim. Acta 555: 157–160.

Chen, F., Zhang, X., Zhang, Q., Du, X., Yang, L., Zu, Y. et al. 2016. Simultaneous synergistic microwave–ultrasonic extraction and hydrolysis for preparation of trans-resveratrol in tree peony seed oil-extracted residues using imidazolium-based ionic liquid. Ind. Crops Prod. 94: 266–280.

Coustets, M., Al-Karablieh, N., Thomsen, C. and Teissié, J. 2013. Flow process for electroextraction of total proteins from microalgae. J. Membr. Biol. 246: 751–760.

Dahmoune, F., Spigno, G., Moussi, K., Remini, H., Cherbal, A. and Madani, K. 2014. Pistacia lentiscus leaves as a source of phenolic compounds: Microwave-assisted extraction optimized and compared with ultrasound-assisted and conventional solvent extraction. Ind. Crops Prod. 61: 31–40.

Dahmoune, F., Nayak, B., Moussi, K., Remini, H. and Madani, K. 2015. Optimization of microwave-assisted extraction of polyphenols from *Myrtus communis* L. leaves. Food Chem. 166: 585–595.

Danh, L.T, Triet, N.D.A., Han, L.T.N., Zhao, J., Mammucari, R. and Foster, N. 2012. Antioxidant activity, yield and chemical composition of lavender essential oil extracted by supercritical CO_2. J. Supercrit. Fluids 70: 27–34.

Danh, L.T., Han, L.T.N., Triet, N.D.A., Zhao, J., Mammucari, R. and Foster N. 2013. Comparison of chemical composition, antioxidant and antimicrobial activity of lavender (*Lavandula angustifolia* L.) essential oils extracted by supercritical CO_2, hexane and hydrodistillation. Food Bioproc. Tech. 6: 3481–3489.

Dauksas, E., Venskutonis, P.R., Povilaityte, V. and Sivik, B. 2001. Rapid screening of antioxidant activity of sage (*Salvia officinalis* L.) extracts obtained by supercritical carbon dioxide at different extraction conditions. Nahrung 45: 338–341.

Day, L., Seymour, R.B., Pitts, K.F., Konczak, I. and Lundin, L. 2009. Incorporation of functional ingredients into foods. Trends Food Sci. Technol. 20: 388–395.

Dima, C., Ifrim, G.A., Coman, G., Alexe, P. and Dima, S. 2016. Supercritical CO_2 extraction and characterization of *Coriandrum sativum* L. essential oil. J. Food Process 39: 204–211.

Eikani, M.H., Golmohammad, F. and Rowshanzamir, S. 2007. Subcritical waterextraction of essential oils from coriander seeds (*Coriandrum sativum* L.). J. Food Eng. 80: 735–740.

El Darra, N., Grimi, N., Maroun, R.G., Louka, N. and Vorobiev, E. 2013. Pulsed electric field, ultrasound, and thermal pretreatments for better phenolic extraction during red fermentation. Eur. Food Res. Technol. 236: 47–56.

Faggian, M., Sut, S., Perissutti, B., Baldan, V., Grabnar, I. and Dall'Acqua, S. 2016. Natural deep eutectic solvents (NADES) as a tool for bioavailability improvement: pharmacokinetics of rutin dissolved in proline/glycine after oral administration in rats: possible application in nutraceuticals. Molecules 21: 1531.

Farhat, A., Fabiano-Tixier, A.S., Visinoni, F., Romdhane, M. and Chemat, F. 2010. A surprising method for green extraction of essential oil from dry spices: Microwave dry-diffusion and gravity. J. Chromatogr. A 1217: 7345–7350.

Ferhat, M.A., Meklati, B.Y. and Chemat, F. 2007. Comparison of different isolation methods of essential oil from Citrus fruits: cold pressing, hydrodistillation and microwave "dry" distillation. Flavour Fragr J. 22: 494–504.

Foegeding, E.A. 2015. Food protein functionality—A new model. J. Food Sci. 80(12): 2670–2677.

Ganzler, K., Salgó, A. and Valkó, K. 1986. Microwave extraction. A novel sample preparation method for chromatography. J. Chromatogr. A. 371: 299–306.

García, A., Rodríguez-Juan, E., Rodríguez-Gutiérrez, G., Rios, J.J. and Fernández-Bolaños, J. 2016. Extraction of phenolic compounds from virgin olive oil by deep eutectic solvents (DESs). Food Chem. 197: 554–561.

Gilbert-López, B., Barranco, A., Herrero, M., Cifuentes, A. and Ibáñez, E. 2017. Development of new green processes for the recovery of bioactives from *Phaeodactylum tricornutum*. Food Res. Int. 99: 1056–1065.

Giner, J., Grouberman, P., Gimeno, V. and Martín, O. 2005. Reduction of pectinesterase activity in a commercial enzyme preparation by pulsed electric fields: comparison of inactivation kinetic models. J. Sci. Food Agric 85: 1613–1621.

Glisic, S., Ivanovic, J., Ristic, M. and Skala, D. 2010. Extraction of sage (*Salvia officinalis* L.) by supercritical CO_2: Kinetic data, chemical composition and selectivity of diterpenes. J. Supercrit. Fluids 52: 62–70.

González, C.G., Mustafa, N.R., Wilson, E.G., Verpoorte, R. and Choi, Y.H. 2017. Application of natural deep eutectic solvents for the "green" extraction of vanillin from vanilla pods. Flavour Fragr. J. 33: 91–96.

Grosso, C., Ferraro, V., Figueiredo, A.C., Barroso, J.G., Coelho, J.A. and Palavra, A.M. 2008. Supercritical carbon dioxide extraction of volatile oil from Italiancoriander seeds. Food Chem. 111: 197–203.

Guan, W., Li, S., Yan, R., Tang, S. and Quan, C. 2007. Comparison of essential oils of clove buds extracted with supercritical carbon dioxide and other three traditional extraction methods. Food Chem. 101: 1558–1564.

Guimarães, J.T., Balthazar, C.F., Scudino, H., Pimentel, T.C., Esmerino, E.A., Ashokkumar, M. et al. 2019. High-intensity ultrasound: A novel technology for the development of probiotic and prebiotic dairy products. Ultrason. Sonochem. 57: 12–21.

Illés, V., Daood, H.G., Perneczki, S., Szokonya, L. and Then, M. 2000. Extraction ofcoriander seed oil by CO_2 and propane at super-and subcritical conditions. J. Supercrit. Fluids 17: 177–186.

Jeong, K.M., Zhao, J., Jin, Y., Heo, S.R., Han, S.Y., Yoo, D.E. et al. 2015. Highly efficient extraction of anthocyanins from grape skin using deep eutectic solvents as green and tunable media. Arch. Pharm. Res. 38: 2143–2152.

Kaderides, K., Papaoikonomou, l., Serafim, M. and Goula, A.M. 2019. Microwave-assisted extraction of phenolics from pomegranate peels: Optimization, kinetics, and comparison with ultrasounds extraction. Chem. Eng. Process. 137: 1–11.

Kalyani Nair, K., Kharb, S. and Thompkinson, D.K. 2010. Inulin dietary fiber with functional and health attributes—a review. Food Rev. Int. 26: 189–203.

Knorr, D. 2003. Impact of non-thermal processing on plant metabolites. J. Food Eng. 56: 131–134.

Kris-Etherton, P.M., Hecker, K.D., Bonanome, A., Coval, S.M., Binkoski, A.E., Hilpert, K.F. et al. 2002. Bioactive compounds in foods: their role in the prevention of cardiovascular disease and cancer. Am. J. Med. 9B: 71–88.

Kulathooran Ramalakshmi, S.T. 2015. Microwave-assisted extraction of inulin from chicory roots using response surface methodology. J. Nutr. Food Sci. 5: 1–6.

Kumoro, A.C., Hasan, M. and Singh, H. 2010. Extraction of sarawak black pepper essential oil using supercritical carbon dioxide. Arab. J. Sci. Eng. 35: 7–16.

Li, S., Wu, Q., Yin, F., Zhu, Z., He, J. and Barba, F. 2018. Development of a combined trifluoroacetic acid hydrolysis and HPLC-ELSD method to identify and quantify inulin recovered from jerusalem artichoke assisted by ultrasound extraction. Appl. Sci. 8: 710.

Liazid, A., Palma, M., Brigui, J. and Barroso, C.G. 2007. Investigation on phenolic compounds stability during microwave-assisted extraction. J. Chromatogr. A 1140: 29–34.

Lin, D.C. 2003. Probiotics as functional foods. Nutr. Clin. Pract. 18: 497–506.

Lingyun, W., Jianhua, W., Xiaodong, Z., Da, T., Yalin, Y., Chenggang, C., Tianhua, F. and Fan, Z. 2007. Studies on the extracting technical conditions of inulin from Jerusalem artichoke tubers. J. Food Eng. 79: 1087–1093.

Loginova, K.V., Shynkaryk, M.V., Lebovka, N.I. and Vorobiev, E. 2010. Acceleration of soluble matter extraction from chicory with pulsed electric fields. J. Food Eng. 96: 374–379.

López, N., Puértolas, E., Condón, S., Raso, J. and Álvarez, I. 2009. Enhancement of the solid-liquid extraction of sucrose from sugar beet (Beta vulgaris) by pulsed electric fields. LWT - Food Sci. Technol. 42: 1674–1680.

Lou, Z., Wang, H., Li, Ji., Zhu, S., Lu, W. and Ma, C. 2011. Effect of simultaneous ultrasonic/microwave assisted extraction on the antioxidant and antibacterial activities of burdock leaves. J. Med. Plant Res. 5: 5370–5377.

Lucchesi, M.E., Smadja, J., Bradshaw, S., Louw, W. and Chemat, F. 2007. Solvent free microwave extraction of *Elletaria cardamomum* L.: A multivariate study of a new technique for the extraction of essential oil. J. Food Eng. 79: 1079–1086.

Mamidipally, P.K. and Liu, S.X. 2004. First approach on rice bran oil extraction using limonene. Eur. J. Lipid Sci. Technol. 106: 122–125.

Martirosyan, D.M. and Singh, J. 2015. A new definition of functional food by FFC: what makes a new definition unique? Funct. Food Health Dis. 5: 209–223.

Marzouki, H., Piras, A., Marongiu, B., Rosa, A. and Dessi, M. 2008. Extraction and separation of volatile and fixed oils from berries of *Laurus nobilis* L. by supercritical CO_2. Molecules 13: 1702–1711.

Medina-Meza, I.G. and Barbosa-Cánovas, G.V. 2015. Assisted extraction of bioactive compounds from plum and grape peels by ultrasonics and pulsed electric fields. J. Food Eng. 166: 268–275.

Mejri, J., Aydi, A., Abderrabba, M. and Mejri, M. 2018. Emerging extraction processes of essential oils: A review. Asian J. Green Chem. 2: 171–280.

Mendes, M.F., Cataldo, L.F., da Silva, C.A., Nogueira, R.I. and Freitas, S.P. 2005. Extraction of the inuline from chicory roots (*Chicorium intybus* L.) using supercritical carbon dioxide. Second Mercosur Congress on Chemical Engineering.

Mhemdi, H., Rodier, E., Kechaou, N. and Fages, J. 2011. A supercritical tuneable process for the selective extraction of fats and essential oil from coriander seeds. J. Food Eng. 105: 609–616.

Milani, E., Koocheki, A. and Golimovahhed, Q.A. 2011. Extraction of inulin from Burdock root (Arctium lappa) using high intensity ultrasound. Int. J. Food Sci. Technol. 46: 1699–1704.

Mnayer, D., Fabiano-Tixier, A.S., Petitcolas, E., Ruiz, K., Hamieh, T. and Chemat, F. 2017. Extraction of green absolute from thyme using ultrasound and sunflower oil. Resource-Efficient Technologies 3: 12–21.

Moradi, S., Fazlali, A. and Hamedi, H. 2018. Microwave-assisted hydro-distillation of essential oil from rosemary: comparison with traditional distillation. Avicenna J. Med. Biotechnol. 10: 22–28.

Mustafa, A. and Turner, C. 2011. Pressurized liquid extraction as a green approach in food and herbal plants extraction: A review. Anal. Chim. Acta 703: 8–18.

Nayak, B., Dahmoune, F., Moussi, K., Remini, H., Dairi, S., Aoun, O. and Khodir, M. 2015. Comparison of microwave, ultrasound and accelerated-assisted solvent extraction for recovery of polyphenols from Citrus sinensis peels. Food Chem. 187: 507–516.

Niness, K.R. 1999. Inulin and oligofructose: What are they? J. Nutr. 129: 1402–1406.

Occhipinti, A., Capuzzo, A., Bossi, S., Milanesi, C. and Maffei, M.E. 2013. Comparative analysis of supercritical CO_2 extracts and essential oils from an *Ocimum basilicum* chemotype particularly rich in T-cadinol. J. Essent. Oil Res. 25: 272–277.

Paini, M., Casazza, A.A., Aliakbarian, B., Perego, P., Binello, A. and Cravotto G. 2016. Influence of ethanol/water ratio in ultrasound and high-pressure/high-temperature phenolic compound extraction from agri-food waste. International Journal of Food Science and Technology 51: 349–358.

Pandey, K.B. and Rizvi, S.I. 2009. Plant polyphenols as dietary antioxidants in human health and disease. Oxid. Med. Cell Longev. 2: 270–278.

Parniakov, O., Barba F.J., Grimi, N., Lebovka, N. and Vorobiev, E. 2014. Impact of pulsed electric fields and high voltage electrical discharges on extraction of high-added value compounds from papaya peels. Food Res. Int. 65: 337–343.

Patrignani, F., Iucci, L., Lanciotti, R., Vallicelli, M., Maina Mathara, J. Holzapfel, W.H. et al. 2007. Effect of high-pressure homogenization, nonfat milk solids, and milkfat on the technological performance of a functional strain for the production of probiotic fermented milks. J. Dairy Sci. 90: 4513–4523.

Pavlić, B., Vidović, S., Vladić, J., Radosavljević, R. and Zeković, Z. 2015. Isolation of coriander (*Coriandrum sativum* L.) essential oil by green extractions versus traditional techniques. J. Supercrit. Fluids 99: 23–28.

Paximada, P., Dimitrakopoulou, E.A., Tsouko, E., Koutinas, A.A., Fasseas, C. and I. Mandala. 2016. Structural modification of bacterial cellulose fibrils under ultrasonic irradiation. Carbohydr. Polym. 150: 5–12.

Perakis, C., Louli, V. and Magoulas, K. 2005. Supercritical fluid extraction of black pepper oil. J. Food Eng. 71: 386–393.

Perez, O.E. and Pilosof, A.M.R. 2004. Pulsed electric fields effects on the molecular structure and gelation of β-lactoglobulin concentrate and egg white. Food Res. Int. 37: 102–110.

Périno-Issartier, S., Ginies, C., Cravotto, G. and Chemat, F. 2013. A comparison of essential oils obtained from lavandin via different extraction processes: Ultrasound, microwave, turbohydrodistillation, steam and hydrodistillation. J. Chromatogr. A 1305: 41–47.

Pongmalai, P., Devahastin, S., Chiewchan, N. and Soponronnarit, S. 2015. Enhancement of microwave-assisted extraction of bioactive compounds from cabbage outer leaves via the application of ultrasonic pretreatment. Sep. Purif. Technol. 144: 37–45.

Pourfarzad, A., Habibi Najafi, M.B., Haddad Khodaparast, M.H. and Hassanzadeh Khayyat, M. 2015. Characterization of fructan extracted from Eremurus spectabilis tubers: a comparative study on different technical conditions. J. Food Sci. Technol. 52: 2657–67.

Pourmortazavi, S.M. and Hajimirsadeghi, S.S. 2007. Supercritical fluid extraction in plant essential and volatile oil analysis. J. Chromatogr. A 1163: 2–24.

Proestos, C. and Komaitis, M. 2008. Application of microwave-assisted extraction to the fast extraction of plant phenolic compounds. LWT - Food Sci. Technol. 41: 652–659.

Qin, B.L., Zhang, Q., Barbosa-Canovas, G.V., Swanson, B.G. and Pedrow, P.D. 1994. Inactivation of microorganisms by pulsed electric fields of different voltage waveforms. IEEE Trans Dielectr. Electr. Insul. 1: 1047–1057.

Radošević, K., Ćurko, N., Gaurina Srček, V., Cvjetko Bubalo, M., Tomašević, M., Kovačević Ganić, K. et al. 2016. Natural deep eutectic solvents as beneficial extractants for enhancement of plant extracts bioactivity. LWT - Food Sci. Technol. 73: 45–51.

Rajha, H.N, Boussetta, N., Louka, N., Maroun, R.G. and Vorobiev, E. 2014. A comparative study of physical pretreatments for the extraction of polyphenols and proteins from vine shoots. Food Res. Int. 65: 462–468.

Rajha, H.N., Boussetta, N., Louka, N., Maroun, R.G. and Vorobiev, E. 2015. Effect of alternative physical pretreatments (pulsed electric field, high voltage electrical discharges and ultrasound) on the dead-end ultrafiltration of vine-shoot extracts. Sep. Purif. Technol. 146: 243–251.

Rezzoug, S.A., Boutekedjiret, C. and Allaf, K. 2005. Optimization of operating conditions of rosemary essential oil extraction by a fast controlled pressure drop process using response surface methodology. J. Food Eng. 71: 9.

Rois Mansur, A., Song, N.E., Won Jang, H., Lim, T.-G., Yoo, M. and Gyu Nam, T. 2019. Optimizing the ultrasound-assisted deep eutectic solvent extraction of flavonoids in common buckwheat sprouts. Food Chem. 293: 438–445.

Roselló-Sotto, E., Barba, F.J., Parniakov, O., Galanakis, C.M., Lebovka, N., Grimi, N. et al. 2015. High voltage electrical discarges, pulsed electric field, and ultrasound assisted extraction of protein and phenolic compounds from olive kernel. Food Bioproc. Tech. 8: 885–894.

Ross, R.P., Desmond, C., Fitzgerald, G.F. and Stanton, C. 2005. Overcoming the technological hurdles in the development of probiotic foods. J. Appl. Microbiol. 98: 1410–1417.

Rubel, I.A., Iraporda, C., Novosad, R., Cabrera, F.A, Genovese D.B. and Manrique, G.D. 2018. Inulin rich carbohydrates extraction from Jerusalem artichoke (*Helianthus tuberosus* L.) tubers and application of different drying methods. Food Res. Int. 103: 226–233

Ruesgas-Ramón, M., Figueroa-Espinoza, M.C. and Durand, E. 2017. Application of deep eutectic solvents (DES) for phenolic compounds extraction: overview, challenges, and opportunities. J. Agric. Food Chem. 65: 3591–3601.

Samaram, S., Mirhosseini, H., Ping Tan, C. and Mohd Ghazali, H. 2014. Ultrasound-assisted extraction and solvent extraction of papaya seed oil: Crystallization and thermal behavior, saturation degree, color and oxidative stability. Ind. Crops Prod. 52: 702–708.

Santos, P.S., Baggio Ribeiro, D.H., Micke, G.A., Vitali, L. and Hense, H. 2019. Extraction of bioactive compounds from feijoa (*Acca sellowiana* (O. Berg) Burret) peel by low and high-pressure techniques. J. Supercrit. Fluids 145: 219–227.

Scalbert, A., Johnson, I.T. and Saltmarsh, M. 2005. Polyphenols: antioxidants and beyond. Am. J. Clin. Nutr. 81: 215–217.

Sereshti, H., Rohanifar, A., Bakhtiari, S. and Samadi, S. 2012. Bifunctional ultrasound assisted extraction and determination of Elettaria cardamomum Maton essential oil. J. Chromatogr. A 1238: 46–53.

Tekin, K., Akalın, M.K. and Şeker, M.G. 2015. Ultrasound bath-assisted extraction of essential oils from clove using central composite design. Ind. Crops Prod. 77: 954–960.

Tewari, S., Ramalakshmi, K., Methre, L. and Rao, L. 2015. Microwave-assisted extraction of inulin from chicory roots using response surface methodology. Food Sci. Nutr. 5: 1–7.

Tiwari, B.K. 2015. Ultrasound: A clean, green extraction technology. Trends Analyt. Chem. 71: 100–109.

Vinderola, C.G., Bailo, N. and Reinheimer, J.A. 2000. Survival of probiotic microflora in Argentinian yoghurts during refrigerated storage. Food Res. Int. 33: 97–102.

Wang, D., Sakakibara, M., Kondoh, N. and Suzuki, K. 1996. Ultrasound-enhanced lactose hydrolysis in milk fermentation with *Lactobacillus bulgaricus*. J. Chem. Technol. Biotechnol. 65: 86–92.

Xiang, B.Y., Ngadi M.O., Ochoa-Martinez, L.A. and Simpson, M.V. 2011. Pulsed electric field-induced structural modification of whey protein isolate. Food Bioprocess Technol. 4: 1341–1348.

Xiao, Z.J., Zhu, D.H., Wang, X.H. and Zhang, M.D. 2013. Study on extraction process of inulin from Helianthus tuberosus. Mod. Food Sci. Technol. 29: 315–318.

Yang, Y. and Zhang, F. 2008. Ultrasound-assisted extraction of rutin and quercetin from *Euonymus alatus* (Thunb.) Sieb. Ultrason Sonochem. 15: 308–313.

Yang, S., Zhang, H. and Wang, W. 2010. The ultrasonic effect on the mechanism of cholesterol oxidase production by Brevibacterium sp. African J. Biotechnol. 9: 2574–2578.

Yeo, S.K. and Liong, M.T. 2011. Effect of ultrasound on the growth of probiotics and bioconversion of isoflavones in prebiotic-supplemented soymilk. J. Agric. Food Chem. 59: 885–897.

Zbinden, M.D.A., Sturm, B.S.M., Nord, R.D., Carey, W.J., Moore, D., Shinogle, H. and Stagg-Williams, S.M. 2013. Pulsed electric field (PEF) as an intensification pretreatment for greener solvent lipid extraction from microalgae. Biotechnol. Bioeng. 110: 1605–1615.

Zhang, R., Grimi, N., Marchal, L., Lebovka, N. and Vorobiev, E. 2019. Effect of ultrasonication, high pressure homogenization and their combination on efficiency of extraction of bio-molecules from microalgae *Parachlorella kessleri*. Algal Research 40: 101524.

Zhu, Z., Bals, O., Grimi, N. and Vorobiev, E. 2012. Pilot scale inulin extraction from chicory roots assisted by pulsed electric fields. Int. J. Food Sci. Technol. 47: 1361–1368.

Zhu, Z., He, J., Liu, G., Barba, F.J., Koubaa, M., Ding, L. et al. 2016. Recent insights for the green recovery of inulin from plant food materials using non-conventional extraction technologies: A review. Innov. Food Sci. Emerg. Technol. 33: 1–9.

CHAPTER 14

Foodomics and Green Analytical Technologies

Ruchi Khare,[1] *Mohammad Yasir,*[2,]* *Alok Kumar Shukla,*[3]
Sivarama Krishna Lakkaboyana[4] *and Rahul Shrivastava*[1,]*

1. Introduction

Food is the direct source of energy and nutrition required for human survival. With the increased population with respect to time, remove food products have been consumed according to their supply, quality, and disease prevention ability for different diseases. Recent developments in food technology have discovered new techniques for designing new food products, increasing shelf life, and gathering molecular level information in food science for the improvement of human nutrition and healthcare (Atalay et al. 2011). Simple food products like pre-biotics and pro-biotics play a major role in maintaining the gut health of humans. Although high throughput approaches like *omics* have connected human nutrition with food for assessment of diseases, metabolic behaviour and their long effects on human health (Balkir et al. 2021). For the analysis of complex biological systems, real-time simulation and computational modelling are performed under the discipline of system biology. All four types of omics categories, i.e., genomics, proteomics, transcriptomics, and metabolomics, have been acquired differently. Foodomics is a new area of nutrition that uses both "dry" and "wet" lab techniques to study the effects of food on the human body.

[1] Department of Biological Science and Engineering, Maulana Azad National Institute of Technology, Bhopal (M.P.), India.
[2] Amity Institute of Pharmacy, Lucknow, Amity University Uttar Pradesh, Noida.
[3] Babu Sunder Singh College of Pharmacy, Nigohan, Lucknow (U.P.), India.
[4] School of Ocean Engineering, Universiti Malaysia Terengganu, 21030, Kuala Nerus, Terengganu Darul Iman, Malaysia.
Emails: ruchi92khare@gmail.com; yasirmohammad6@gmail.com; svurams@gmail.com
* Corresponding author: shrivastavarm1972@gmail.com

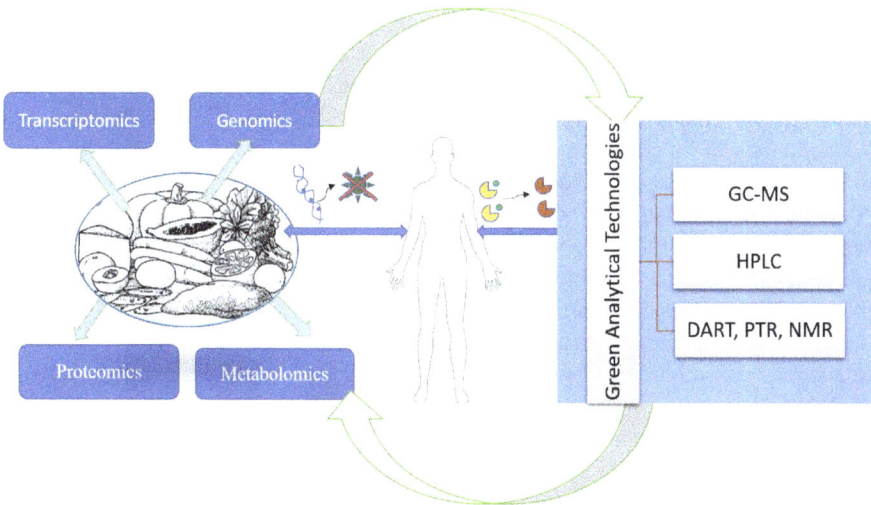

Figure 1. The schematic diagram describes the correlation between green analytical technologies, human nutrition, and foodomics; for prevention of diseases and better survival achieved via manipulation of food at various levels.

The foundation of foodomics as a separate discipline was first introduced by professor Alejandro Cifuentes from the Institute of Food Science Research (CIAL) of the National Research Council, Spain, in 2009 (Zheng and Chen 2014). This foodomics approach also allows the researchers to look forward to the nutritional values of plants, fermented food products, and vegetables to insert specific genes to increase their nutritional values. The techniques like NMR (Nuclear Magnetic Resonance) and MS (Mass Spectroscopy) offer molecular profiling of the whole metabolome of wild food crops and GMO's (Genetically Modified Organisms), as seen in Figure 1.

The genomics offers comprehensive study of retorts to different foods by analysing DNA and its functions triggering different responses.

Proteomics offers information related to food toxicity and food safety, which may result in allergic or non-allergic diseases.

Transcriptomics can help us to reduce the stress related to molecular actions of food nutrients, which may allow modulations in gene expression of proteins related to disease prevention.

Metabolomics allows us to study biochemical interactions interfering with specific cellular pathways (Cifuentes 2012). However, the role of system biology in the development of sustainable foodomics methodologies is discussed further.

By exploiting different methodologies of system biology, new sources of essential food supplements, their suitability, the behaviour of extracted food components, and their effect on the human body at a molecular level can be predicted. Such approaches also allow the healthcare system for specific treatment of genetic and non-genetic lethal diseases which reduces the mortality rate and enhance the quality of human life. With a growing population, problems like an improper supply of food, intensive farming, and desertification have aroused. Environmental sustainability is the key factor to be considered for food supply, food production systems and the development of new food products. The excessive load on environmental parameters can be balanced by opting for green technologies for the foodomics discipline. For designing and supply of functional foods for the prevention of diseases, the extraction of bioactive compounds, their sample preparation, sustainability, and molecular mode of action have been examined by following sustainable GAC (Green Analytical Chemistry) methodologies (Gilbert-López et al. 2017). Many tools for separation and analysis of food constituents have been developed, like LC (Liquid Chromatography) was a simple technique but with enhancement. In techniques, many products related to specific columns and programs have been developed for natural and synthetic biological samples. Likewise, nano-chips have been developed to quantify the amount of DNA, protein and metabolites in sample extracts (Wilson et al. 2015).

The tools developed for foodomics help in detecting food adulteration, the authenticity of the product, foodborne pathogens, allergens, etc. Food safety is one of the major aspects with respect to its nutritional value (Piñeiro et al. 2015, Mansor et al. 2020). System biology provides a platform for the development of new software and techniques that enabled food safety and nutrition value measurements worldwide. The development of foodomics in parallel to system biology has enabled FDA to set parameters for instrumentation, handling, and storage of food samples (Andjelković et al. 2017).

2. Applications and Drawbacks

The foodomics branch has connected human health with food nutrition. It has enabled researchers to gather information about food compositions and their effect on the physiological and psychological health of human beings. The increase in research regarding foodomics and the green technology supporting real-time monitoring, extraction, separation, and storage techniques have generated a huge amount of data. This strategy, however, has limited the use of radiochemical and thermal approaches. Other methods, including spectroscopy, ultra-pressure liquid chromatography, and Fourier transform infrared spectroscopy, were developed, and high-throughput technology was developed to store, forecast, and disseminate such data (García-Cañas et al. 2012). To provide the best techniques and methodologies, the GAC and system biology emerged as an interdisciplinary subject with food sciences. Collaboration of such three different disciplines has created a huge impact on the living standards of people. Some of the other applications are given below:

- Healthy individuals are seeking more nutritional food products like pre-biotics and pro-biotics, and diseased individuals are having specific food which has enhanced nutritional values (Jiménez-Pranteda et al. 2015). These techniques especially genomics and transcriptomics have helped the researchers to develop plant and plant products with more nutritional value that serve people having malnutrition issues within geographical boundaries. Other than malnutrition, genomics has served people by regulating specific protein/amino-acid diets, which decreased the susceptibility of humans to certain diseases (Pérez-Massot et al. 2013).

- Proteomics has become a major tool when it comes to biological sample analysis. Proteomics has enabled the research to investigate wild variety and genetically modified food varieties and their protein profiling. It also focused on the large-scale protein product development given to patients of different ages (Gong and Wang 2013).
- Metabolomics helped in investigating the role of food supplements in mitigating harmful toxic substances in the body. The tracing of toxic substances in the food, their permitted concentration in food and their deleterious effects on the body can be assessed with the help of interdisciplinary studies.
- Animal and plant-derived foods are susceptible to damage, oxidation and putrefaction during storage and transport. It is a major issue of food safety and food storage. The tools developed by foodomics can help in assessment quality maintenance by developing biological markers for the identification of freshness of food and fabricating and maintaining specific refrigerators (Viant et al. 2019).

There are major challenges related to foodomics and system biology. The foodborne illness is one of the major issues in countries that are dependent upon industrialized fermented food products. With the change in climate, the micro-organisms are mutating periodically which is affecting the global standards ruled in for food safety. The technologies have to be exploited more for tracing such micro-organisms and viruses inoculated in the environment. After globalization, a need was felt for the development of similar tools and protocols to analyse food products marking their quality, nutritional values and monetary values.

3. Green Analytical Technologies

Qualitative qualities of foods may be discerned by screening procedures or instrumental procedures, and this has been vital to the food business (Gallo and Ferranti 2016). The essential ideas of green chemistry in foodomics will be explained in this chapter.

Enormous leaps in the study of biology have occurred due to omics technology. GC-MS and HPLC are the most often used and are costliest of the chromatography-related techniques, although the results are long and time-consuming to process (Lynch 2017). So, to achieve green chemistry's objective, which is to utilise less or to replace hazardous materials (Ratti 2020).

MS-based procedures (such as DART (Direct Analysis in Real Time), PTR (Proton Transfer Reaction), and IM (Ion Mobility) among others) and NMR (Nuclear Magnetic Resonance) are commonly used in Foodomics for direct analysis (Mendiola et al. 2013). Other techniques, such as spectroscopic analysis, are available. NIR (Near Infrared) and Raman spectroscopy are examples of techniques that can help. Because they are quite versatile, chromatography and electrophoresis are particularly useful for isolating, profiling, and quantifying materials and are very popular in the food industry (Cifuentes 2012).

The subject of foodomics, especially for the growth of green processing techniques for extracting useful components from natural sources, has gained popularity. Using traditional extraction procedures (such as the process that includes heating, the sonication process, and the solid-liquid extraction process) will result in lengthy extraction durations and significant quantities of samples. Because of these attributes, the extraction yields will be poor and the extraction results will not be very selective. Because of this, huge volumes of organic solvent waste will be produced, which has the potential to negatively impact the environment (Tzanova et al. 2020).

In this regard, ultrasound-assisted extraction (UAE) and microwave-assisted extraction (MAE) are flexible approaches to green extraction procedures because they enable the use of multiple solvents with different polarities, allowing for quick extractions while reducing the number of solvents used (Fomo et al. 2020).

Modern superheated extraction methods, on the other hand, such as supercritical fluid extraction (SFE), pressurised liquid extraction (PLE), or pressurised hot water extraction (PHWE), also known as subcritical water extraction (SWE), adhere to green chemistry and engineering principles and have proven to be very valuable (Haghighi and Khajenoori 2013).

A wide range of innovative analytical methods combined with recent developments in microelectronics and miniaturisation provide the means for improved process and pollution management. With improvements in chemometrics, analytical systems have also been miniaturised, bringing them closer to customers (Kricka 1998). The growth of chemometrics has benefited the progress of solvent-free techniques that concentrate on the mathematical treatment of signals generated by direct measurements on untreated solid or liquid materials.

4. Green Separation Techniques

4.1 Gas Chromatography

Chromatographic procedures entail significant environmental damage because of the sample preparation step, which is frequently the most ecologically hazardous section of the overall chromatographic analysis process. The removal of the sample preparation stage is typically easier with gas chromatographic. The invention of on-column injection is a breakthrough. Halogenated and hydrocarbons containing compounds in samples are detected by coupling with flame ionisation detection or mass spectrometric detection (Regmi and Agah 2018). The ideas of green chemistry may be integrated into several ways into gas chromatography.

- The choice of carrier gas.
- Reduction in analysis time.
- Utilization of low thermal mass technology.
- Making use of direct resistive heating of nickel-clad fused silica GC columns.
- Comprehensive use of two-dimensional gas chromatography combined with time-of-flight mass spectrometry (TOFMS) results in significant time and reagent saving.
- Use of consumable-free modulator (CFM). It cools itself by utilising ambient airflow. This type of modulator does not require the use of cryogenic agents, making it environmentally friendly.

4.2 Supercritical Fluid Chromatography (SFC)

The bioactive components should be extracted appropriately to provide the highest extraction efficiency and lowest degradation of the active component, while still being economical and having little environmental effect (Khan et al. 2019). The most prominent ways of achieving this green approach are pressurised liquid extraction (PLE), supercritical fluid extraction (SFE), and gas-expanded liquid extraction (GEL) (GXLs). SFC is a technology that uses supercritical fluid (the most popular of which is carbon dioxide) and could vary the solvent strength. Even though SFC offers several important potential advantages, such as high resolution and performance, it has not been developed (Dean 1993).

4.3 Capillary Electrophoresis (CE)

Capillary electrophoresis is a flexible isolation technique that can be applied to a wide variety of materials. This involves the movement of electrically charged species present in an electrolytic solution within a capillary to which an electric field is applied, resulting in the generation of a current in the capillary's interior (Voeten et al. 2018). The analysis of several pharmaceutical and biopharmaceuticals involves capillary electrophoresis. The CE technique is often used for research and determining impurities and analyses of protein, glycoproteins, complex carbohydrates, liposaccharides, DNA and virus particles, small molecular products, excipients, and counterion ions in pharmaceuticals (Pedroso et al. 2019).

Capillary electrophoresis-mass spectrometry (CE-MS) incorporates the benefits of both CE and MS to provide high separation efficiency and molecular mass knowledge in a single study. It has high resolving power and sensitivity, so it only takes up a small amount of space and can analyse data quickly. The combination of CE's high performance and speed with MS detection's tailored properties is extremely appealing. This hyphenation has a broad range of uses, including in biomolecules, pharmaceuticals, metabolites, food analysis, environmental analyses and technical substance investigations (Drouin et al. 2020).

4.4 Microfluidic and Lab on a Chip

In the food industry, microfluidic applications can be used in both food preparation and food diagnostics. The principle of combining optics, fluidics, electronics, and biosensors in extremely narrow channels is cutting-edge technology. Many advantages will definitely be achieved by using very small fluid concentrations (micro- to pico-litres) of solvents and chemicals, including high throughput, high power, decreased waste production, and comparatively low energy consumption (Atalay et al. 2011).

Microfluidic Capillary electrophoresis (MCE) also has a lot of potential in food quality research, such as measuring organoleptic properties (colour, odour, flavour, texture, and so on), as well as undesirable additives and other harmful compounds formed during the manufacturing process. For the identification of trace phenolic compounds, an MCE-based analytical method with on-chip pre-concentration has been developed (Shiddiky et al. 2006, Gao and Zhong 2021).

5. Conclusion

Modern foodomics have many stages that may be changed to produce ecological procedures. Modern foodomics has many stages that may be changed to produce ecological procedures. So, it's important for food chemists, analytical chemists, biochemists, microbiologists, molecular biologists, food technologists, clinical scientists, and human biologists to work together on food chemistry research using omic methods with the ultimate goal of enhancing population wellbeing.

References

Andjelković, U., Šrajer Gajdošik, M., Gašo-Sokač, D., Martinović, T. and Josić, D. 2017. Foodomics and food safety: where we are. Food Technology and Biotechnology 55: 290–307.

Atalay, Y.T., Vermeir, S., Witters, D., Vergauwe, N., Verbruggen, B., Verboven, P., Nicolaï, B.M. and Lammertyn, J. 2011. Microfluidic analytical systems for food analysis. Trends in Food Science and Technology 22: 386–404.

Balkir, P., Kemahlioglu, K. and Yucel, U. 2021. Foodomics: A new approach in food quality and safety. Trends in Food Science and Technology 108: 49–57.

Cifuentes, A. 2012. Food analysis: present, future, and foodomics. ISRN Analytical Chemistry 2012: 1–16.

Dean, J.R. 1993. Applications of supercritical fluids in food science. pp. 130–158. *In*: John R. Dean (ed.). Applications of Supercritical Fluids in Industrial Analysis. Springer Netherlands: Dordrecht.

Drouin, N., van Mever, M., Zhang, W., Tobolkina, E., Ferre, S., Servais, A.-C., Gou, M.-J., Nyssen, L., Fillet, M., Lageveen-Kammeijer, G.S.M., Nouta, J., Chetwynd, A.J., Lynch, I., Thorn, J.A., Meixner, J., Lößner, C., Taverna, M., Liu, S., Tran, N.T., Francois, Y., Lechner, A., Nehmé, R., Al Hamoui Dit Banni, G., Nasreddine, R., Colas, C., Lindner, H.H., Faserl, K., Neusüß, C., Nelke, M., Lämmerer, S., Perrin, C., Bich-Muracciole, C., Barbas, C., Gonzálvez, Á.L., Guttman, A., Szigeti, M., Britz-McKibbin, P., Kroezen, Z., Shanmuganathan, M., Nemes, P., Portero, E.P., Hankemeier, T., Codesido, S., González-Ruiz, V., Rudaz, S. and Ramautar, R. 2020. Capillary electrophoresis-mass spectrometry at trial by metabo-ring: effective electrophoretic mobility for reproducible and robust compound annotation. Analytical Chemistry 92: 14103–14112.

Fomo, G., Madzimbamuto, T.N. and Ojumu, T.V. 2020. Applications of nonconventional green extraction technologies in process industries: challenges, limitations and perspectives. Sustainability 12: 5244.

Gallo, M. and Ferranti, P. 2016. The evolution of analytical chemistry methods in foodomics. Journal of Chromatography A 1428: 3–15.

Gao, Z. and Zhong, W. 2021. Recent (2018–2020) development in capillary electrophoresis. Analytical and Bioanalytical Chemistry.

García-Cañas, V., Simó, C., Herrero, M., Ibáñez, E. and Cifuentes, A. 2012. Present and future challenges in food analysis: foodomics. Analytical Chemistry 84: 10150–10159.

Gilbert-López, B., Mendiola, J.A. and Ibáñez, E. 2017. Green foodomics. Towards a cleaner scientific discipline. TrAC Trends in Analytical Chemistry 96: 31–41.

Gong, C.Y. and Wang, T. 2013. Proteomic evaluation of genetically modified crops: current status and challenges. Frontiers in Plant Science 4: 41.

Haghighi, A. and Khajenoori, M. 2013. Subcritical water extraction. *In*: Nakajima, H. (ed.). Mass Transfer—Advances in Sustainable Energy and Environment Oriented Numerical Modeling. InTech: Rijeka, p.

Jiménez-Pranteda, M.L., Pérez-Davó, A., Monteoliva-Sánchez, M., Ramos-Cormenzana, A. and Aguilera, M. 2015. Food omics validation: towards understanding key features for gut microbiota, probiotics and human health. Food Analytical Methods 8: 272–289.

Khan, S.A., Aslam, R. and Makroo, H.A. 2019. High pressure extraction and its application in the extraction of bio-active compounds: A review. Journal of Food Process Engineering 42: e12896.

Kricka, L.J. 1998. Miniaturization of analytical systems. Clinical Chemistry 44: 2008–2014.

Lynch, K.L. 2017. Toxicology: liquid chromatography mass spectrometry. pp. 109–130. *In*: Mass Spectrometry for the Clinical Laboratory. Elsevier.

Mansor, M., Al-Obaidi, J.R., Jaafar, N.N., Ismail, I.H., Zakaria, A.F., Abidin, M.A.Z., Selamat, J., Radu, S. and Jambari, N.N. 2020. Optimization of protein extraction method for 2DE proteomics of goat's milk. Molecules (Basel, Switzerland) 25: 2625.

Mendiola, J.A., Castro-Puyana, M., Herrero, M. and Ibáñez, E. 2013. Green foodomics. pp. 471–506. *In*: Foodomics. John Wiley and Sons, Inc.: Hoboken, NJ, USA.

Pedroso, T.M., Schepdael, A. Van and Salgado, H.R.N. 2019. Application of the principles of green chemistry for the development of a new and sensitive method for analysis of ertapenem sodium by capillary electrophoresis. International Journal of Analytical Chemistry 2019: 1–11.

Pérez-Massot, E., Banakar, R., Gómez-Galera, S., Zorrilla-López, U., Sanahuja, G., Arjó, G., Miralpeix, B., Vamvaka, E., Farré, G., Rivera, S.M., Dashevskaya, S., Berman, J., Sabalza, M., Yuan, D., Bai, C., Bassie, L., Twyman, R.M., Capell, T., Christou, P. and Zhu, C. 2013. The contribution of transgenic plants to better health through improved nutrition: opportunities and constraints. Genes and Nutrition 8: 29–41.

Piñeiro, C., Carrera, M., Cañas, B., Lekube, X. and Martinez, I. 2015. Proteomics and Food Analysis: Principles, Techniques, and Applications. pp. 369–391.

Ratti, R. 2020. Industrial applications of green chemistry: Status, challenges and prospects. SN Applied Sciences 2: 263.

Regmi, B.P. and Agah, M. 2018. Micro gas chromatography: an overview of critical components and their integration. Analytical Chemistry 90: 13133–13150.

Shiddiky, M.J.A., Park, H. and Shim, Y.-B. 2006. Direct analysis of trace phenolics with a microchip: in-channel sample preconcentration, separation, and electrochemical detection. Analytical Chemistry 78: 6809–6817.

Tzanova, M., Atanasov, V., Yaneva, Z., Ivanova, D. and Dinev, T. 2020. Selectivity of current extraction techniques for flavonoids from plant materials. Processes 8: 1222.

Viant, M.R., Ebbels, T.M.D., Beger, R.D., Ekman, D.R., Epps, D.J.T., Kamp, H., Leonards, P.E.G., Loizou, G.D., MacRae, J.I., van Ravenzwaay, B., Rocca-Serra, P., Salek, R.M., Walk, T. and Weber, R.J.M. 2019. Use cases, best practice and reporting standards for metabolomics in regulatory toxicology. Nature Communications 10: 3041.

Voeten, R.L.C., Ventouri, I.K., Haselberg, R. and Somsen, G.W. 2018. Capillary electrophoresis: trends and recent advances. Analytical Chemistry 90: 1464–1481.

Wilson, S., Vehus, T. and Berg, H.S. 2015. Nano-LC in proteomics: recent advances and approaches. Bioanalysis 7.

Zheng, C. and Chen, A. 2014. System biological research on food quality for personalised nutrition and health using foodomics techniques: a review. Journal of Food and Nutrition Research 2: 608–616.

Index

Biographies

Dr. Vinay Kumar completed his Ph.D. in Biotechnology from the School of Biotechnology, Jawaharlal Nehru University, New Delhi, India. He has received prestigious awards including CSIR-JRF, DBT-JRF, CSIR-NET, GATE, and ICAR-NET. In addition, he completed a research grant from SERB as Principal Investigator at IIT Roorkee. He worked in various countries as a postdoctoral fellow. He has more than twelve years of research experience working on environmental engineering, nanobiotechnology and nanotoxicology. He has published several outstanding research papers in prestigious journals. Moreover, he has research collaboration with prominent scientists from Finland, Mexico, Malaysia, Thailand, China, Switzerland, Chile, Brazil, Ireland, Greece, etc. Currently, he is working as Professor at the Department of Community Medicine, Saveetha Medical College, Saveetha Institute of Medical and Technical Sciences, Chennai, India.

Dr. Kleopatra Tsatsaragkou is a Chemical Engineer who graduated from the National Technical University of Athens. She holds a Ph.D. in Food Engineering Lab at Agricultural University of Athens, focused on Physical Properties of Foods. She has worked as a Research Engineer at the Department of Food and Nutritional Sciences at the University of Reading. Currently, she works as a Chemical Engineer at the Oil and Gas Industry.

Dr. Nilofar Asim obtained her BSc in Applied Chemistry (1995) and MSc in Inorganic Chemistry (1999) from Tehran University, and her Ph.D. in Materials Science (2008) from the National University of Malaysia. She is presently working as an Associate professor at the Solar Energy Research Institute, a center of excellence for the research and development of solar energy technology, National University of Malaysia (UKM). She has been involved in the field of materials science for more than 24 years. Her main research interests fall into the fields of circular economy, especially in the areas of recycling and sustainable manufacturing including the development of new products, new fabrication routes, and enhancement of existing products. She is experienced in the valorisation of different agricultural and industrial waste materials for fabrication of value-added materials with various applications such as adsorbents for decontamination and desiccant materials, the synthesis of various metal oxides and metal oxide nanocomposites to be utilized in solar cells, renewable system technologies, catalysts, and depollution as well as Design & synthesis of functional geopolymers for cooling systems as well as depollution. As a material scientist, her focus expands to set on the synthesis of various types of functional, advanced, and environmentally sustainable materials from varieties of resources for different applications. Considering various environmental issues her focus was on applying the green chemistry concept to overcome these issues as much as possible by utilizing various agricultural and industrial wastes as resource materials. She has published more than 120 research papers in journals, conferences, and book chapters.